Oils and Fats as Raw Materials for Industry

Scrivener Publishing
100 Cummings Center, Suite 541J
Beverly, MA 01915-6106

Publishers at Scrivener
Martin Scrivener (martin@scrivenerpublishing.com)
Phillip Carmical (pcarmical@scrivenerpublishing.com)

Oils and Fats as Raw Materials for Industry

Edited by
Divya Bajpai Tripathy
Anjali Gupta
Pooja Agarwal
Anuradha Mishra
and
Arvind Kumar Jain

Scrivener Publishing

WILEY

This edition first published 2024 by John Wiley & Sons, Inc., 111 River Street, Hoboken, NJ 07030, USA and Scrivener Publishing LLC, 100 Cummings Center, Suite 541J, Beverly, MA 01915, USA

For more information about Scrivener publications please visit www.scrivenerpublishing.com.

Wiley Global Headquarters
111 River Street, Hoboken, NJ 07030, USA

For details of our global editorial offices, customer services, and more information about Wiley products visit us at www.wiley.com.

Library of Congress Cataloging-in-Publication Data

ISBN 9781119910411

Cover image: Pixabay.com
Cover design by Russell Richardson

Set in size of 11pt and Minion Pro by Manila Typesetting Company, Makati, Philippines

Printed in the USA

10 9 8 7 6 5 4 3 2 1

Contents

Preface

In today's world, oils and fats play a crucial role in various industries and applications. From the food industry to cosmetics, from pharmaceuticals to coating industries, these versatile substances have become indispensable raw materials. With increasing concerns about sustainability and environmental impact, there is a growing need to explore alternative sources and utilize biotechnology to harness the potential of oils and fats. This book is divided into 12 chapters, discussing the potential of oil and fats as raw materials for various industrial sectors.

In chapter 1, fundamentals of oils and fats has been covered. Discussion has been made on their various types, ranging from vegetable oils to animal fats, on their chemistry and properties. Understanding the characteristics of different oils and fats is essential to comprehend their applications across industries.

The second chapter of the book is focuses on the advancements in biotechnology that have revolutionized the production and utilization of oils and fats. We explore the role of biotechnology in enhancing yield, improving quality, and developing sustainable processes. The chapter highlights the potential of genetic engineering, enzyme technology, and other biotechnological tools in the field of oils and fats.

Chapter 3 of the book is based on the sustainability of oils and fats over petrochemicals. This chapter insights the concerns about the environmental impact of petrochemicals rise and examines the sustainability of oils and fats as an alternative. Comparison on the ecological footprint of oils and fats with that of petrochemicals and discuss the advantages and challenges associated with transitioning to renewable sources has been done. The chapter also addresses the importance of sustainable sourcing and production practices.

The fourth chapter explores the extensive use of oils and fats in the food industry. This discusses their role in cooking, flavouring, and texture enhancement, and their impact on product quality. Additionally, the chapter examines the health considerations associated with consuming different types of oils and fats, shedding light on the ongoing debates surrounding their effects on human health.

Chapter 5 investigates the potential of oils and fats as raw materials for surfactants and laundry detergents. Traditional petrochemical-based surfactants can have adverse environmental effects, and this chapter explores the benefits of utilizing renewable sources to produce eco-friendly alternatives. Challenges and opportunities in this emerging field has also been covered.

Cosmetics have long relied on oils and fats for their moisturizing, emollient, and functional properties. Chapter 6 examines the use of oils and fats in the cosmetic industry, ranging from skincare to haircare products and delve into the formulation aspects and discuss the growing demand for natural and sustainable ingredients.

Coatings are essential for protection and enhancement in numerous industries. In seventh chapter, applications of oils and fats as raw materials for coatings has been explored. In addition, their role in providing durability, adhesion, and corrosion resistance, while also considering the sustainability aspects of such coatings have also been discussed.

The eighth chapter focuses on the utilization of oils and fats as raw materials for corrosion inhibitors and biolubricants. We delve into their effectiveness in protecting metal surfaces from degradation and their potential as environmentally friendly lubricants. The chapter discusses the challenges in implementing these alternatives and their benefits for various industries.

The pharmaceutical industry constantly seeks new and innovative solutions, and vegetable oils have emerged as promising candidates. This chapter delves into the unique properties of vegetable oils and their applications in pharmaceutical formulations. From drug delivery systems to excipients, vegetable oils offer numerous advantages, such as improved bioavailability and enhanced stability. We explore the latest research and developments in this exciting field.

As the world strives towards a greener future, the search for sustainable energy sources becomes increasingly crucial. In this chapter, we explore the utilization of non-edible oils as biodiesel. These oils, derived from plants that are not traditionally used for food, offer a renewable and environmentally friendly alternative to conventional fossil fuels. Chapter 10 delves into the production methods, properties, and challenges associated with non-edible oils as biodiesel, shedding light on their potential impact on energy sustainability.

Oils and fats play a significant role in both ecological and economic realms. Chapter 11 examines the intricate relationship between oil and fat production, consumption, and their environmental impact and explores the topics such as deforestation, land use, and the carbon footprint associated with oil production, as well as the economic implications of oil markets and trade. By understanding the ecological and economic aspects of oils and fats, decisions to foster a sustainable and balanced future can be made.

Corrosion poses a significant challenge across various industries, leading to economic losses and safety concerns. In chapter 12, potential of oils and fats as raw materials for corrosion inhibitors has been explored. This last chapter of the book delve into the unique properties of these substances that make them effective in preventing or reducing corrosion. By examining their mechanisms of action and applications, the possibilities of harnessing oils and fats to mitigate the damaging effects of corrosion have been uncovered.

Throughout this book, editors and authors aim to provide a comprehensive understanding of oils and fats, showcasing their versatility and potential in different fields. From pharmaceutical applications to energy sustainability and environmental considerations, the chapters ahead will shed light on the multifaceted nature of oils and fats. This exploration will inspire further research and innovation, pushing the boundaries of what oils and fats can achieve.

Divya Bajpai Tripathy
Anjali Gupta

1

Oil and Fats as Raw Materials for Industry: An Introduction

Sonali Kesarwani[1], Mukul Kumar[1], Divya Bajpai Tripathy[2*], Anjali Gupta[2] and Suneet Kumar[3]

[1]*Division of Forensic Science, School of Basic and Applied Sciences, Galgotias University, Greater Noida, Uttar Pradesh, India*
[2]*Division of Chemistry, School of Basic and Applied Sciences, Galgotias University, Greater Noida, Uttar Pradesh, India*
[3]*Forensic Science Laboratory, Moradabad, Uttar Pradesh, India*

Abstract
Worldwide, fats and oils occupy major roles in nearly all industries like food, pharmaceutical, surfactants, biodiesel, agriculture, etc. The major renewable raw materials for the chemical industry are oils and fats, which are lipids. When ingested in sufficient quantities, they serve as a vital component of a healthy diet. Triacylglycerols (TAGs), also referred to as triglycerides, are essentially ester-linked glycerol moieties connected with three fatty acids that make up oils and fats. These fatty acids provide fats with useful properties. Because of the modification in lipids with improved nutritional and functional qualities, scientists and researchers have recently developed a growing interest in fats and oils. Triacylglycerols have a variety of properties and use depending on their structure and lipid composition. Therefore, altering the TAGs of lipids is a method for producing personalized lipids and expanding the range of their applications. Due to the detrimental health effects of trans fats, which are mostly brought on through partial hydrogenation, interesterification has become the dominant alternative strategy. This chapter provides a wide range of unprocessed fats and oils, along with information on their characteristics, classification, and historical evolution. In addition, the chapter emphasises the applications and difficulties in this field.

***Keywords*:** Fats, oils, raw materials, saturated, unsaturated, applications, triglycerides

**Corresponding author*: divyabaj@gmail.com

Divya Bajpai Tripathy, Anjali Gupta, Pooja Agarwal, Anuradha Mishra, and Arvind Kumar Jain (eds.) Oils and Fats as Raw Materials for Industry, (1–32) © 2024 Scrivener Publishing LLC

1.1 Introduction

"Fatty acids" (FAs) are molecules with long carbon chains in which the hydrogen atoms are connected to carbon atom bonds along with chains bearing carboxyl (-CO) as a functional group. Fatty acids are the building blocks of "fats" and "oils. "Saturated" fats are simple straight chains of carbon (C) atoms having alkanes, i.e., only single bonds between C-atoms, whereas unsaturated fats have alkenes or alkynes, i.e., double or more bonds. As a result, polyunsaturated fatty acids have more than two double bonds between C-atoms while monounsaturated fatty acids only have one double bond [1].

Natural fats and oils are mainly composed of triacylglycerols (TAGs), and the composition and molecular structure of TAGs differ from the physico-chemical and determines functional and nutritional properties of fatty acids, i.e., saturated, mono or polyunsaturated [2]. The suitability of fats and oils are usually determined by their physico-chemical properties. The use of the words "oil and fat" in most languages recognizes the property of being liquid or solid at ambient temperature as the basis for distinguishing between these two classes of products [3]. The molecular structure, melting and crystallization behaviour, and rheological properties of fats and oils are important for the rational design of lipid materials [4]. It is well known that the properties of fats and oils mainly depend on their triacylglycerol composition, i.e., the unsaturation degree and carbon chain length of the TAGs component fatty acid (FA) molecules [5].

Oils and fats are organic compounds with carboxylic acids as active elements. As a renewable source, oils and fats from plant seeds and animal tissues are used to manufacture petroleum products in industries [6]. Fats and oils are a "great source of energy (9 kcal/g) and essential fatty acids such as linoleic acid (18:2), linolenic acid (18:3), EPA (20:5) and DHA (22:6) to humans" [7]. Fats have a vital role in meat products, influencing organoleptic like palatability, texture and flavour, and technological aspects like cooking loss, emulsion stability, water holding capacity and rheological properties. In comparison to animal fats, vegetable oils have a higher ratio of unsaturated to saturated fatty acids and do not include cholesterol [8]. Being a source of important fatty acids and fat-soluble bioactive substances, vegetable oils and fats are a crucial part of the human diet.

Palm oil, sunflower oil, rapeseed oil and soybean oil are the four edible vegetable oils that are traded the most on a global scale. Interest in alternate sources of vegetable oils has developed due to the current demand for oils and fats for human consumption, with an emphasis on nutritious eating and new industrial uses [9]. Studies on the chemical and physical characteristics of oils and fats produced from distinctive oil plants have been done in order to identify substitute lipid sources with high nutritional value as well as to enhance commercial applications [10]. Due to the presence of carotenoids and chlorophyll pigments, vegetable oils are often clear and have a yellow or green colour [7]. Dietary sources of fats that play significant and varied roles in biological processes include vegetable oils and animal fats. Triacylglycerols are important components of oils and fats, and the profiles of various oils and fats varies significantly. For nutritional evaluation, quality assurance, and ensuring the safety of oils and fats, TAG profiles are required [11].

According to O'Brien (2009), "edible oils and fats consisting of esters of glycerine and fatty acids (>90%) are structurally distinguished from triglycerol based on their chain length, the position of the double and cis/trans bonds as well as the relative ratios of saturated and unsaturated fatty acids, i.e., number and position of double bonds" [9]. In human nutrition, meats, fats, edible oils, fish, dairy products and nuts are the main sources of lipids in the human diet. Researchers have found that "dietary fatty acids play an important role in cholesterol metabolism and thus may be associated with cardio-vascular disease (CVD)" [12]. As we gain more knowledge about the relation between dietary fats and fatty acids and chronic diseases, particularly coronary artery disease (CAD), the significance of edible fats and oils in health and illness continues to develop [13].

In total, India grows nine different oil seed crops, seven of which are used to make edible oils, including sunflower, safflower, oilseed rape (mustard), niger, sesame, soybean, and peanut. Linseed and Castor, on the other hand, are non-edible oil seeds. Significant amounts of edible oil are also produced by maize germ, cotton seed, and rice bran [13]. Fats and oils are widely used in domestic and industrial settings, as well as being the most frequent contaminants in wastewater. Because they are typically removed by physical processes in sewage treatment plants, these substances are left behind in large quantities that are challenging to discard and process [13]. Fats and oils are significant organic components of municipal and industrial wastewaters, although it is unclear how they behave exactly during the treatment process [14]. Although recent advances in this sector have made it possible to remove a sizable part of oil from wastewater, biological treatment of oily wastewater is not well developed [15]. Anaerobic digesters of various designs have been researched for the treatment of sludge and oil simultaneously. For instance, anaerobic co-digestion of food waste, oils and fats with sewage sludge [16], long-chain fatty acids of oils and fats with municipal sludge [17], and aerobic co-digestion of sludge with oils and fats were also examined [18, 19].

1.2 Classification of Oils and Fats

Essential nutrients in a good diet include oils and fats. The terms oil and fat differ by the states of triglycerides, i.e., at room temperature, liquid triglyceride is termed as oils whereas at solid state it is fat [20]. Most commonly used oils and fats have long chain fatty acids, i.e., typically 8 to 20 carbon atoms, mostly having an even number of carbons; however, an odd number of carbon long chains is found mostly in animal fats. Such fatty acids contain -CH_3 (methyl) group at one end and -COOH (carboxylic acid) group at other end. An ester bond of triacylglycerol is formed by the reaction of this carboxylic group with the hydroxyl (-OH) group present on the glycerol molecule [20].

- Fats and oils
  - Long chain
    - Long chain saturated 14-18
      - Myristic acid
      - Palmitic acid
      - Stearic acid
        - Butter lard
        - Tallow palm
        - Salmon
    - Mono-unsaturated
      - Omega-9
        - Olive
    - Poly-unsaturated
      - Omega-6
        - Corn Soyabean, Sunflower
      - Omega-3
        - Linseed and Fish
  - Medium chain
    - Medium-chain saturated 8-12
      - Caprylic acid
      - Caproic acid
      - Capric acid
      - Lauric acid
        - Coconut

Figure 1.1 Classification of fats and oils.

The three main groups of fatty acids (Figure 1.1) can be divided according to chain length:

1. Long-chain fatty acid
2. Medium-chain fatty acid
3. Short-chain fatty acid

1.2.1 Long-Chain Fatty Acid

Long-chain fatty acids (LCFAs) comprise 13 to 21 carbon atoms having an aliphatic end, as a result of hydrolysis of neutral triglycerides. Palmitate, stearate, and oleate are the three most abundant LCFAs present in the system where digestion takes place in the absence of oxygen termed as anaerobic digesters [21].

1.2.1.1 Long-Chain Saturated Fatty Acid

Saturated fats are solid at room temperature, which is why they are also called solid fat. It is primarily present in "foods of animal origin, like milk, cheese, and meat. Poultry and fish contain less saturated fat than red meat [22]. Saturated fats are also found in tropical oils, such as coconut oil, palm oil, and cocoa butter. Tropical oils like saturated fatty acids are also found in many snacks and non-dairy foods, such as coffee creamer and whipped toppings. Foods made with butter, shortening (pastries and cakes, cookies, and other similar desserts) or margarine, have higher saturated fats which can elevate cholesterol levels. A healthy diet contains less than 10% of daily calories from saturated fat" [22].

Below are the most common long-chain saturated fatty acids.

1.2.1.1.1 Stearic Acid (18 Carbon Atoms)

Stearic acid is a long-chain saturated fatty acid with an 18-C framework and molecular formula $C_{18}H_{36}O_2$ (Octadecanoic acid), Figure 1.2 [23]. It is waxy-solid in texture. Stearic acid is present in "fats and oils of various plants and animals and is one of the useful types of saturated fatty acids, and a major constituent of cocoa butter and shea butter" [23].

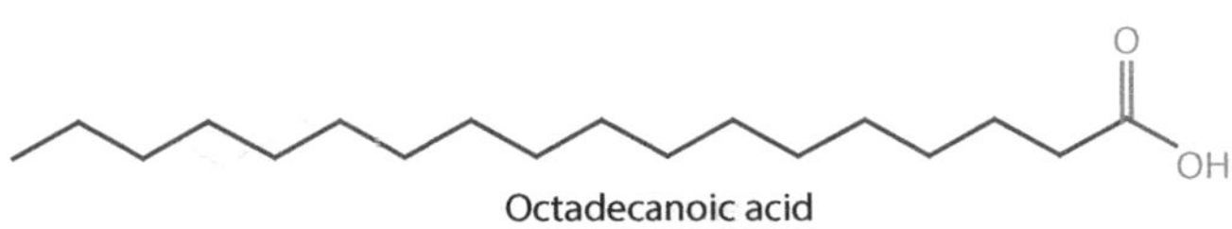

Figure 1.2 Structure of Octadecanoic acid.

1.2.1.1.2 Palmitic Acids (16 Carbon Atoms)

Palmitic acid is a long-chain saturated fatty acid with a 16-C framework and molecular formula $C_{16}H_{32}O_2$ (Hexadecanoic acid), Figure 1.3. Palmitic acid is naturally present in "palm oil and palm kernel oil, milk, butter, cheese, and meat" [24].

1.2.1.1.3 Myristic Acid (14 Carbon Atoms)

The saturated long-chain fatty acid having 14-C framework forms a myristic acid with molecular formula $C_{14}H_{28}O_2$ (Tetradecanoic acid), Figure 1.4. The sources of myristic acid are palm oil, coconut oil, and butter fat [25, 26].

1.2.1.2 Unsaturated Fatty Acid

At normal temperature, fats and unsaturated fatty acids are liquid. This type of fatty acid is generally present in oils from plants and helps to improve cholesterol levels [27]. Unsaturated fats are further classified as Monounsaturated fat and Polyunsaturated fat.

1.2.1.2.1 Monounsaturated Fat

The sources of monounsaturated fats are "avocados, nuts, and vegetable oils, like olive, peanut oil, and canola. Eating foods that are high in such fat can help lower your 'bad' low density lipoprotein (LDL) cholesterol. Thus fat can also maintain 'good' high density lipoprotein (HDL) cholesterol levels high. However, by only consuming higher unsaturated fat without reducing saturated fat may not lower your cholesterol" [27].

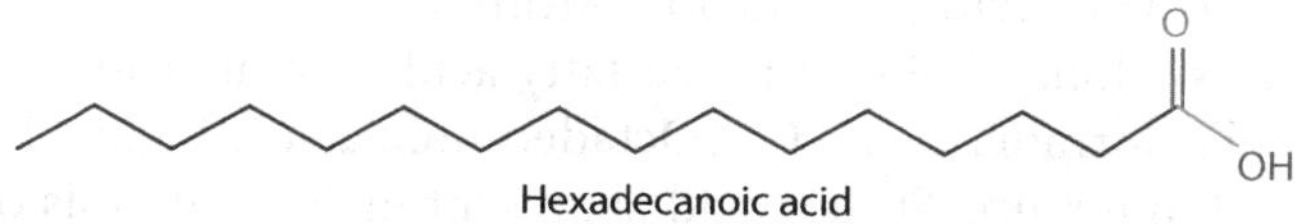

Figure 1.3 Structure of Hexadecanoic acid.

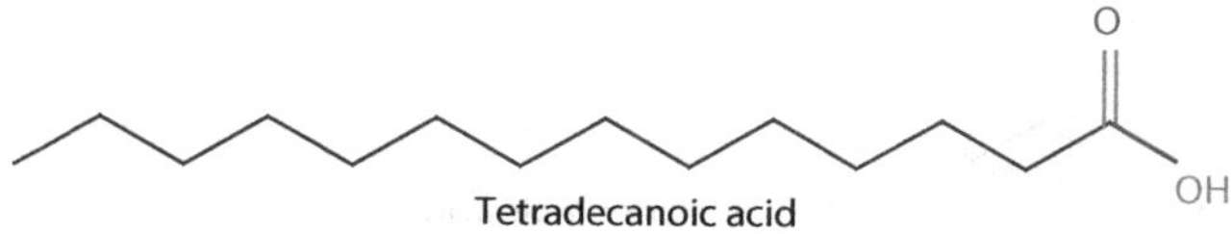

Figure 1.4 Structure of Tetradecanoic acid.

Example: **Omega-9** (or n-9) fatty acids, that contain first double bond at 9th C-atom and include mainly oleic acid [27].

1.2.1.2.2 Polyunsaturated Fat

Polyunsaturated fat is mainly stored in "vegetable oils such as safflower, corn oils, soybean, sesame and sunflower. It is also the main fat present in seafood" [28]. Low-density lipoprotein cholesterol may be decreased by polyunsaturated fats rather than saturated fats. Omega-6, omega-9, and omega-3 fatty acids are three common polyunsaturated fats [27, 28].

Examples:

Omega-3 fatty acids are found in "foods from plants like soybean oil, canola oil, walnuts, and flaxseed. They are also found in fatty fish and shellfish as Eicosapentaenoic acid (EPA) and Docosahexaenoic acid (DHA)" [22, 27, 29].

Omega-6 fatty acids are found mainly in "liquid vegetable oils like corn oil, safflower oil, and soybean oil" [27].

An unsaturated fatty acid is further categorized in two forms, i.e., bent form "*cis*" and straight form "*trans*", based on the position of H-bond, i.e., whether it is on the same, or the opposite side of the molecule. Mostly unsaturated fatty acids are in the *cis* form [30].

1.2.1.3 Cis-Unsaturated Fatty Acid

Cis-unsaturated fatty acids are fatty acids having carboxylic acids attached with long aliphatic carbon chains, being two H-atoms linked by double bond on the same side of C-chain. This type of configuration is known as "cis configuration of unsaturated fatty acids" [31, 32].

Sources of cis-unsaturated fatty acids include "palm oil, nuts and seeds, avocados, and animal fats. Some vegetable oils, such as olive oil (75%), mid-oleic sunflower oil (70%), and rapeseed oil (65%), contain more than 50% of cis-MUFA" [30–32].

1.2.1.4 Trans-Unsaturated Fatty Acid

This fat has undergone a procedure known as hydrogenation. This method makes fat tougher at room temperature and extends the shelf life of the fat. Crispier crackers and flakier pie crusts are produced by harder fat [31]. Consuming more trans-fat increases the cholesterol level, hence it is advisable to consume a lesser amount of it.

Processed foods, "foods prepared with butter, some margarine, and partially hydrogenated oils, snacks including chips and crackers, cookies, certain margarine, and salad dressings are all sources of trans fatty acids" [32].

There are two categories of trans fatty acids (TFA): naturally occurring TFA (ruminant) and industrially produced artificial TFA.

Artificial trans-fats: During the hydrogenation of vegetable oils, which produces partially hydrogenated vegetable oils, artificially or industrially generated trans-fats (iTFAs) are produced. Epidemiologic and biochemical evidence is mounting that iTFAs negatively impact a number of cardio-vascular risk factors, raising the cases of coronary heart disease (CHD) globally [33].

Natural trans fats (ruminant): However, ruminant TFA is formed naturally by the enzymes of microorganisms found in the rumen of animals. This process, known as biohydrogenation, is a complicated one that results in the isomerization, hydration, or hydrogenation of dietary fatty acids that have not been esterified. After being formed, ruminant TFA is ingested and integrated into lipids found in milk and bodily tissue [34–36].

1.2.2 Medium-Chain Fatty Acid

Milk fat and coconut oil both naturally contain medium-chain fatty acids (MCFA). Coconut oil is used to make medium-chain triglycerides (MCT). In comparison to triglycerides containing fatty acids with 16 or 18 C atoms, short- or medium-chain fatty acid triglycerides have a higher fraction of carbon and hydrogen in the molecule and a lower energy density [37]. Here are a few illustrations of medium-chain fatty acids:

1.2.2.1 Lauric Acid (12 Carbon Atoms)

The medium chain saturated fatty acid having 12-C framework and molecular formula $C_{12}H_{24}O_2$ (Dodecanoic Acid), Figure 1.5 is termed as Lauric acid. "It is found naturally in various plant and animal fats and oils and is a major component of coconut oil and palm kernel oil [38]. It is an inexpensive,

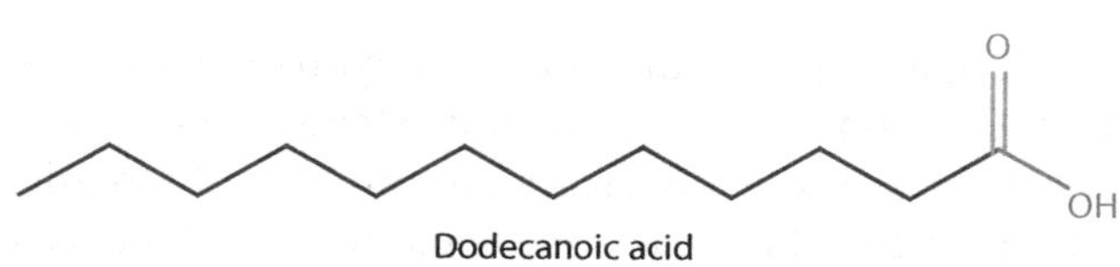

Figure 1.5 Structure of Dodecanoic acid.

non-toxic and safe-to-handle compound often used in laboratory investigations of melting-point depression. It is a solid at room temperature but melts easily in boiling water, so liquid lauric acid can be treated with various solutes and used to determine their molecular masses" [38].

1.2.2.2 Capric Acid (10 Carbon Atoms)

Medium-chain straight saturated fatty acid having 10-C framework and molecular formula $C_{10}H_{20}O_2$ (Decanoic acid), as seen in Figure 1.6, is termed as capric acid. It has a melting point of 31.5 °C and is a white, crystalline solid with a sour smell [40].

1.2.2.3 Caprylic Acid (8 Carbon Atoms)

Medium-chain saturated fatty acid having 8-C backbone and the chemical formula $C_8H_{16}O_2$ (Octanoic acid), as seen in Figure 1.7, forms caprylic acid. "Numerous mammals' milk naturally contains octanoic acid, which is also a minor component of coconut oil and palm kernel oil" [39].

1.2.3 Short-Chain Fatty Acids

Saturated fatty acids that comprise less than 6-C atoms are known as "short-chain fatty acids (SCFAs). The primary metabolic by-products of anaerobic bacterial fermentation in the intestine are short-chain fatty acids. SCFAs influence various processes in the gastrointestinal (GI) tract as well as in other tissues like adipose and immunological tissues, in addition to playing a crucial role as fuel for intestinal epithelial cells" [41].

Here are some of the most significant SCFAs:

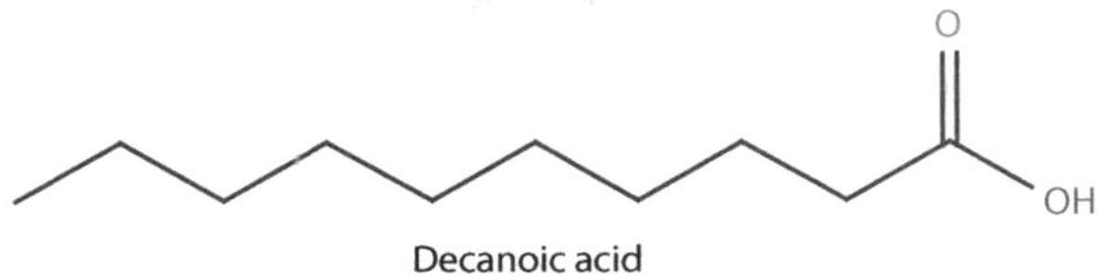

Figure 1.6 Structure of Decanoic acid.

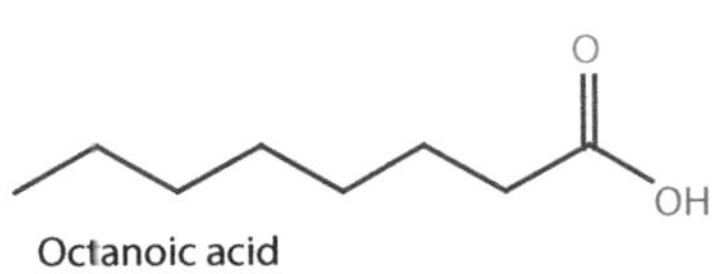

Figure 1.7 Structure of Octanoic acid.

1.2.3.1 Butyric Acid (4 Carbon Atoms)

A short-chain fatty acid having 4-C framework and molecular formula $C_4H_8O_2$(Butanoic acid) as seen in Figure 1.8, forms butyric acid. It appears as a colourless liquid with a strong, unpleasant odour. It causes corrosion to metals and tissues. It has a 70 °F flash point and 8.0 lbs/gallon density [42].

1.2.3.2 Propionic Acid (3 Carbon Atoms)

Propionic acid, with the chemical formula $C_3H_6O_2$ (Propanoic acid), is a short-chain saturated fatty acid that has ethane linked to the carbon of a carboxy group (Figure 1.9). It is a colourless liquid that emits an irritating vapour and has a strong rancid odour. It functions as an antifungal medication. It is the propionate conjugate acid [43].

1.2.3.3 Acetic Acid (2 Carbon Atoms)

Acetic acid, also referred as Ethanoic acid ($C_2H_4O_2$), is a simple monocarboxylic acid with a 2-C backbone, as seen in Figure 1.10. "Protic solvent, food acidity regulator, antibacterial food preservative, and daphnia magna metabolite" are some of its functions. It is conjugate acid of acetate [44].

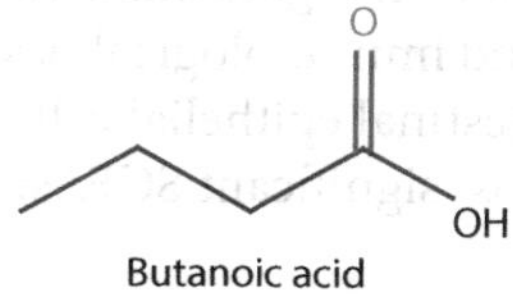

Figure 1.8 Structure of Butanoic acid.

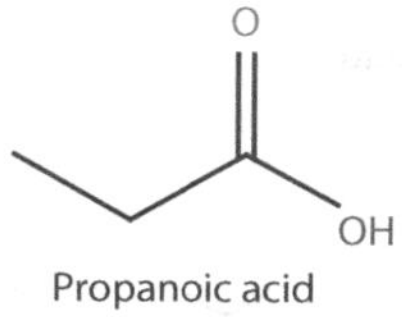

Figure 1.9 Structure of Propanoic acid.

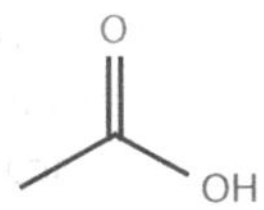

Figure 1.10 Structure of Ethanoic acid.

1.3 Chronology of the Development of Oil and Fats for Industry

Fats and lipids have been regarded as being of utmost importance throughout the history of humankind due to their worth in "food, cosmetics, and natural medicine, as well as many other household applications (such as cooking and candle wax). Vegetable oils and animal fats were used for the first time in Mesopotamia (7000 BC) and ancient Egypt (5000 BC)" [45]. They were employed in cosmetic products like body lotions and oils. These peoples began creating scented oils for mummification, personal cleanliness, medicine, and cosmetics around 2000 BC. "A novel method based on the maceration in oils of flowers, leaves, spices, resins, and occasionally pigments were invented by them [45]. There are surprisingly many different sources of oils, fats, and waxes described in contemporary records. It is evident that even in more prehistoric times, lipid knowledge was more advanced than we think because of the abundance of very common seeds like linseed and poppy seeds, native trees like cedar and palm, fruits like olives and avocados, fish, and even some remarkable animal oils like hippopotamus or crocodile oils [45]. The great Mediterranean societies of the ancient Greeks and Romans adhered to this practice. They introduced new methods for making oils and lotions, like seed pressing and distillation [46]. While the globe was obsessed with the Dark Ages between 400 and 1000 A.D., significant advancements in the use of oils and fats, particularly when applied to medicine and alchemy, were made in various regions of Europe, China, Japan, and North America" [47, 48].

European pharmacopoeia from the medieval and seventeenth centuries both attested to the therapeutic qualities of oils and fats during the Middle Ages [49]. The Industrial Revolution, the development of organic and lipid chemistry, and other factors drastically altered the way that products were produced [50]. Saponification is an illustration of the first novel technique that was widely used throughout the seventeenth and eighteenth century. "The simplest known method of producing soap was heating animal fats or oils with a powerful alkali, adding salts, and then separating the fatty acid salts and glycerol from the final combination [48–50]. The production of soap grew to be a significant industrial-scale business in the second half of the eighteenth century" [50].

As chemistry as a whole developed during the "eighteenth and nineteenth-centuries, so did our understanding of the biological characteristics and uses of fats. As a result, new industrial uses and applications for fatty acids were gradually developed". As a result, there was a constant quest for fresh and interesting natural sources of fatty acids [46]. At that time, "men began hunting sperm whales as it was termed later, a highly rich supply of fats and an industry of oils at its day. They were the largest toothed whales at the period, distinguished by their enormous skull, which was covered in a waxy material known as spermaceti [51]. This extremely valuable oil is derived from the blubber and acoustic fat bodies, which are now understood to be crucial for whale signal transmission" [52].

At the start of the nineteenth century, the spermaceti oil was attractive to the fats industry because of its adaptability and abundance; candles, soaps, and cosmetics were all made with the congealed form [51, 52]. "By the end of 1958, over 20,000 sperm whales were killed annually, and their waxy oil was used to make cattle fodder, dog food, vitamins, supplements, glue, leather preservatives, and brake fluids [51]. The whaling industry experienced a decline from 1880 until 1925 before increasing once more during World War II. Because of the sharp decline in whale populations, whaling was banned in 1982. The number of sperm whales has dropped from approximately one million to barely a few hundred thousand since whaling began. Because whales are crucial for creating phytoplankton, which recycles CO_2 from the atmosphere, this resulted in a tremendous loss for the marine ecosystem" [53].

The introduction of new technologies to the industrial manufacturing of fats and oils resulted from the development of knowledge regarding the chemistry of fats and oils in France and Germany in the late 1800s. Three substantial modifications were employed in this sector. Firstly, "switching from oil extraction to chemical solvent (hexane) extraction via mechanical crushing" was a key advancement in this field [50–53]. Secondly, to "improve storage by reducing rancidification of extracted oils by the addition of antioxidants or refrigeration" and lastly, "building on the second advancement by learning how to use hydrogenation procedures to oils, which transformed polyunsaturated oils into the texture of natural fats" [54].

Despite the large production and the export of a variety of polyunsaturated vegetable oils and fats, people of different cultures and nations still consume large amounts of saturated fat on a regular basis. The Tibetans regularly use yak butter. Due to the historical influence of Ayurvedic treatment, coconut oil, butter, and ghee are still consumed throughout the Indian subcontinent. Despite the availability of processed vegetable oils with a modern Western flavour that are less expensive, Chinese people who live in rural areas still prefer lard and other pork-based foods with saturated fats [54]. For the production of traditional soap, vegetable oils were employed. New processes for refining, bleaching, and deodorising oils as well as high-heat oil extraction procedures enabled the "creation of oils with a neutral flavour that could be utilised for a variety of cooking applications as well as soap making in the late 1800s to early 1900s. Due to their low production costs, the new industrially derived oils were extremely easily adapted for usage in domestic and commercial cooking, frying, and baking [50]. Furthermore, the low-cost margarine that was created by hydrogenation rapidly took the place of natural fats like butter, lard, and tallow in cooking and baking "recipes" [51]. Alongside these developments, better techniques for protecting and storing the novel polyunsaturated vegetable oils were created [54]. Hydrogenated oil and fat can be produced to resemble natural products in terms of look, texture, and flavour while still being inexpensive. Compared to spreading chilled butter on bread, spreading chilled margarine was significantly simpler. These characteristics made hydrogenated goods appealing, but they also appeared to have certain negative impacts on human organ systems and tissues during the course of the twentieth century [54]. The utilisation of fatty acid sources has changed recently: polyunsaturated fatty acids (PUFAs), which have been found to have positive effects on human health [55], are of particular interest to the cosmetics, pharmaceutical and food industries. "Fatty acids have also been recently valued as an innovative and green source for the production of biofuel and feedstock [55, 56]. PUFAs are mainly produced by marine phytoplankton and contained in fish and seafood, but climate change has dramatically affected the marine ecosystem. This is due to the high level of carbon dioxide emission and ultraviolet (UV) irradiation, both of which have resulted in a decrease in the growth of marine sources and reduced synthesis of PUFAs and slowed down the growth of marine source [57]. Vegetable oils are insufficient to fill the existing gap, and microbial synthesis is too expensive [58, 59]. Fish and vegetable oil sources are insufficient to meet industrial needs, so research in this area has begun to examine the genetic engineering of algae, bacteria, yeasts, seeds, and plants as bioengineered factories to produce PUFAs in greater quantities consistent with the increased global demand" [60].

1.4 Chemistry of Oil and Fats

Tachen, in the late seventeenth century, Scheele, d'Arcet, and Berthollet, in the following eras, and Michel Eugène Chevreul, in the 1823 publication "Chemical Research on Fats of Animal Origin," were the first to provide a comprehensive understanding of the chemistry of fats and oils [61, 62]. Chevreul demonstrated that "the excess mass produced by the saponification of fatty acids and glycerine was actually the consequence of the combination of water and the elements rather than the fixation of oxygen" [63]. He came to this conclusion by describing triacylglycerols, which are completely formed acids, as a combination of anhydrous glycerine and oils and fats [61, 63].

As stated earlier, fats and oils are structurally the esters of glycerol having three FAs, hence named triglycerides in the food industry, despite their scientific name of triacylglycerols. Although "oils" and "fats" are sometimes used synonymously, they differ between triglycerides in their liquid state at room temperature (referred to as "oils") and those in their solid condition (referred to as "fats") [64]. Triacylglycerols are the basic units of countless potential combinations because they are "esters of fatty acids with the trihydric alcohol glycerol (propane-1,2,3-triol)" [65, 66]. As seen in Figure 1.11, their position, i.e., R_1, R_2, and R_3 and the carbon chains they comprise in the FAs are what essentially define their chemical and physical characteristics.

The number or location of the double bonds in the fatty acids, as well as their chain length, serve as primary distinguishing factors. As a result, the same fat can include a variety of FAs, and different fats can have the same FAs [67]. Generally, they fall into one of three categories:

a) Major FAs, which are the "five most prevalent naturally occurring FAs (palmitic, stearic, oleic, linoleic, and linolenic), present in large quantities and making up nearly 95%

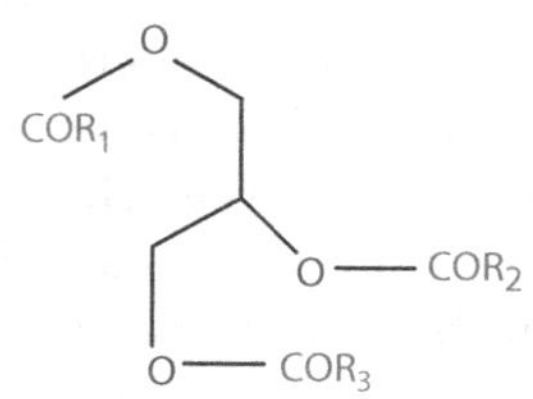

Figure 1.11 Schematic representation of triacylglycerol structure having three fatty acid molecules (-COR_1, -COR_2, and -COR_3) attached with a glycerin molecule [65].

of the FAs of oils and fats used in food products or industrial applications";

b) Minor FAs, which are "either homologous or isologues of the major ones and are present in dietary or industrial fats as secondary constituents";

c) Uncommon FAs, which are "extracted from infrequent sources and are the primary constituents" [67].

The variety of FA molecules is inversely correlated with the number of TAG species, and natural oils and fats are combinations of various forms of triacylglycerols. For instance, cocoa butter only contains three TAG species while having an FA profile that includes three major FA types [68]. In contrast, "the TAG profile of bovine milk fat, which is made up of a wide variety of FAs, contains more than one hundred species [69]. Although TAG molecules make up 95% or more of crude fats and oils, on average, they also contain a variety of minor ingredients that can have an impact on their physico-chemical properties. The greatest amounts are found in free fatty acids and mono- and di-acylglycerols (MAG and DAG, respectively)" [70].

Additionally, non-glyceride molecules often make up 2% or more of crude vegetable oils [66]. Therefore, the physical characteristics of the real-fat systems' primary constituents and the phase behaviour of the mixtures these compounds created are partly responsible for the "complicated melting, crystallization, crystal morphology, and aggregation behaviour of these systems" [71].

Polymorphism and molecular interactions may both disclose the crystallization characteristics of lipids. Lipid crystal polymorphism involves crystallization and subsequent transformation, which are its macroscopic features. Basically, the solid-solid transition temperatures and three polymorphic forms constitutes a TAG molecule [72]. Thus, the sensory properties of items made from fats are significantly influenced by the polymorphism of triglycerides. For instance, due to their significance in explaining the phase behaviour of cocoa butter, the polymorphism of particular TAG types in applications like chocolate has been examined in greater detail [70].

1.5 Properties of Oils and Fats

The sensory qualities and nutritional value of food can be impacted by the physical and chemical properties of oils and fats [74]. The physical and chemical characteristics of fats and oils are covered below.

1.5.1 Physical Properties

Contrary to popular belief, pure fats and oils have no flavour, odour, or colour. Foreign compounds that are "lipid soluble and have been absorbed by these lipids are what give some of them their distinctive colours, smells, and aromas. For instance, the presence of the pigment carotene gives butter its yellow colour, while the chemicals diacetyl and 3-hydroxy-2-butanone (Figure 1.12), which are formed by bacteria in the maturing cream used to make butter, give it its flavour" [75].

Water is heavier than fats and oils, which have densities of roughly 0.8 g/cm3. They act as great "insulators for the body, limiting the loss of heat via the skin since they are poor heat- and electricity-conductors" [73, 75].

1.5.2 Chemical Properties

Oils and fats can participate in a wide range of chemical processes. Due to being "triglycerides as esters, can get hydrolyzed in the presence of an acid, a base, or certain enzymes called lipases. Making soap is a process known as saponification, which involves hydrolyzing fats and oils in the presence of a base", as shown in Figure 1.13 [75].

Today, the majority of soaps are made by hydrolyzing triglycerides (typically derived from coconut oil, tallow or both) with water at a high pressure of 700 lb/in2 (about 50 atm) and temperature of 200°C [75]. Then, using sodium carbonate or sodium hydroxide, the fatty acids are transformed into their sodium salts (soap molecules):

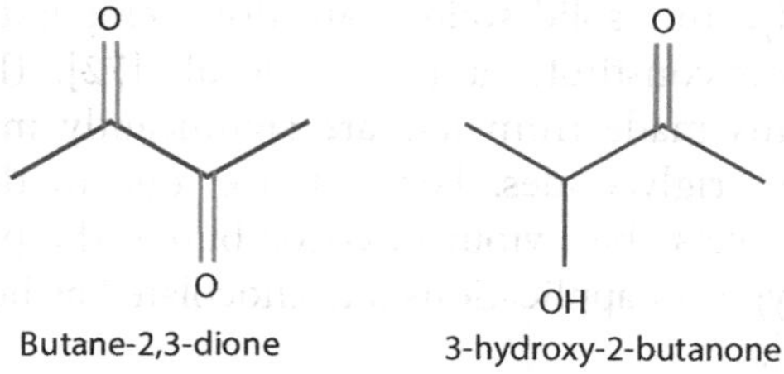

Figure 1.12 Structure of Butane-2,3-dione and 3-hydroxy-2-butanone.

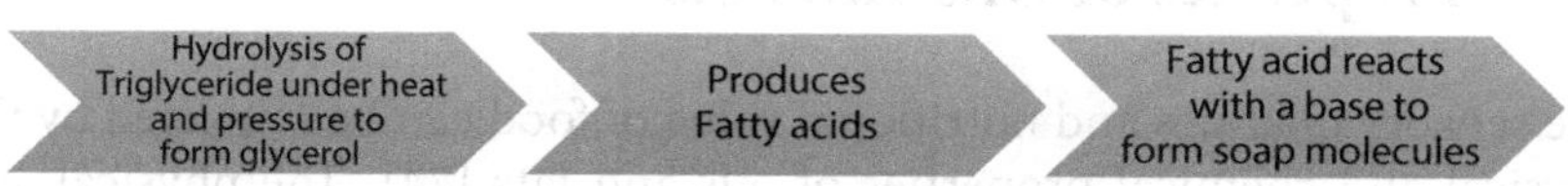

Figure 1.13 Process of Saponification.

One of the earliest organic syntheses used by humans, ordinary soap is a "combination of the sodium salts of several fatty acids (second only to the fermentation of sugars to produce ethyl alcohol). Romans and Phoenicians (600 BCE) both used animal fat and wood ash to make soap. Even so, it wasn't until the 1700s that soap started to be produced widely. In the past, vast open vats of molten lard or tallow were treated with a tiny overabundance of alkali to create soap. Steam was emitted from the heated mixture as it was being stirred" [73].

After the process of saponification was finished, sodium chloride (NaCl) was added to the mixture to precipitate the soap, which was then removed by filtration and repeatedly rinsed with water [73–75]. Then, extra NaCl was used to reprecipitate it after it had been dissolved in water. The aqueous wash solutions were also used to recover the glycerol that was created during the process. To make scouring soap, pumice or sand are added, and to make fragrant, coloured soaps, scents or dyes are added. A floating soap is created by blowing air through molten soap. Although more expensive, soft soaps manufactured with potassium salts have a finer lather and are more soluble. Shaving creams, shampoos, and liquid soaps all include them [75]. Typically, "body oils, cooking fats, lubricating greases, and other chemicals that function as glues combine with dirt and grime to attach to skin, clothing, and other surfaces. These compounds cannot be removed by washing with water alone since water does not dissolve them. However, soap eliminates them because of the dual nature of soap molecules. The tail dissolves in oils, while the head, which carries an ionic charge (a carboxylate anion), carries an ionic charge and dissolves in water [70–73]. The ionic heads stay in the aqueous phase, the hydrocarbon tails break down in the soil, and the soap transforms the oil into micelles, which are tiny droplets of oil that disseminate throughout the solution. Due to the charged surfaces of droplets, they repel one another and do not agglomerate. The soap-enclosed dirt can be readily rinsed away since the oil is no longer 'gluing' the dirt to the soiled surface (skin, fabric, or dish)" [75].

In industrial procedures, the quantity of hydrogenated double bonds is meticulously regulated to provide fats with the correct consistency (pliable and soft). Thus, "canola, corn, and soybean oils-cheap and plentiful-are converted into margarine and cooking fats. For instance, to make margarine, partially hydrogenated oils are combined with water, salt, nonfat dry milk, flavourings, colorings, and vitamins A and D to mimic the appearance, flavour, and nutritional benefits of butter (preservatives and antioxidants are also added) [75]. To keep it from splitting out, the peanut oil in most commercial peanut butter has been partially hydrogenated. By using the original, unprocessed oils on their dishes, consumers might reduce the quantity of saturated fat in their diet, but most people would rather spread margarine on their toast than pour oil on it" [75].

1.6 Applications of Oils and Fats

1.6.1 Oils and Fats in Surfactants Industry

Surfactants are used in virtually every step of oil exploration, exploitation, and processing [76]. An organic chemical known as surfactant, sometimes known as a "surface active agent," is amphiphilic and contains both hydrophilic groups (polar heads) and hydrophobic groups (nonpolar tails). Both molecules can build up between fluid phases like oil/water or air/water, lowering the surface and interfacial tensions and generating micelles or emulsions. This is one of these compounds' special qualities and adaptability [77].

Surfactants are used in the formulation of drilling muds that are water- or oil-based, in hydraulic fracturing and cementing fluids, in acidizing oil wells, in increased oil recovery, in preventing corrosion, in oil-water-gas separation, and in the transportation of crude oil. Surfactants have a variety of roles in the extraction of oil and gas, as well as in the formulation of petroleum products downstream and in the clean-up of oil pollution for environmental protection [76]. Natural surfactants are found in crude oil, and their physical and chemical characteristics have both benefits and downsides. On the other hand, petroleum hydrocarbons continue to provide the majority of the hydrophobes, to be functionalized to amphiphilic compounds, as well as ethylene and propylene oxides, necessary to the synthesis of alkoxylated non-ionic [76, 78]. Carbohydrates and fats are an important source of raw materials for the manufacture of surfactants.

1.6.2 Oils and Fats in Food Industry

For many "food products, including confectionery, bakery, ice creams, emulsions and sauces, shortenings, margarine, and other specially crafted items, fats and oils are crucial raw materials and functional ingredients [79]. Fats and oils add flavour to foods and affect the order in which flavour components are released when foods are eaten, in addition to having a lubricating effect and creating a sensation of moistness in the mouth" [80]. They also contribute to tenderness in shortened cakes and by airing batter, fats help to establish texture in cakes.

Oil helps food retain heat, emulsifies or thickens sauces, prevents sticking, and gives food a crisp texture. Olive oil and butter both have a 9 calorie (kcal)/gm content. Reducing the quantity of fat in baked goods lowers calories while also reducing the amount of fat [81, 82]. The texture, stability, spreadability, and mouthfeel of fat crystals are significantly influenced by their melting profiles, as well as other characteristics. The concentration, shape, and interactions of fat crystals impact the texture of goods like chocolate, shortenings, and especially butter. Shortenings are fats that provide pies, bread, pasta, and other foods particular functional qualities (structural integrity, increased shelf-life, softness, texture, mouthfeel, heat transfer, and air incorporation) (McClements & Decker, 2010) [79].

1.6.3 Oils and Fats in Pharmaceuticals

The role that fats and oils, also known as lipids, play in metabolic processes and cell membrane structure, particularly as precursors of signalling eicosanoids and cell membrane transport agents, is one of the most crucial. "In a variety of delivery forms, such as tablets, capsules, suppositories, emulsions (enteral/parenteral), ointments, creams, and lotions, several lipids are used as fillers, binders, lubricants, solubilizers, emulsifiers, and emollients when highly purified and refined to meet pharmaceutical specifications [83]. Bioactive lipids and lipid-soluble micronutrients like fat-soluble vitamins are important in the prevention of disease" [85]. The human body uses sterols and steroids to generate vital hormones like progesterone and testosterone as well as for various therapeutic applications (Pelletier *et al.*, 1995) [84]. Sterols and steroids also have significant effects on cellular processes. The majority of popular vegetable oils include phytosterols. For instance, one of the most popular oils, soybean oil, is said to contain about 0.124% tocopherols and 0.36% sterols. The potential prevention of intestinal reabsorption of circulating cholesterol has been suggested as the mechanism by which phytosterols reduce blood cholesterol. Since stanols, the saturated form of sterols, are said to be more easily digested, they have been used commercially in food products like margarines and as nutritional supplements [83]. Sitostanol, a saturated sitosterol derivative, is the main stanol described for this purpose. Diagnostic imaging, artificial blood, gene delivery, and medical equipment are among more non-direct applications (Hernandez, 2005) [85].

1.6.4 Oils and Fats in Cosmetics

Natural lipids and oils are employed in cosmetics for a variety of reasons. The combination of skin and cosmetic lipids results in emolliency, the imparting of softness and flexibility. If emulsifiers are utilised, moisture may be retained or even added as a result. Other significant advantages of employing lipids in cosmetics include lubricity, adherence, gloss, and colouring [86]. There may occasionally be biochemical advantages. For instance, animals are unable to manufacture critical fatty acids like linoleic acid, which is then transformed to gamma linoleic acid, an anti-inflammatory drug [87].

1.6.5 Oils and Fats in Agriculture (Pesticide/Herbicide Adjuvants)

In the agriculture industry, methyl esters and plant lipids can be used in place of petroleum-based components in herbicide adjuvant formulations [88]. These ingredients are "biodegradable, non-volatile, low in toxicity, and safer for workers while enhancing herbicide activity during application and reducing the overall pesticide burden on the environment" [86].

1.6.6 Oils and Fats as Biodiesel

Vegetable oil was proposed as a possible fuel source in 1980 [87]. The idea of using edible oil as fuel implies that petroleum will be the alternative fuel instead of alcohol and vegetable oil and that non-renewable resources must be replaced by some type of renewable energy [88].

Animal fats, used cooking oil, and plant and vegetable oils can all be used to make biodiesel. Likewise, fungus, microalgae, and algae can be used to make biodiesel. However, oil-bearing plants have received the majority of attention. The selection of feedstock is the initial stage. More than 350 oil-bearing plants have been identified as possible sources for biodiesel production worldwide [89]. The accessibility of a variety of feedstocks is crucial for the manufacture of biodiesel [88, 90, 91]. Low production costs and mass production are the two primary criteria that the feedstock should meet. Biodiesel feedstock production and availability are influenced by climatic factors, regional soil characteristics, geographic locations, and agricultural methods. Four categories of biodiesel feedstocks are distinguished [88, 92–96]:

i. Waste or recycled oil.
ii. Edible vegetable oils such as rapeseed, palm, sunflower, peanut, coconut oil and soybean.
iii. Animal fats, such as by-products from fish oils, cattle tallow, yellow grease, and chicken fat [97].
iv. Vegetable oils that are not edible, including halophytes, algae, sea mango, jatropha, and karanja [95, 98].

1.6.7 Oils and Fats in Coating and Polymers

Oils and fats form the component of paints and dyes. Because of the "capacity to cross-link or polymerize on surfaces to form water-proof coatings, conjugated oils like linseed, tung, and some fish oils are referred to as drying oils". These substances are utilised as paints and varnishes when mixed with pigments [89]. The amount of fatty acids still employed in these applications is estimated to be around 450 million lbs today [90, 99].

To create reactive hydroxyl groups in epoxides, "highly unsaturated fatty acids and triglycerides are oxygenated. Epoxides are then reacted with isocyanates or other cross-linking agents to create polymeric coatings, polymers, and foams" [90].

1.6.8 Oils and Fats in Lubricants

Natural plant oils have been used as lubricants since ancient times. Although these materials are biodegradable and have great lubricating qualities, they also have poor chemical stability. Triglyceride oils are "reactive in the presence of oxygen and water at high temperatures due to the existence of numerous double bonds and ester linkages, which has sparked the development of synthetic lubricants" [86].

1.7 Challenges

Despite the many benefits and uses of fats and oils in various industries, both on the plus and minus sides, those industries face numerous difficulties. For a variety of reasons, fat is today a nutrient with great visibility. The most recent labelling suggestion for trans fatty acids comes first. Second, the highly charged argument around biotechnology revolves around lipids and oils [99]. Additionally, there is a lot of interest in small lipid components, particularly phytosterols, which have lately started to appear on the market as functional food ingredients and demonstrate specific biological affects [100, 101].

1.7.1 The Challenge for Trans-Fatty-Acids

Recent clinical investigations have demonstrated a link "between dietary trans fatty acids (TFAs) and elevated levels of total cholesterol and low-density lipoprotein (LDL) cholesterol, as well as unaltered or decreased levels of HDL cholesterol in comparison to control diets [101]. These studies' trans intakes ranged from 4–11% of total energy. In fact, there is disagreement over what "average intake" of TFA means. "On average, the U.S. population eats a mean of 5.3 grams trans-fatty acids per day - roughly 2.6% of their overall energy and 7.4% of their fat energy," according to Allison *et al.* (1999) [102].

The scientists state that this study also supports past findings showing that "saturated fatty acid (SFA) intake is still high and, at 12.5% of total energy, is more concerning than trans fatty acid intake [103]. It is crucial to remember that most trans isomers produced during partial hydrogenation of vegetable oils and fats are situated along the fatty-acid backbone at different locations (mostly elaidic) than those that naturally exist in animal fats (vaccenic). 20–25% of the TFA intake is apparently made up of fats from ruminants. However, there may be differences between the relationships between trans-fat from vegetable and animal sources and risk factors for heart disease. Many nutritionists advise against substituting saturates for TFAs in the diet [104]. The American Heart Association (AHA) suggests that, instead, "naturally occurring un-hydrogenated oil be utilized when possible and attempts made to substitute un-hydrogenated oil for hydrogenated or saturated fat in processed meals. Additionally, it seems fair to suggest using softer margarines and cooking fats instead of tougher ones. By asterisking the total and listing the grammes of TFAs at the bottom of the panel, the FDA suggests adding TFAs in the saturated fatty acid line of the Nutrition Facts panel" [104]. Foods containing "more than 4 gms of combined saturated and trans-fat per serving would not be allowed to make health claims" [101]. Additionally, the EPA suggests trans limitations for goods that make nutritional content claims [102–104]:

- "Low saturated fat" claims are only valid when a serving contains no more than 0.5 grams of trans fat in addition to the present restriction of 1 grammes or less of saturated fat.
- Only when there is a combined reduction of at least 25% in saturated and trans fats, in addition to the present minimum of 25% less saturated fat, are "reduced saturated fat" claims allowed [104].
- Instead of the current requirement of 2 gm or less saturated fat, cholesterol claims will only be allowed when the total amount of saturated and trans fat is 2 gm or less.
- Only foods that meet the current limits for total fat and cholesterol, as well as having less than 4.5 grammes of saturated and trans fats combined, are allowed to make "lean" claims [103].

- "Extra lean" claims are only accepted for foods that meet the current requirements for total fat and cholesterol, plus fewer than 2 grams of saturated and trans fats combined.
- The proposal includes a new "trans-free fat" claim that could be used on products that have less than 0.5 g of trans fat and less than 0.5 g of saturated fat per serving [101, 104].

Additionally, "the food industry has actively reduced the amount of saturated fats in processed foods, but this change has not been straightforward because solid fats are crucial for the shelf-life and acceptability of meals. For instance, the solid fat content of dairy, meat and some plant meals, like chocolate, greatly influences their textural qualities. The sensation of solid fat melting away and turning into liquid oil in the mouth is especially imparted by solid fats [102]. Solid fats are crucial in baking because they improve the incorporation of air into products, resulting in fluffy textures (such as cookies), and they prevent the production of gluten, resulting in flaky textures. Simple substitutions of unsaturated for saturated fats also present issues since the rancidity of unsaturated fats results in food waste. A long-term trend to shift the kind of fats in foods is reflected in the current decrease in saturated fats in the diet [105]. For instance, prior to the widespread availability of vegetable oils, animal fats like lard, butter and tallow were used to manufacture the majority of foods (the original fat in fast food French fries).

However, nutritionists believed in the late 1950s and early 1960s that consuming cholesterol was linked to heart disease, which led the food industry to reduce the amount of animal fats in many meals [106]. Animal fats had to be replaced because they serve so many crucial functions in diet, and in this case, tropical oils predominated (coconut, palm kernel and palm). Because they include a lot of saturated fatty acids (coconut oil has 85% of them, for example), tropical fats are solid. The food industry moved to partially hydrogenated fats, though, in the early 1980s as a result of new research that revealed saturated fats were strongly linked to heart disease [102–104]. Because partially hydrogenated fats are solid at room temperature but can be modified to have high levels of unsaturated fatty acids, they have a number of benefits. For example, partial hydrogenation can be used to remove linolenic acid from soybean oil, making it much less likely to become rancid. Later studies revealed that trans fatty acids produced during hydrogenation increased LDL cholesterol in a manner similar to that of saturated fatty acids while simultaneously lowering "good" HDL cholesterol, potentially making them worse than saturated fatty acids [105–108].

Trans fatty acids must now be listed on the nutrition fact panel by law starting in 2006 due to the potential health risk they bring. As a result, food producers drastically cut back on the use of hydrogenated fats in their products. This measure caused a significant drop in trans-fat consumption (>60%) in a relatively short period of time. To further reduce trans fatty acids in the food supply, the FDA eliminated the "generally regarded as safe" designation from partly hydrogenated fat in 2013 [107]. This modification was made because certain items continued to utilise partly hydrogenated oils in spite of the labelling law. Unless businesses have FDA approval, partly hydrogenated oil from food must be removed by 2016 [108]. According to the FDA, further banning trans fats might avert an additional 20,000 heart attacks and 7,000 heart-related deaths annually. Sadly, there is no proof that eliminating the extremely low amounts of trans from the diet will have a positive impact on health.

1.8 Conclusion

An essential component of a healthy diet is fat and oil. Perhaps more than any other nutrient, fats are misunderstood, and this misperception may lead people to make eating decisions that either limit their access to beneficial sources of fats or overconsume bad types. As was already mentioned, the proportions of various fats in the human diet have distinct effects. "Trans-fatty acids were shown to be the most detrimental of all the fatty acids in a study because they increased the levels of bad LDL cholesterol and decreased the levels of good HDL cholesterol. Furthermore, because a significant portion of those saturated fatty acids are stearic acid, it can be inferred from the effects of individual saturated fatty acids on blood cholesterol levels that fats like cocoa butter and shea butter, although saturated, are not as bad for cardiovascular health as would be predicted from their saturated fatty acid levels [109]. In contrast, cis-unsaturated fatty acids raise the good HDL cholesterol and lower the bad LDL cholesterol in the blood, which is helpful for blood cholesterol levels. But now that partly hydrogenated fat is prohibited from being used in food, the food sector needs to look for a new fat substitute. High oleic oils and interesterified fats are probably the ones to cause this.

References

1. Roccisano, D., Kumaratilake, J., Saniotis, A. & Henneberg, M. (2016). Dietary Fats and Oils: Some Evolutionary and Historical Perspectives Concerning

Edible Lipids for Human Consumption. *Food and Nutrition Sciences*, 7, 689-702. doi: 10.4236/fns.2016.78070].
2. Sivakanthan, S., & Madhujith, T. (2020). *Current trends in applications of enzymatic interesterification of fats and oils: A review. LWT, 109880.* doi: 10.1016/j.lwt.2020.109880].
3. Timms, R.E. (1985). Physical properties of oils and mixtures of oils, *J. Am. Oil Chem. Soc.* 62(2), 241-249, https://doi.org/10.1007/BF02541385.
4. Marangoni, A.G., Acevedo, N., Maleky, F., Co, E., Peyronel, F., Mazzanti, G., Quinn, B., Pink, D. (2012). Structure and functionality of edible fats, *Soft Matter*, 8, 1275e1300, https://doi.org/10.1039/C1SM06234D.
5. Lawson, H. W. (1995). *Food oils and fats: technology, utilization and nutrition.* Springer Science & Business Media.
6. Tripathy, D. B., Yadav, A., & Mishra, A. (2017). Applications of petrochemicals: A mini review. *Recent Adv. Petrochem. Sci, 2*, 67-69.
7. Endo, Y. (2018). Analytical Methods to Evaluate the Quality of Edible Fats and Oils: The JOCS Standard Methods for Analysis of Fats, Oils and Related Materials (2013) and Advanced Methods. *Journal of Oleo Science,* 67(1), 1-10. doi:10.5650/jos.ess17130.
8. Selani, M. M., Shirado, G. A., Margiotta, G. B., Saldana, E., Spada, F. P., Piedade, S. M. & Canniatti-Brazaca, S. G. (2016). Effects of pineapple by-product and canola oil as fat replacers on physicochemical and sensory qualities of a low-fat beef burger. *Meat Science*, 112, 69-76.
9. Dorni, C., Sharma, P., Saikia, G., & Longvah, T. (2018). Fatty acid profile of edible oils and fats consumed in India. *Food Chemistry*, 238, 9-15. doi: 10.1016/j.foodchem.2017.05.07.
10. Serra, Lima, J. *et al.* (2019). Alternative sources of oils and fats from Amazonian plants: Fatty acids, methyl tocols, total carotenoids, and chemical composition. *Food Research International* 116: 12-19.
11. Xu, Shu-ling, *et al.* (2018). Research advances based on mass spectrometry for profiling of triacylglycerols in oils and fats and their applications. *Electrophoresis* 39.13: 1558-1568.
12. Huth, P. J., Fulgoni III, V. L., & Larson, B. T. (2015). A systematic review of high-oleic vegetable oil substitutions for other fats and oils on cardiovascular disease risk factors: implications for novel high-oleic soybean oils. *Advances in Nutrition*, 6(6), 674-693.
13. Dorni, C., Sharma, P., Saikia, G., & Longvah, T. (2017). Fatty acid profile of edible oils and fats consumed in India. *Food Chemistry,* S0308814617308634. doi:10.101. foodchem.2017.05.072.
14. Chipasa, K. & Medrzycka, K. (2006). Behavior of lipids in biological wastewater treatment processes. *J. Ind. Microbiol. Biotechnol*, 33, 635-645.
15. Jamaly, S., Giwa, A. & Hasan, S. (2015). Recent improvements in oily wastewater treatment: Progress, challenges and future, opportunities. *J. Environ. Sci.*, 37, 15-30.

16. Li, C., Champagne, P., Anderson, B. (2015). Enhanced biogas production from anaerobic co-digestion of municipal wastewater treatment sludge and fat, oil, and grease (FOG) by a modified two-stage thermophilic digester system with selected thermo-chemical pre-treatment. *Renew. Energy.*, 83, 474-482.
17. Awe, O., Zhao, Y., Zhou, A., Minh, P. & Lyczko, N. (2018). Anaerobic co-digestion of food waste and FOG, with sewage sludge – realizing its potential in Ireland. *Int. J. Environ. Stud.*, 75, 496-517.
18. Pintor, A., Vilar, V., Botelho, C. & Boaventura, R. (2016). Oil and grease removal from wastewaters: Sorption treatment as an alternative to state-of-the-art technologies. A critical review. *Chem. Eng. J.* 297, 229-255.
19. Ziels, R., Karlsson, A., Beck, D., Ejlertsson, J., Yekta, S., Bjorn, A. & Svensson, B. Microbial community adaptation influences long-chain fatty acid conversion during anaerobic co-digestion of fats, oils, and grease with municipal sludge. *Water Res.*, 103, 372-382.
20. Oils and Fats (2017). Institute of Food Science + Technology. Available at: https://www.ifst.org/resources/information-statements/oils-and-fats
21. Labatut, R. A., & Pronto, J. L. (2018). Sustainable Waste-to-Energy Technologies: Anaerobic Digestion. In *Sustainable Food Waste-to-Energy Systems*, 47-67. doi:10.1016/b978-0-12-811157-4.00004-8.
22. U.S. Department of Health and Human Services, U.S. Department of Agriculture (2015). 2015-2020 Dietary Guidelines for Americans, 8th ed. http://health.gov/dietaryguidelines/2015/guidelines/. Accessed January 12, 2016.
23. National Center for Biotechnology Information (2022). PubChem Compound Summary for CID 5281, Stearic acid. Available at: https://pubchem.ncbi.nlm.nih.gov/compound/Stearic-acid#section=2D-Structure
24. National Center for Biotechnology Information (2022). PubChem Compound Summary for CID 985, Palmitic acid.
25. National Center for Biotechnology Information (2022). PubChem Compound Summary for CID 11005, Myristic acid.
26. Moll, J. (2022). Saturated vs. Unsaturated Fats. *Nutrition*. Available at: https://www.verywellhealth.com/difference-between-saturated-fats-and-unsaturated-fats-697517.
27. Romito, K. & O'Brien, R. (2019). Types of Fats, available at: https://wa.kaiser-permanente.org/kbase/topic.jhtml?docId=aa160619
28. Functions, Classification, and Characteristics of Fats (2014). *Food facts for healthy choices*. Available at: https://www.eufic.org/en/whats-in-food/article/facts-on-fats-the-basics#:~:text=in%20the%20body).,Classification%20of%20unsaturated%20fatty%20acids%20(cis%20and%20trans),are%20found%20in%20cis%20form.
29. U.S. Department of Health and Human Services, U.S. Department of Agriculture (2010). Dietary Guidelines for Americans. 7th ed. Washington,

DC: U.S. Government Printing Office. Available at: http://health.gov/dietaryguidelines/2010.asp

30. Joris, P.J. & Mensink, R.P. (2016). Role of cis-Monounsaturated Fatty Acids in the Prevention of Coronary Heart Disease. *CurrAtheroscler Rep.* 18 (7):38. doi: 10.1007/s11883-016-0597-y. PMID: 27221500; PMCID: PMC4879159.
31. Madhu(2018),DifferencebetweenCisandTransfattyacids,Availableat: https://www.differencebetween.com/difference-between-cis-and-trans-fatty-acids/
32. "The Kinds of Fats." Indiana University Bloomington, The Lilly Library, Indiana University.
33. Firdaus, M. A., & Mishra, S. (2021). Industrial Trans-Fatty Acid Intake Associated with Coronary Heart Disease Risk: A Review. *J Nutr Food Sci.* 11: 806.
34. Bessa, R. J. B., Alves, S. P. & Santos-Silva, J. (2015). Constraints and potentials for the nutritional modulation of the fatty acid composition of ruminant meat. *Eur. J. Lipid Sci. Technol.* 117, 1325-1344.
35. Mapiye, C., Vahmani, P., Mlambo, V., Muchenje, V., Dzama, K., Hoffman, L.C. & Dugan, M.E.R. (2015). The trans-octadecenoic fatty acid profile of beef: implications for global food and nutrition security. *Food Res. Int.* 76, 992-1000.
36. Verneque, B.J.F., Machado, A.M., Silva, L. de A., Lopes, A.C.S. & Duarte, C. K. (2022). Ruminant and industrial trans-fatty acids consumption and cardiometabolic risk markers: A systematic review. *Critical Reviews in Food Science and Nutrition* 62.8: 2050-2060.
37. De Vrese, M., Pfeuffer, M., Roos, N., Scholz-Ahrens, K., & Schrezenmeir, J. (2010). The health aspects of milk. Improving the Safety and Quality of Milk, 28–73. doi:10.1533/9781845699437.1.28
38. National Center for Biotechnology Information. PubChem Compound Summary for CID 3893, Lauric acid. https://pubchem.ncbi.nlm.nih.gov/compound/Lauric-acid.
39. National Center for Biotechnology Information. PubChem Compound Summary for CID 379, Octanoic acid. https://pubchem.ncbi.nlm.nih.gov/compound/Octanoic-acid.
40. National Center for Biotechnology Information. PubChem Compound Summary for CID 2969, Decanoic acid. https://pubchem.ncbi.nlm.nih.gov/compound/Decanoic-acid.
41. Vinolo, M. A. R., Rodrigues, H. G., Nachbar, R. T., & Curi, R. (2013). Modulation of inflammatory and immune responses by short-chain fatty acids. Diet, Immunity and Inflammation, 435-458. doi:10.1533/9780857095749.3.435.
42. National Center for Biotechnology Information (2022). PubChem Compound Summary for CID 264, Butyric acid.
43. National Center for Biotechnology Information (2022). PubChem Compound Summary for CID 1032, Propionic acid.

44. National Center for Biotechnology Information (2022). PubChem Compound Summary for CID 176, Acetic Acid.
45. DeNavarre M.G. (1978). Oils and fats, the historical cosmetics. *J Am Oil Chem Soc.* 55:435-7. 10.1007/BF02911908.
46. Gidez L.I. (1984). The lore of lipids. *J Lipid Res.* 25:1430-6. Doi: 10.1016/S0022-2275(20)34415-1.
47. Gwei-Djen, L. & Needham, J. (1964). Medieval preparations of urinary steroid hormones. Med Hist. 8:101-21. Doi: 10.1017/S0025727300029355
48. Wu, L.C., Davis, T.L. & Po-Yang, W. (1932). An ancient Chinese treatise on alchemy entitled Ts'anT'ung Ch'i. *Isis.* 18:210-89. 10.1086/346707.
49. Brockbank, W. (1964). Sovereign remedies: a critical depreciation of the 17th-century London pharmacopeia. *Med Hist.* 8:1-14. Doi: 10.1017/S0025727300029057.
50. Stahlman W.D. (1961). A short history of technology from the earliest times to A.D. 1900. *Science.* 134:183–3. 10.1126/science.134.3473.183.
51. Wilson, D.E. & Ruff, S. (1999). American Society of Mammalogists Smithsonian Institution. *The Smithsonian Book of North American Mammals.* Vancouver, BC: UBC Press.
52. Koopman, H.N. (2018). Function and evolution of specialized endogenous lipids in toothed whales. J Exp Biol. 221 (Suppl. 1): jeb161471. 10.1242/jeb.161471.
53. Pitman, R.L., Balance, L.T., Mesnick, S.I. & Chivers, S.J. (2001). Killer whale predation on sperm whales: observations and implications. Marine Mammal Sci. 17:494-507. 10.1111/j.1748-7692.2001.tb01000.x.
54. Roccisano, D., Kumaratilake, J., Saniotis, A. & Henneberg, M. (2016). Dietary Fats and Oils: Some Evolutionary and Historical Perspectives Concerning Edible Lipids for Human Consumption. *Food and Nutrition Sciences*, 7, 689-702. doi: 10.4236/fns.2016.78070
55. Yashodhara, B.M., Umakanth, S., Pappachan, J.M., Bhat, S.K., Kamath, R. & Choo, B.H. (2009). Omega-3 fatty acids: a comprehensive review of their role in health and disease. *Postgrad Med J.* 85:84-90. 10.1136/pgmj.2008.073338.
56. Lestari, S., Maki-Arvela, P., Beltramini, J., Lu, G.M. & Murzin, D.Y. (2009). Transforming Triglycerides and fatty acids into biofuels. *Chem Sus Chem.* 2:1109-19. 10.1002/cssc.200900107.
57. Kang, J.X. (2011). Omega-3: a link between global climate change and human health. *Biotechnol Adv.* 29:388-90. 10.1016/j.biotechadv.2011.02.003.
58. Kojima, Y., Parcell, J., Cain, J., Kojima, Y., Parcell, J. & Cain, J. (2016). *A Global Demand Analysis of Vegetable Oils for Food and Industrial Use: A Cross-Country Panel Data Analysis with Spatial Econometrics.* Available at: https://ageconsearch.umn.edu/record/235744.
59. Christophe, G., Kumar, V., Nouaille, R., Gaudet, G., Fontanille, P., Pandey, A., *et al.* (2012). Recent developments in microbial oils production: a possible alternative to vegetable oils for biodiesel without competition with human food? *Braz Arch Biol Technol.* 55:29-46. 10.1590/S1516-89132012000100004.

60. Lee, J., Lee, H., Kang, S. & Park, W. (2016). Fatty acid desaturases, polyunsaturated fatty acid regulation, and biotechnological advances. *Nutrients.* 8:23. 10.3390/nu8010023.
61. List, G.R., Kenar, J.A. & Moser, B.R. (2017). History of fatty acids chemistry, in: M.U. Ahmad (Ed.), *Fat. Acids, Elsevier*, pp. 1e22, https://doi.org/10.1016/B978-0-12-809521-8.00001-5.
62. Klere, J. (1996). The history of oils and fats, *Oils Fats Man. A Compr. Treatise d Prop. Prod. Appl.*, Lavoisier Publishing, Paris, p. 1527.
63. Chevreul, M.E. (1823). Rechercheschimiques sur les corps gras d'origineanimale, Levrault, Paris, 1923-1997.
64. Ogori, A.F. (2020).Extraction and Constituents of Fats and Oils, *Journal of Food Science & Nutrition,* doi:10.24966/FSN-1076/100060
65. Bockish, M. (1998). *Fats, and Oils Handbook*, Elsevier, 145-157,available at: https://doi.org/10.1016/ C2015-0-02417-0.
66. O'Brien, R.D. (2009). *Fats and Oils: Formulating and Processing for Applications*, Third Ed., CRC Press, Boca Raton, 184-192.
67. Naudet, M. (1996). Main chemical constituents of fats, in A. Karleskind, K.-P. Wolff (Eds.), Oils Fats Man. *A Compr. Treatise d Prop. Prod. Appl.*, Lavoisier Publishing, Paris, p. 1527.
68. Ghazani, S.M. & Marangoni, A.G. (2019). The ternary solid-state phase behavior of triclinic POP, POS, and SOS and its relationship to CB and CBE properties, *Cryst. Growth Des.* 19. 704e713, https://doi.org/10.1021/acs.cgd.8b01273.
69. Mottram, H.R. & Evershed, R.P. (2001). Elucidation of the composition of bovine milk fat triacylglycerols using high-performance liquid chromatography atmospheric pressure chemical ionization mass spectrometry, *J. Chromatogr., A.* 926. 239e253, https://doi.org/10.1016/S0021-9673(01)01048-2.
70. Wesdorp, L., van Meeteren, J., de Jong, S., van der Giessen, R., Overbosch, P., Grootscholten, P., Struik, M., Royers, E. Don, A., de Loos, T., Peters, C. & Gandasasmita, I. (2012). Liquid multiple solid phase equilibria in fats. *Struct. Prop. Fat Cryst. Networks* 2nd ed., CRC Press, Boca Raton, pp. 241-418, https://doi.org/10.1201/b12883-10.
71. Sato, K. (2001). Crystallization behavior of fats and lipids d a review, *Chem. Eng. Sci.* 56. 2255e2265. https://doi.org/10.1016/S0009-2509(00)00458-9.
72. Sato, K., Bay_es-García, L., Calvet, T., Cuevas-Diarte, M._A. & Ueno, S. (2013). External factors affecting polymorphic crystallization of lipids, *Eur. J. Lipid Sci. Technol.* 115. 1224-1238, https://doi.org/10.1002/ejlt.201300049.
73. Sato, K. & Ueno, S. (2001). Molecular interactions and phase behavior of polymorphic fats, in N. Garti, K. Sato (Eds.), *Cryst. Process. Fats Lipid Syst.*, Marcel Dekker Inc., New York, pp. 177e210.
74. Endo, Y. (2018). Analytical Methods to Evaluate the Quality of Edible Fats and Oils: The JOCS Standard Methods for Analysis of Fats, Oils and

Related Materials and Advanced Methods. *Journal of Oleo Science,* 67(1), 1-10. doi:10.5650/jos.ess17130.

75. Fats and Oils: Structure of Fats and Oils, *Introductory, Conceptual, and GOB Chemistry,* Book Chapter 17. *Bookshelves.* Available at: https://chem.libretexts.org/Bookshelves/Introductory_Chemistry/Basics_of_General_Organic_and_Biological_Chemistry_(Ball_et_al.)/17%3A_Lipids/17.02%3A_Fats_and_Oils
76. Tripathy, D. B., & Mishra, A. (2016). Renewable plant-based raw materials for industry. *Sustainable Inorganic Chemistry,* 353.
77. Abe, M. (1997). Microemulsions in Enhanced Oil Recovery: middle-phase microemulsion formation with some typical anionic surfactants, in Industrial applications of microemulsions, *Surfactant Science series,* vol. 66, pp. 279-303, C. Solans and H. Kunieda eds., Marcel Dekker, New York.
78. Rios, R. V., Pessanha, M. D. F., Almeida, P. F. de, Viana, C. L., & Lannes, S. C. da S. (2014). Application of fats in some food products. *Food Science and Technology (Campinas),* 34(1), 3–15.
79. Kumari, R., Jeyarani, T., Soumya, C., & Indrani, D. (2011). Use of vegetable oils, emulsifiers and hydrocolloids on rheological, fatty acid profile and quality characteristics of pound cake. *Journal of Texture Studies, 42,* 377-386. http://dx.doi.org/10.1111/j.1745-4603.2011.00297.
80. Guarino, A. M. (2011). "Recipe Substitutes". *Human Nutrition and Food,* Louisiana State University, Louisiana Cooperative Extension.
81. Peterson, L. (2016). Functions of Fats in the Kitchen. Available at: https://extension.illinois.edu/blogs/live-well-eatwell/2016-02-22-functions-fats-kitchen.
82. Hernandez, E. M. & Eldin, A.K. (2013). Cosmetic and Pharmaceutical Properties of Fats and Oils. *Processing and Nutrition of Fats and Oils,* 223–243. doi:10.1002/9781118528761.ch13.
83. Pelletier, X., Belbraouet, S., Mirabel, D., Mordret, F., Perrin, J.L., Pages, X., & Debry, G. (1995). A diet moderately enriched in phytosterols lowers plasma cholesterol concentrations in normocholesterolemic humans. *Annals of Nutrition and Metabolism,* 39, 291–295.
84. Hernandez, E. (2005). Pharmaceutical and cosmetic use of lipids. In: *Bailey's Industrial Oils and Fat Products* (ed. F. Shahidi), vol. 6, Chapter 12, John Wiley & Sons, Inc., New York.
85. Tao, B. Y. (2007). Industrial Applications for Plant Oils and Lipids. Bioprocessing for Value-Added Products from Renewable Resources, 611-627. doi:10.1016/b978-044452114-9/50025-6
86. Spitz, L. (2004). SODEOPEC Soaps, Detergents, Oleochemicals, and Personal Care Products, ed., AOCS Press.
87. Bist, S. (2004). Development of vegetable lipids derived fatty acid methyl esters as aviation turbine fuel extender, MS thesis, School of Agricultural and Biological Engineering, Purdue University, W. Lafayette, IN.
88. Kwatra, H. S., Caruthers, J. M., & Tao, B. Y. (1992). Synthesis of long chain fatty acids esterified onto cellulose via the vacuum-acid chloride process.

Industrial & Engineering Chemistry Research, 31(12), 2647-2651. doi:10.1021/ie00012a004.

89. P. Wang and B. Y. Tao, *Appl. Polymer Sci.*, 52, (1994) 755.
90. Ma, F., Hanna, M.A. (1999). Biodiesel production: a review. *Bioresour Technol.*70:1-15.
91. Bartholomew, D. (1981). Vegetable oil fuel. *J Am Oil Chem Sot.* 58:286A-288A.
92. Atabani, A.E., Silitongac, A.S., Badruddin, I.A., *et al.* (2012). A comprehensive review on biodiesel as an alternative energy resource and its characteristics. *Renew Sustain Energy Rev.*16:2070-2093.
93. Shahid, E.M. & Jamal, J. (2011). Production of biodiesel: a technical review. *Renew Sustain Energy Rev.* 15(9):4732-4745.
94. Atadashi, I.M., Aroua, M.K. & Abdul Aziz, A. (2010). High-quality biodiesel and its diesel engine application: a review. *Renew Sustain Energy Rev.* 14(7):1999-2008.
95. Janaun, J. & Ellis, N. (2010). Perspectives on biodiesel as a sustainable fuel. *Renew Sustain Energy Rev.* 14(4):1312-1320.
96. Srivastava, A. & Prasad, R. (2000). Triglycerides-based diesel fuels. *Renew Sust Energ Rev.* 4:111-133.
97. Balat, M. & Balat, H. (2008). A critical review of biodiesel as a vehicular fuel. *Energy Convers Manage.* 49:2727-2741.
98. Balat, M. & Balat, H. (2010). Progress in biodiesel processing. *Appl Energy.* 87(6):1815-1835.
99. Demirbas, A. (2008). Biodiesel: a realistic fuel alternative for diesel engines. Springer- Verlag London Limited. doi: 10.1007/978-1-84628-995-8.
100. Lin, L., Cunshan, Z., Vittayapadung, S., Xiangqian, S. *et al.* (2011). Opportunities and challenges for biodiesel fuel. *Appl Energy.* 88(4):1020-1031.
101. Allison, David, B., *et al.* (1999). Estimated intakes of trans fatty and other fatty acids in the US population. Journal of the *Academy of Nutrition and Dietetics Association*, vol. 99, no. 2, pp. 166-74. https://www.jandonline.org/article/S0002-8223(99)00041-3/fulltext.
102. Wainwright, B. (2000), Fats and Oils 2000: Challenges and Opportunities, available at: https://www.naturalproductsinsider.com/regulatory/fats-and-oils-2000-challenges-and-opportunities
103. Vieira, S.A., McClements, D.J. & Decker, E.A. (2015). Challenges of utilizing healthy fats in foods.*Adv Nutr.* 6(3):309S-17S.
104. Mensink, R.P., Zock, P.L., Kester, A.D.M. & Katan, M.B. (2003). Effects of dietary fatty acids and carbohydrates on the ratio of serum total to HDL cholesterol and on serum lipids and apolipoproteins: a meta-analysis of 60 controlled trials. *Am J. Clin. Nutr.* 77. 1146-1155.
105. McClements, D. J. and Decker, E. A. (2008). Lipids. In: *Fennema's Food Chemistry*, 4th ed., pp. 155–216. Srinivasan, D., Parkin, K. L. and Fennema, O. R., Eds., CRC Press/Taylor & Francis, Boca Raton, FL.
106. Decker, E. (2016). Fatty Acids. *Health and Nutrition.* Vol. 24. Available at: https://www.soyconnection.com/resources/health-nutrition-newsletter/

newsletter-article-list/fats-and-oils-many-challenges-faced-by-industry-consumers
107. Kritchevsky, D. (1998). History of recommendations to the public about dietary fat. *J Nutr.* 128(2):449S-52S.
108. European Union (2000). European Parliament and Council Directive 2000/36/EC (23 June 2000) relating to cocoa and chocolate products intended for human consumption. *Official Journal of the European Communities* 43(L197), 19.
109. Sacks, F.M. *et al.* (2017). Dietary fats and cardiovascular disease. *Circulation.* 135:00–00. DOI: 10.1161/CIR.0000000000000510.

2

Biotechnology for Oil and Fat

Nabya Nehal[1] and Priyanka Singh[2]*

[1]Department of Biotechnology, Faculty of Engineering and Technology, Rama University, Kanpur, Uttar Pradesh, India
[2]Institute of Allied Medical Science and Technology, NIMS, University, Rajasthan, India

Abstract

Oilseed crops are currently potent sources for secretion of edible oil enriched with protein, nutrients, and bioactive compounds. Vegetable oils with improved nutritional properties are being generally formulated for oilseed crops by selection and breeding of fatty acid biosynthetic mutants. These mutant varieties of oilseed crops have altered amounts of saturated and unsaturated fatty acids. The recent trends of biotechnological tools have enabled the generation of genetically modified oilseeds with ability for secreting high-quality oil enriched with nutritional values. Genome editing tools have currently improved the oil quality to a greater extent in comparison to traditional breeding techniques. Pretreatment methodologies are also employed to improve quality of edible oil by changing the physiochemical properties of oilseed crop. In this chapter, the composition of edible oil and their specific role as a nutritive component has been highlighted. Genome editing and other molecular tools are also discussed for improving the quality of oil from oilseed crops. The different pretreatment methodologies have been subsequently applied to reduce the oxidative property and anti-nutritional component in edible oils. Their application for development of varieties of oilseeds with expanded utility for food applications with respect to higher nutritional values has also been discussed in detail.

Keywords: Oilseed crop, edible oil, recombinant method, genome editing, pretreatment method

**Corresponding author*: priyay20@gmail.com

Divya Bajpai Tripathy, Anjali Gupta, Pooja Agarwal, Anuradha Mishra, and Arvind Kumar Jain (eds.) Oils and Fats as Raw Materials for Industry, (33–64) © 2024 Scrivener Publishing LLC

2.1 Introduction

The requirements of dietary fats and oils increase day by day due to the worldwide increase in population diversity; hence, the rate of resources starts decreasing very fast, and in coming years, they will vanish. The considerable development of dietary fats is an energy-driven nutrient and hence provides essential fatty acids (FAs) (Carreiro and Buhman, 2019). Nowadays, the major use of vegetable oils and fats is in food products such as salad shortenings, margarines and cooking oils, and can serve other large-scale biotechnological applications. Continued development of new commercial processes or industrial products ensures that fats and oils will be able to maintain their global market share. For applications such as chemical additives and lubricants, soybean oil and cosmetics are viscous and reactive toward atmospheric oxygen. For other uses including detergents, flavors, polymers, coatings, pharmaceuticals and reactivity of vegetable oils and agricultural chemicals needs to be enhanced by introducing additional functionalities or cleaving the fatty acid molecules (Figure 2.1).

Dietary fats are source of fatty acids, fat-soluble vitamins (A, D, E and K), macronutrients that play diverse role in metabolic pathways (activation, inhibition) in living organisms. They can form structural components of the cell membranes, can be a source of energy in the form of Adenosine

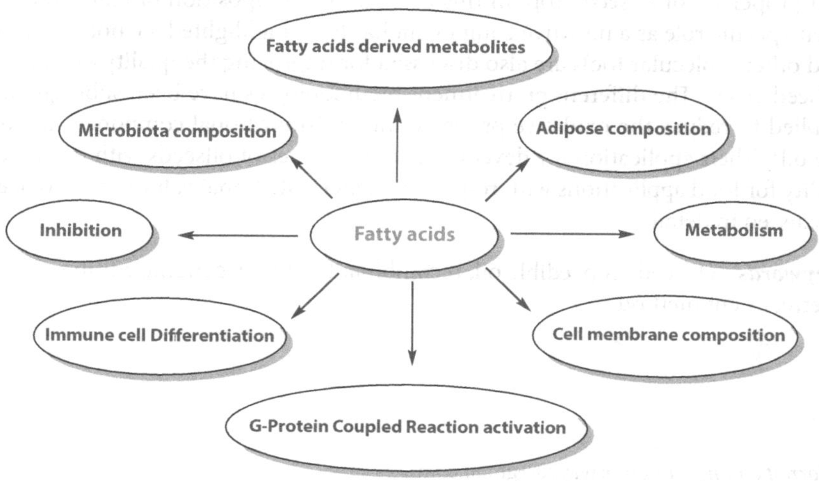

Figure 2.1 Biological application of fatty acids in different industrial sectors.

Triphosphate (ATP) and can trigger and mediate signals (Xiang *et al.*, 2019).

Seeds for Oilseed crops are primarily grown for the processing of edible oils. Peanuts, soybeans, rapeseed, sesame and sunflower are considered as potential oilseed crops for production of high yield of edible oil. The vegetable oil has been obtained by extracting the oil from pressed seeds. The cake or meal obtained in the extraction is a rich source of proteins, minerals, vitamins and therefore is added in diets fed to livestock to provide protein and energy. Oilseeds crops like cotton, soybeans, and sunflower provide higher energy concentrations from oil in addition to protein. Genetic diversity in oilseed crop germplasm has been generally identified using different molecular markers like randomly amplified polymorphic DNA (RAPDs), simple-sequence repeats (SSR) and inter-simple sequence repeats (ISSRs) (Khan *et al.*, 2015; Amini *et al.*, 2008; Tripathy and Mishra, 2016). Flowers of some oilseed crops have been commercially used for treatment of several chronic diseases and therefore are preferred for formulation of herbal medicines (Dajue and Mündel, 1996). Their leaves and shoots are also being used as salad in human diet due to enrichment of iron, calcium, phosphorus, vitamins A and E (Nimbkar, 2002). Seeds of these crops are being used to get high-quality edible oil rich in polyunsaturated fatty acid that reduced the level of blood cholesterol. These oilseed crops secrete oil containing a higher percentage of unsaturated fatty acid like oleic acid and linoleic acid and a small percentage of saturated fatty acids, i.e., steric acid and palmitic acid (Velasco and Fernandez-Martinez, 2001). Safflower oil consists of a higher percentage of linoleic acid (70-75%) compared to cotton seed, soybean, corn, peanut, or olive oils. Oleic acid as monounsaturated fatty acid has high therapeutic value for reducing low-density lipoprotein in the blood, and therefore it is considered as ideal oil (Weiss, 1971; Smith, 1996). These oilseed crops have also anti-nutritional components which reduce their nutritional values due to formation of indigestible complexes with minerals and proteins (Francis *et al.*, 2002; Lonnerdal, 2002). Therefore, it is necessary to reduce the levels of anti-nutritional factors in edible crops to prevent toxicity and associated health problems (Gemede and Ratta, 2014). From the last few decades, these anti-nutritional factors can be reduced by use of traditional food preparation methods such as roasting, soaking, boiling, microwaving, autoclaving, and fermentation to improve the biological value of vegetable oils (Handa *et al.*, 2017; Jaybhaye and Srivastav, 2015). In oilseeds processing, pretreatment has been used as a rapid and efficient oil extraction process including roasting, boiling, pressing and steaming. All oilseeds required their specific method of pretreatment. By heat treatment, the coating of seeds would be weakened and oil secreting glands would be exposed to the moisture and heat released during oil extraction. These factors lowered its viscosity; coagulate the protein for improved diffusion during extraction (Carr, 1993). Roasting method has been also used for the enhancement of oil content and anti-oxidant activity in oilseeds at optimum temperature (120-180°C) without affecting the phenolic compounds (Taha and Mätthaus, 2018). After oil extraction, the processed oilseed has been mostly used as a protein-enriched fodder for animal feeding (Nagaraj, 1995; Oelke *et al.*, 1992).

In this chapter, the composition of different nutritional components in oilseed crops has been primarily discussed. The various pretreatment methodologies have been also discussed to lower the percentage of anti-nutritional factors and improve the quality of edible oils.

2.2 Review of Literature

2.2.1 Development of Dietary Fats and Vegetable Oils

The concept of using plants, microorganisms and animals as sources of oils and fats has a long history. In the mid-1870s, microbial lipids got attention as they became the source of dietary fats and oils (Kyle and Ratledge, 1992). However, considerable developments in investigations of lipid-producing organisms were discovered and identified in Germany (1920 to 1945) (Hesse, 1949). However, it has been realized that the microorganisms-derived oils would never be able to compete in terms of price with the bulk commodity oils, such as sunflower oil, soybean oil and canola oil. In the 20th century, not only the rate of fat consumption by individuals increased but also the quality of fats consumed by the individuals dramatically increased (Xiang *et al.*, 2019). The studies investigated that the consumption rate of trans-saturated fat and omega-6 are dramatically increasing while the rate of omega-3 long-chain fatty acids are decreasingly prevalent in the diet (Cordain *et al.*, 2005). This lifestyle may cause diseases such as nonalcoholic steato-hepatitis, cardiovascular disease, and type 2 diabetes, and typically lead to poorer health outcomes, inflammation and immune dysfunction (Jakobsen *et al.*, 2009).

Lipids as biological molecules composed of fats and oils, highly solubilize in non-polar solvents and are less soluble in polar liquid (water). Lipids serve as food reservoirs in animal and plant tissues and are referred to as dietary fat (Sabikhi and Kumar, 2012). Triglycerides are composed of fatty acids and backbone of glycerol; three molecules of fatty acids are

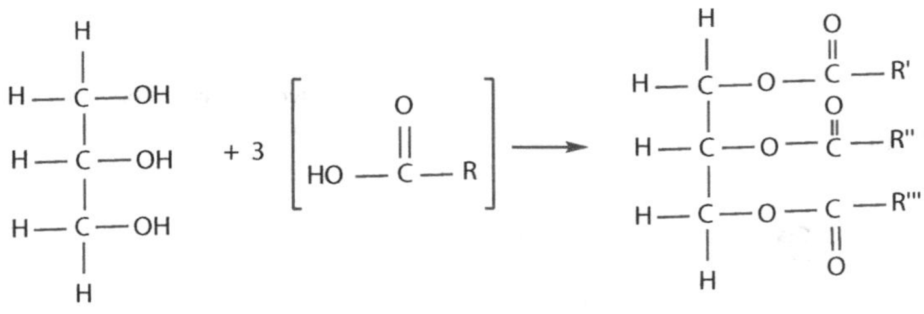

Figure 2.2 Chemical process for synthesis of triglyceride from glycerol and fatty acids.

attached with glycerol moieties (Figure 2.2) and can be non-identical (mixed glycerides) or identical (simple triglycerides).

Fatty acids are long hydrocarbon chains, length varying from 2-20 carbon molecules or more, with acid group at one end and methyl group at the other end of the chain. Fats are solid at room temperature (RT) while oils are liquid at room temperature. Fatty acids are building blocks of fat and constitute 90% of the dietary fats (Anthea *et al.*, 1993). Free fatty acids, mono- and di-glycerides, phosphatides, fatty alcohols, sterols, fat-soluble vitamins, chlorophyll, tocopherols and carotenoids (Strayer, 2006) are some minor components whereas fats and oils are major components. Animal fats and fish oils are the main source of cholesterol, whereas plant oils usually contain phyto-sterols (Griffin and Cunnane, 2009).

2.2.2 Classification of Fatty Acids

On the basis of the number of double bonds, dietary fatty acids are classified as saturated and unsaturated fatty acids, respectively. Saturated fatty acids (SFA) contain no double bonds whereas unsaturated fatty acids (USFA) contain hydrocarbon chains containing at least one double bond. Furthermore, it is classified into two categories: Mono-unsaturated fatty acid (MUFA) and Poly-unsaturated fatty acid (PUFA). MUFA contains one double bond whereas PUFA contains two or more. Saturated fatty acids have higher melting points than unsaturated fatty acids (Masurkar *et al.*, 2015). Kinks are formed in the double bonds between carbon atoms in FA and prevent molecules from stacking together. Double bonds between carbon atoms are arranged in cis and trans forms and can give rise to geometric isomers (Sabikhi and Kumar, 2012). In cis unsaturated fatty acids, double bond form on the same side and result in kinked geometry whereas

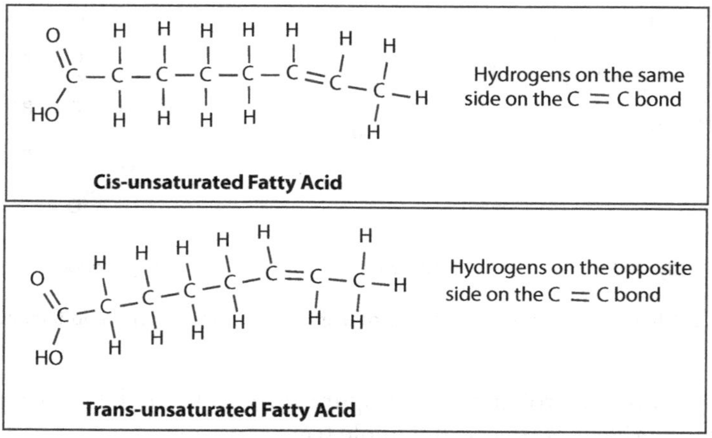

Figure 2.3 Structure of different configuration of unsaturated fatty acid.

in trans unsaturated fatty acids, double bond forms the opposite side, hence, arrangement appears in a straight line (Figure 2.3).

2.2.2.1 *Saturated Fatty Acids (SFA)*

General formula of SFA is R-COOH, C-C atoms are connected together with a single bond. On the basis of chain length, SFA are further categorized into the following classes:

- **Short-chain SFA** consists of 3 to 7 carbon atoms
- **Medium-chain SFA** consists of 8 to 13 carbon atoms
- **Long-chain SFA** consists of 14 to 20 carbon atoms
- **Very long-chain SFA** consists of 21 or more carbon atoms

Butyric acid (C4:0), caprylic acid (C8:0), caproic acid (C6:0), capric acid (C10:0), myristic acid (C14:0), lauric acid (C12:0), palmitic acid (C16:0), arachidic acid (C20:0), stearic acid (C18:0), lignoceric acid (C24:0) and behenic acid (C22:0) are some examples of SFA (Strayer *et al.*, 2006).

2.2.2.2 *Unsaturated Fatty Acid*

On the basis of chain length, UFA are further categorized into the following classes:

- **Short-chain UFA** consists of 19 carbon atoms

- **Long-chain UFA** consists of 20 to 24 carbon atoms
- **Very long-chain UFA** consists of 25 or more carbon atoms

MUFA- Carbon- carbon atoms are linked together with single double bonds. They are found in nuts, red meat, high fats fruits such as olive oil and avocados. Palmitoleic acid (C16:1 ω-9), erucic acid (C22:1 ω-9) and

Table 2.1 List of food items enriched with different types of fatty acid.

Types of lipids	**Source**
Saturated Fatty Acid (SFA)	
Short-chain FA	Dairy fat
Medium-chain FA	Coconut fat, dairy fat and palm kernel oil
Long-chain FA	Animal fat, peanut oil, dairy fat and peanut oil
Very long-chain FA	Peanut oil
Unsaturated Fatty Acid (UFA)	
Monounsaturated FA (MUFA)	Canola oils and olive oils
Linoleic acid	Safflower, sunflower, soybean and corn oils
α-Linoleic acid	Canola, walnuts, flaxseed oil and soybean oil
γ-Linoleic acid	Blackcurrant seed oil, borage oil and primrose oil
Arachidonic acid	Egg lipids, Lean meat, fish and organ lipids
Stearidonic acid	Blackcurrant seed oil, fish oil and hemp oil
Docosahexaenoic acid and Eicosa-pentaenoic acid	Shellfish and oily fish (herring, salmon, mackerel, smelt and anchovy)
trans Fatty Acids	Hydrogenated oils and fats
Other lipid constituents	
Phytosterols	Olive oil and soya products
Cholesterol	Organ meats, shellfish and eggs
Phospholipids	Soybean, lean animal meat and leafy plants

oleic acid (C18:1 ω-9) are some examples of MUFA. In oleic acid, the first double bond is present at the ninth position of carbon atoms (Nivetha *et al.*, 2021).

PUFA- Carbon- carbon atoms are linked together with two or more double bonds. Significantly found in walnuts, fish, seeds including sesame seeds and sunflower (FAO, 2010). Linoleic acid (LA) (C18:2 ω-6), α-linolenic acid (ALA) (C18:3 ω-3), arachidonic acid (AA) (C20:4 ω-6), γ-linolenic acid (GLA) (C18:3 ω-6), eicosa-pentanoic acid (EPA) (C20:5 ω-3), docosa-hexaenoic acid (DHA) (C22:6 ω-3) and docosa-pentanoic acid (DPA) (C22:5 ω-3) (Spector and Kim, 2015) listed in Table 2.1. In ALA, EPA and DHA, the first bond is present at the third position of C atoms whereas in case of AA and LA, the first double bond is present at the sixth position of carbon atoms as shown in Figure 2.4.

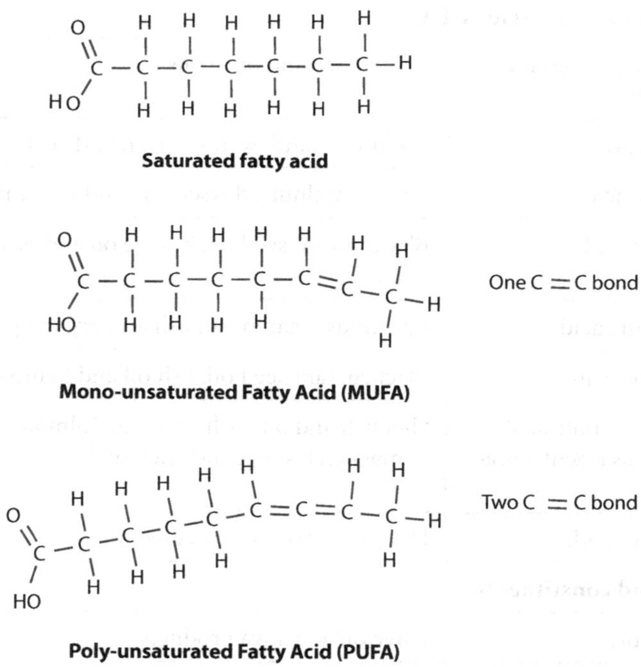

Figure 2.4 Structure configuration for mono and polyunsaturated fatty acid.

2.2.3 Sources of Dietary Fats

2.2.3.1 Microorganisms as Source of Lipids

Microbial lipids produced economically got attention as acceptance rate by individuals increased. In the 1980s, yeasts were selected to produce cocoa butter enriched with fats (oleate, stearate and palmitate) are esterified to glycerol and equal amounts of triacylglycerol (TAG). Microorganisms such as bacteria, fungus (mold and yeast) and algae are not ubiquitous, can accumulate high level of dietary fats in cells. Approximately 600 species of yeast have been isolated and identified, out of which only 30 to 40 can accumulate 20% of oil in cells. These species are known as oleaginous microorganisms (Thorpe and Ratledge, 1972). Synthesis of fatty acids occurs in every phase of growth; phospholipids are continuously synthesized as they are incorporated into cell membranes. Quality and quantity of fats may vary from organism to organism (Table 2.2). Yeasts species such as *Lipomyces* and *Rhodotorula* accumulate more than 70% of cell mass as lipid whereas some species such as *Candida utilis* and *Sacchromyces cerevisiae* accumulate less than 10% of biomass as lipid, respectively (Rattray, 1988).

2.2.3.2 Conventional Oilseeds as Source of Dietary Fats

Oilseed crops are defined as seeds that contain a large amount of oils; they are the main source of nutrition and play a major role in industries. Oilseeds include soybean, olive, cotton seed, sunflower seed, palm kernel, rapeseed, safflower seed, linseed, sesame seed and groundnut. Palm oil is one of the major sources of oils and fats. The palm oil is extracted from mesocarp region of the fruit whereas kernel oil is extracted from kernel. Palm oil consists of 44% of palmitic acid (SFA), 39% of oleic (MUFA), 11% of linoleic acids (PUFA), 5% of stearic and 1% of myristic acids, whereas palm kernel oil consists of 44-51% of lauric, 7-10% of palmitic, 15-17% of myristic, 3-5% of capric, 12-19% of oleic, 2-3% of stearic, 1-2% of linoleic acid and 3-5% caprylic, respectively (Khosla, 2006; Gervajio, 2005). Rapeseed are extracted from the family of *Brassica* including cabbage, turnips and mustard. Forty percent of rapeseed contains carcinogenic erucic acid (inedible oil). New varieties of rapeseed replace erucic acid with oleic acid and were referred to as Canola oil, basically originated in Canada and USA (Bailey and Hui, 1996; McKevith, 2005). Sunflower seed extracted from sunflower is primarily a source of oil; the seed is eaten as an edible snack nut. Compositions of fatty acids depend on the climatic conditions. A warmer climate produces oleic acid (MUFA) whereas a cooler climate

Table 2.2 Distribution of fatty acid content in different types of living organism.

	Content of different form of fat (% w/w)									
Yeasts	**Lipid**	**Myristic acid**	**Palmitic acid**	**Palmitoleic acid**	**Stearic acid**	**Oleic acid**	**Linoleic acid**	**α-linolenic acid**	**Arachidonic acid**	**Docosa-hexaenoic acid**
Cryptococcus curvatus	58	-	32	-	15	44	8	-	-	-
Rhodotorula glutinis	72	-	37	1	3	47	8	-	-	-
Lipomyces starkeyi	63	-	34	6	5	51	3	-	-	-
Rhodosporidium toruloides	66	-	18	3	3	66	-	-	-	-
Rhodotorula graminis	36	-	30	2	12	36	15	4	-	-
Yarrowia lipolytica	36	-	11	6	1	28	51	1	-	-
Cryptococcus albidus	65	-	12	1	3	73	12	-	-	-
Waltomyces lipofer	64	-	37	4	7	48	3	-	-	-
Candida didensiae	37	-	19	3	5	45	17	5	-	-

(*Continued*)

Table 2.2 Distribution of fatty acid content in different types of living organism. (*Continued*)

Molds	Majority residues of fatty acids (% w/w)									
	Lipid	**Myristic acid**	**Palmitic acid**	**Palmitoleic acid**	**Stearic acid**	**Oleic acid**	**Linoleic acid**	**α-linolenic acid**	**Arachidonic acid**	**Docosa-hexaenoic acid**
Cunninghamella japonica	60	-	16	-	14	48	14	8	-	-
Entomophthora coronata	43	31	9	-	2	14	2	4	4	-
Pythium ultimum	48	7	15	-	2	20	16	-	11	14
Mucor circinelloides	25	-	22	1	5	38	10	15	-	-
Mortierella alpina	50	-	19	-	8	28	9	8	21	-
Aspergillus terreus	57	2	23	-	-	14	40	21	-	-
Cunninghamella echinulata	24	-	13	1	2	46	16	19.5	-	-
Mortierella alpina	50	-	8	-	11	14	7	14	49	-
Pythium irregulare	>27	-	17	7	2	14	18	-	11	-

(*Continued*)

Table 2.2 Distribution of fatty acid content in different types of living organism. (*Continued*)

Microalgae	**Majority residues of fatty acids (w/w)**									
	Lipid	**Myristic acid**	**Palmitic acid**	**Palmitoleic acid**	**Stearic acid**	**Oleic acid**	**Linoleic acid**	**α-linolenic acid**	**Arachidonic acid**	**Docosa-hexaenoic acid**
Cryptocodinium cohnii	40	16	16	1	-	1	-	-	-	40
Thraustochytrium aureum	15	3	8	-	-	16	-	-	3	52
Schizochytrium limacinum	50	4	56	-	1	-	-	-	-	30
Isochrysis galbana	22-30	12	1	11	3	2	-	-	-	11
Nannochloropsis oculata	30-35	4	15	22	-	3	1	-	4	-
Schizochytrium sp.	40	8	22	<0.5	0.5	1	-	-	-	41
Porphyridium cruentum	10	-	30	5	-	<1	-	-	-	-

Table 2.3 Fatty acid profile of selected conventional oilseeds (a) Saturated fatty acids (b) Monounsaturated and polyunsaturated fatty acids.

Conventional oilseeds	Saturated fatty acids								
	Caproic (C6:0)	Caprylic (C8:0)	Capric (C10:0)	Lauric (C12:0)	Myristic (C14:0)	Palmitic (C16:0)	Stearic (18:0)	Arachidic (C20:0)	Behenic (C22:0)
Palm	-	-	-	0.2	1.1	44.0	4.5	0.4	-
Soybean	-	-	-	0.1	0.2	10.7	3.9	0.2	-
Rapeseed	-	-	-	-	0.2	3.5	1.5	0.6	0.3
Sunflower	0.1	0.1	0.1	0.1	-	6.0	4.0	0.4	0.7
Palm Kernel	0.3	4.4	3.7	48.3	15.6	7.8	2.0	-	-
Groundnut	-	-	-	-	-	8.1	1.5	1.1	2.1
Cottonseed	0.1	0.1	0.1	0.1	0.8	27.3	2.0	0.3	-
Coconut	0.5	7.0	6.7	47.5	18.1	8.8	-	0.1	-
Olive	-	0	0	0.1	0.1	10.1	3.0	0.4	0.1

(*Continued*)

Table 2.3 Fatty acid profile of selected conventional oilseeds (a) Saturated fatty acids (b) Monounsaturated and polyunsaturated fatty acids. (*Continued*)

Conventional oilseeds	Monounsaturated fatty acid				Polyunsaturated fatty acid		
	Palmitoleic (C16:1)	Oleic (C18:1)	Gadoleic (C20:1)	Erucic (C22:1)	Linoleic (C18:2)	γ-Linoleic (C18:3)	α-Linoleic (C18:3)
Palm	0.1	39.2	-	-	10.1	0	0.3
Soybean	0.3	22.8	-	-	50.8	0	7.3
Rapeseed	0.2	60.1	1.4	0.2	20.1	0	9.6
Sunflower	0.1	16.5	0.1	-	72.4	0	0.5
Palm Kernel	-	22.1	-	-	3.3	-	-
Groundnut	Trace	49.9	0.9	-	35.4	0	0
Cottonseed	0.8	18.3	-	-	50.5	0	0.1
Coconut	2.6	6.2	Trace	-	1.6	-	-
Olive	0.7	71.9	-	-	7.5	0	0.7

produces ω-6 linoleic acid (PUFA), respectively (Morrison *et al.*, 1995). Peanut oil is also referred to as groundnut oil, mainly originated from India and China. Cottonseed is an end-product of cotton, primarily used as vegetable oil. Copra (coconut kernel), dried meat kernel, about 64% of oils are produced. Coconut oil consists of 46-50% lauric acid, 8-10% palmitic acid, 17-19% of myristic acid, 6-9% of caprylic acid, 5-7% of oleic acid, 2-3% of stearic acid, 1-2.5% of linoleic and 6-10% of capric acid, respectively, as shown in Table 2.3 (Bruinsma, 2003). It is primarily used as toilet soap and edible oil in several countries (South Asian, Caribbean and the Far East). Olive contains 15-20% of oil, seed contains 2-4% whereas pulp contains 96-98% of oils. Olive oil, naturally producing vitamin E, PUFA and MUFA, primarily consists of 78-83% of oleic acid, 6-9% of linoleic acid, 8-15% of palmitic acid, and 1.5-3% of stearic acid (Mailer, 2006). Corn oil obtained from the germ of the seed is primarily used as cooking oil. The approximate compositions of the oil are as follows: 58% of ω-6 linoleic acid, 12% of palmitic acid, 27% of oleic, 2% of stearic acid and 1% of ALA, respectively, as shown in Table 2.3 (Black *et al.*, 1994). Sesame seed originated from tropical countries; the seed is eaten as an edible snack nut and used to produce oil and paste (tahini). It primarily consists of less than 20% of SFA, 4.8-6.1%% of stearic acids and 7.9-12% of palmitic acids whereas 80% of total fatty acids consist of linoleic acid and oleic acid. 41.5-47.9% of linoleic acid and 35.9-42.3% of oleic acid are roughly found in sesame seed (Hwang, 2005). Flaxseed, also referred to as linseed, originated from Babylon in 3000 B.C. It primarily consists of 35-60% of ω-3 ALA and 19% of LA, whereas lowest level of SFA occur. Safflower seed is an annual crop, principally originated in India, U.S., Middle East and Mexico. Generally, unroasted safflower oil consists of 5.53% of palmitic acid, 11% of oleic acid, 81.5% of linoleic acid, 1.62% of stearic acid, 0.40% of linolenic acid, whereas roasted safflower seeds contain 1.69% of stearic, 4.99% of palmitic acid, 82.0% of linoleic acid, 10.9% of oleic acid and 0.41% of linoleic acid, respectively (Lee *et al.*, 2004).

2.2.3.3 Unconventional Oilseeds as Source of Dietary Dats

Cucumber and melon are unconventional sources of fatty acids (Datta, 1976). About 51% of oil contents were found in *Citrullus vulgaris* (watermelon) and 49.3% of oil found in *Cucumis melo* (muskmelon kernel), 41.4% of *Cucumis sativus* (cucumber), 44.4% of *Cucurbita moschata* (squash), 43.9% of *Cucurbita pepo* (pumpkin) and 41.8% of oil in *Luffa acutangula* (sponge gourd) (de Mello *et al.*, 2001), 50% of *Telfairia occidentalis* (fluted pumpkin), respectively (Esuoso *et al.*, 1998) as shown in Table 2.4.

Cucumis amaris originated from North of Ivory Coast region; the fruit is inedible but seeds are used in cuisine (Sorho *et al.*, 2006). Primarily, 65.2% of linoleic acid, 10.21% of palmitic acid, 12.68% of oleic acid and 10.2% of stearic acids were present (de Mello *et al.*, 2001). Some vegetable seeds have less oil contents as compared to melon seeds such as 38.6% in *Brassica oleracea* var. *botrytis* (cauliflower), 20% in *Capsicum frutescens* (sweet chillies), 16.3% in *Hibiscus esculentus* (okra), 19.8% in *Solanum melongena* (eggplant), 34% in *Brassica oleracea* var. *caulorapa* (Kohlrabi) and 35% in *Brassica oleracea* var. *capitata* (cabbage) respectively as listed in Table 2.4 (Nzikou *et al.*, 2006).

Moringa olifera (drumstick) originated from different parts of Africa and Indian subcontinents. Table 2.4 shows that seeds of drumstick had 77.6% of oleic acid, 4.05% of linoleic acid, 5.1% of palmitic acid, 5.7% of behenic acid, 2.5% of arachidic acid and 4% of stearic acid, respectively

Table 2.4 Fatty acid profile in selected seed of unconventional plant.

Plant seeds	Palmitic (C16:0)	Stearic (C18:1)	Oleic (C18:1)	Arachidic (C20:0)	Linoleic (C18:2)
Bassia butyracea	60.8	3.2	30.9	-	4.9
Neem (Refined oil)	16.2	16.6	56.6	3.4	9.0
Azadirachta indica	14.9	14.4	61.9	1.3	7.5
***Madhuca* sp.**	23.7 to 24.7	19.9 to 29.9	36.3 to 43.3	-	11.6 to 15.8
Mangifera indica	5.4	44.0	43.7	3.2	3.7
Sal	4.5	44.2	42.2	6.3	2.8
Garcinia indica	3.4	67.4	28.1	0.3	0.6
Pongamia glabra	10.6	6.8	49.0	4.1	19.0
Vateria indica	9	46.9	41.4	1.4	1.3

(Anwar *et al.*, 2007). In 2012, Jorge and Luzia widely recognized fruits of *Pachira aquatic* (Guiana chestnut, water chestnut and money tree) originated from Amazon region, primarily produce oil and were used in pharmaceutical, cosmetics and food industries. Fatty acids comprised 44.93% of palmitic acid, followed by 11.35% of linoleic acid and 39.27% of oleic acid (Jorge and Luzia, 2012). In 2009, Ariffin *et al.* recognized dragon fruit seed as source of FAs. Red flesh and white flesh are two different varieties of fruit seed have varied composition of fatty acids such as red flesh, comprises 23.6% of SFA, 25.6% of MUFA and 50.8% of PUFA whereas white flesh comprises 21.7% of SFA, 27.2% of MUFA and 51.1% of PUFA, respectively (Ariffin *et al.*, 2009). Seeds of *Garcinia mangostana* (Mangosteen) originated from tropical region and prevalent FAs were amounted as 59.6% of SFA, 35.3% of USFA, 34% of oleic and 49.5% of palmitic (Ajayi *et al.*, 2007). In 2010, Wannes and colleagues evaluated the seeds of myrtle fruit (*Myrtus communis* var. *italic*) were prevalent FAs were 79.2% of linoleic acid, 9.5% of palmitic acid, 8% of oleic acid and 0.5% of α-Linolenic acid (Wannes *et al.*, 2010). In 2003, Ramadan and Morsel recognized that the concentrations and compositions of fats and oils may vary from parts of plants (pulp, peel and seeds) (Ramadan *et al.*, 2003a and 2003b).

In 2006, researchers found that the cactus pear is a rich source of PUFA; it contained 59% of oil, fatty acid contents were in g/100g (palmitic acid (9.32), palmitoleic acid (1.42), stearic acid (3.11), oleic acid (16.77) and linoleic acid (70.29) with total unsaturated FA content of 88.48g/100g, respectively (Ennouri *et al.*, 2006). Fruit samples of *Olea cuspidate* (wild olive) were extracted and quantified and were used as edible oil for human consumption (Gulfraz *et al.*, 2009) and contained 32.1-38.6% of oil. Seeds of Hound's berry (*Solanum nigrum*) originated from European countries and contained 37.12% of seed oil, 67.77% of linoleic acid, 14.59% of elaidic acid, 12.46% of palmitic acid, 0.64% of linolenic acid and 4.31% of stearic acid (Nzikou *et al.*, 2007).

In 2001, Garcia-Lopez *et al.* examined high concentration of oil content in wild variety of Legumes (lupin). Originated in Mexico, oil content contained 57-115g/kg, palmitic acid ranged from 175-287g/kg, linoleic acid (324-517g/kg), respectively (Garcıa-López *et al.*, 2001). Mahua (*Madhuca longifolia*) also known as buttercup, originated in the North and Central Indian plains, produce 58% of crude fat, about 37.4% of MUFA, 16.5% of PUFA and 46% of SFA, respectively (Ramadan *et al.*, 2006). Mahua were used as edible flowers, in confectionary products and chocolates, candles, shortenings, pharmaceuticals and cosmetics, in dairy products (ice-cream, whipping cream, cheese and coffee cream).

2.2.4 Applications of Fatty Acids

Omega-3 fatty acids implementations are essential as they prevent a number of cardiovascular and neuro-degenerative diseases mainly, anti-inflammatory, mental retardation, anti-atherosclerosis, endothelial relaxation, anti-fibrotic, memory loss, anti-arrhythmic and antithrombotic, strokes and trauma and other neuro-degenerative diseases such as dementia, Alzheimer's and Parkinson's disease and so on.

2.2.4.1 Fatty Acids can Modulate Inflammation

Vivo and *in vitro* elucidation conducted that the molecular and cellular pathways link fatty acids that can modulate inflammation (Norde *et al.*, 2019). Accordingly, interventions that SFAs act as precursors contribute to endoplasmic reticulum stress (Cao *et al.*, 2016) and ceramides (Chaurasia and Summers, 2015) and can generate reactive oxygen species (Harijith *et al.*, 2014). Signals trigger cellular inflammatory pathways and increase the secretion and production of pro-inflammatory cytokines in adipocytes (Alhusaini *et al.*, 2010), immune cells (Namgaladze and Brüne, 2016), pancreatic beta cells (Diakogiannaki *et al.*, 2008), myocytes (Barazzoni *et al.*, 2012) and hepatocytes (Nakamura *et al.*, 2009).

Saturated fatty acid (SFA), monounsaturated fatty acid (MUFA), trans-unsaturated fatty acid (TFA), and polyunsaturated fatty acid (PUFA) are myriad of fatty acids effects on inflammation (Norde *et al.*, 2019).

2.2.4.2 Treatment of Neuro-Degenerative Diseases (NDDs)

Omega-3 FAs influence human health conditions to decelerate neuro-degenerative diseases. During the embryonic development stage, PUFAs are prominently essential for neuronal differentiation and cell proliferation. Studies suggest the brain needs all the essential lipids associated with 30-35% of omega-3 and omega-6 FAs, which help in neurotransmitter signaling. Low concentration of FAs may lead to disorganized thought processes, memory loss, mental retardation and trauma, stroke, dementia and Alzheimer's disease. High levels of FAs may lead to modulate immune and vascular functions and can direct the central nervous system (CNS) (Nagy and Tiuca, 2017). FAs (DHA and EPA) are involved in reducing the risk of neuro-psychiatric diseases (Young and Conquer, 2005) and known to strengthen the functional and structural approach by improving elasticity, membrane organization and permeability of ions and thus improving metabolic pathways. Phospholipid eater of omega-3 FAs associated in the basal membrane aids in the synthesis of neurons (neurogenesis) and neuro-transmission, enhance the integrity and flexibility and improve the speed of synaptic transmissions and hence protect the brain from oxidative stress (Nagy and Tiuca 2017). High concentration of DHA may lead to reduced risk of dementia (Nivetha *et al.*, 2021) while EPA possesses anti-inflammatory and anti-oxidative properties. Thus, use of DHA and EPA may reduce the risk of diseases, such as bipolar disorder, chronic mental disorder, depression and attention deficit hyperactivity disorder (ADHD). Depression is a mental health condition in which a person shows mood swings and fluctuations, shows loss of interest in daily activities, and may often express melancholy and anger. High concentration of omega-6 and omega-3 induces the rate of inflammation and depression. From studies, researchers concluded that the excess intake of arachidonic acid and docosahexaenoic acid form an effective biomarker for depression (Cryż *et al.*, 2016).

Loss of essential PUFAs in diet may lead to ADHD in children and adults. ADHD typically refers to learning and behavioral patterns. Elevation in blood pressure level within arteries due to impairment of endothelial functioning may refer to cause hypertension. Sufficient consumption of α-linolenic acid may decelerate the effect of hypertension (Das, 2010). Alzheimer's, referred to as a common form of dementia, deficit consumption of PUFAs may lead to loss of efficient cognitive function and synapses (Dyall and Michael-Titus, 2008).

2.2.5 Pretreatments Methodologies for Improving Oil Yield

The marketing strategy and production of edible oil has been greatly affected by oxidative deterioration of oils under environmental stress condition. Lipid act as greater importance in oil content due to providing energy during cellular metabolism process and are also used as texture agent, and flavor agent in food sectors (Shahidi and Zhong, 2010). The oxidation of lipid in oil content would result in the loss of nutritive values, reduction of shelf life, and production of off-flavoring agents in edible oil. The oxidative process for these oil content results in the secretion of ketones and aldehydes as secondary oxidative products, which will cause health issues (Min and Jeffrey, 2002; Vieira *et al.*, 2017). The nutritive value of edible oil will be maintained by applying physico chemical pretreatment process during processing of oil (Grosshagauer *et al.*, 2019). The processing technique for extraction of oil from oilseed crop will lead to the formation of Maillard reaction products (MRPs), which can inhibit lipid oxidation (Liu *et al.*, 2020; Shrestha and De Meulenaer,

2014). In the oil industry, physiochemical extraction methodologies commonly involve screw pressing, solvent extraction, supercritical fluid extraction and enzyme-assisted aqueous extraction (Divine Bup Nde, and Anuanwen Claris Foncha, 2020). Oil processing technique based on supercritical fluid and enzyme-assisted extraction are highly efficient and biocompatible (Shankar *et al.*, 1997; Temelli, 2009), but these processes are not being preferred by the oil industry due to high cost of equipment and reagent (enzyme) costs (Rosenthal *et al.*, 1996; Rombaut *et al.*, 2015). Solvent extraction or pressing methods are most commonly used for extraction of oil from oilseed crops. Oil seed crops are being commercially used by human society for multipurpose uses. Their nutraceutical application has been limited by the availability of anti-nutritional factors (ANFs) in oilseed crops. Many processing methodologies have been used for minimizing the level of anti-nutritional factors to improve the quality of oil from these oilseed crops.

Over the last few decades, roasting, microwave, and boiling have been used as the most common thermal pretreatment process for extraction of oil from oilseeds (Ozdemir and Devres, 2000; Zhang *et al.*, 2020; Waszkowiak *et al.*, 2018). The treatment of oilseed at high temperature induces changes in the microstructure of seed, organoleptic properties, physicochemical properties, and oxidative properties of the oil. These thermal treatments improved the oil yield by reducing moisture content (Mohamed *et al.*, 2020); and stimulating the production of polyphenol and tocopherol (Yang *et al.*, 2018). Their antioxidant property improved by formation of melanoidin (MRPs) secondary product for enhancing their stability during storage and cooking (Oracz *et al.*, 2014; Kaur *et al.*, 2017). Boiling of oilseed crop at 85-90°C for short duration has caused the degradation of seed coat of oilseed crop, inactivation of enzyme and coagulation of protein with lower degree of oil viscosity to improve the secretion of oil from their oil glands. The boiling of oil seed for a longer time will result in the inactivation of nutritive components and finally decrease the quality of oil (Wakelyn and Wan, 2005). The secretion of oil from oil-seed crops will be enhanced by disrupting boiled oilseed with appropriate physical cell disruption methodology. Solvent extraction methodology causes the coagulation of seed proteins and boiling at 105-130°C for 30-120 minutes has resulted in the coalescence of oil droplets with increased level of permeability of seed coat to secrete oil. Boiling methodology has caused the expansion of the oil glands due to rupture of the cell wall and finally resulted in the secretion of oil from these oil glands (Shukla *et al.*, 1992). The combinatorial boiling pretreatment methodology associated with soaking and germination will effectively reduce the level of the anti-nutritional factor at a higher extent in comparison to individual methodology. These combinatorial pretreatment methodologies have reduced the concentration of trypsin, phytic acid, and tannin in oilseed (Khokhar and Chauhan, 1986; TV *et al.*, 2018). Recently, specific hydrothermal processing methodologies have also been used for significant removal of phytic acid, saponins, trypsin inhibitor and tannin from oilseed crops (Ojha *et al.*, 2018).

2.2.5.1 Microwave Pretreatment Process

Microwave pretreatment methodology has been also used for the extraction of high-quality vegetable oils from seeds enriched with tocopherols, phytosterols, flavonoid, phenolic components. These contents have improved the antioxidant properties and enhanced shelf life of edible oils (Azadmard-Damirchi *et al.*, 2011; Kittiphoom and Sutasinee, 2015). This methodology has not allowed the degradation of quality of edible oil due to maintaining nutritional components and is considered as a highly efficient pretreatment methodology for extraction of oil from oil seed crops. The high-quality oil has been extracted from seed of oil seed crop of *Nigella sativa* L. by using microwave pretreatment technique (Mazaheri *et al.*, 2019). Roasting pretreatment has degraded the quality of edible oil by affecting anti-oxidant activity and fatty acid composition in oilseed crops (Taha and Matthaus, 2018).

2.2.5.2 Biological Pretreatment Techniques

The extraction of oil from kernels of oilseed crop has been effectively increased with subsequent treatment of heat processed oilseed with mixed enzymatic mixture. Enzymatic treatment with cellulase has disrupted the cell wall for loosening the structure of inner cotyledons of seed coat. Pectinase enzyme has denatured the pectin layer for diffusion of oil granules after expansion of cotyledon cells (Danso-Boateng, 2011). The enzyme assisted-aqueous oil-extraction process (EAEP) has been also used for recovery of soybean oil. The higher yield of oil has been obtained after enzymatic treatment of soybean slurry with ohmic heat at specific electric fields (Pare *et al.*, 2014). This electroporation has caused thermal softening of seed coat tissues for higher rate of secretion of oil from oilseed crop (Pare *et al.*, 2014). These enzymatic treatment methodologies have been accomplished by treating seeds with different concentrations of commercial enzymes like cellulase, pectinase and viscozyme (Bisht *et al.*, 2015; Oberoi *et al.*, 2012). The pretreatment with physicochemical methodologies has effectively reduced the percentage of the anti-nutrient factors in oil seed crop (Bora, 2014). Among these methodologies, boiling minimized the percentage of anti-nutrients at a greater extent in the processed oil-seed (Piorreck *et al.*, 1984). The content of phytate and tannin has been reduced after germination during biological pretreatment process (Bulbula and Urga, 2018). In rapeseeds, steam treatment has improved the quality of seed cake which could be used as good animal feed supplement (Van Bruggen *et al.*, 1993). The oil of some safflower seeds has been also used for production of biodiesel in some countries. These crude oils have been recovered by solvent extraction followed by a trans-esterification process. The viscosity of crude oil for biodiesel has been reported as closer to that of petroleum derived diesel (Hamamci *et al.*, 2011).

2.2.5.3 Genetically Improved Oil Seed Crop

The nutritional composition of oilseed crops has generally comprised fatty acids, proteins, flavonoids, alkaloids, lignans, steroids, phenolics, polysaccharides, carboxylic acids, minerals, glycosides, quinone and kaempferol derivatives (Sato *et al.*, 1985; Nagatsu *et al.*, 1988; Chang *et al.*, 1999; Li *et al.*, 2002; Roh *et al.*, 2004; Yoo *et al.*, 2006; Elarbi *et al.*, 2009 and Ren *et al.*, 2016).

Genetically modified oil seed crops have been developed by genetic divergence of eleven clusters for phenotypic traits like plant height, oil content, number of capitula, seed weight and seed yield (Atole *et al.*, 2018). The mutant varieties have been studied by genetic breeding program using specific molecular markers like ISSR (Inter Simple Sequence Repeats), RAPD (Random Amplified Polymorphic DNA) and AFLP (Amplified Fragment Length Polymorphism) (Sujatha, 2008). The quality of oil in some mutant oil seed crops has been modified by studying linkage characters between constricted appearance of seed hull and oil percentages (Ceron-Diaz and Rubis, 1970). Genes have been mapped for 61 quantitative trait loci (QTL) in safflower oilseeds for morphological traits of flower color and spine nature of leaves (Pearl *et al.*, 2014). Their genetic diversity has been also estimated using polymorphic start codon targeted (SCoT) and polymorphic sequence-related amplified polymorphic marker (SRAPs) markers (Golkar and Mokhtari, 2018).

Genome editing has been currently used as advanced molecular technique to facilitate efficient and targeted modifications at specific gene loci (Chen and Gao, 2013; Gao, 2015). In the last two decades, genome editing has been carried out using zinc-finger nucleases (ZFNs) (Kim *et al.*, 1996; Zhang *et al.*, 2018) and transcription activator-like effector nucleases

(TALENs) (Christian *et al.*, 2010). In recent years, clustered regularly interspaced short palindromic repeats (CRISPR)/Cas systems have provided ease of targeted gene editing (Jinek *et al.*, 2012; Zhang *et al.*, 2018). This CRISPR/Cas system uses sequence-specific nucleases (SSNs) to induce recognition of specific DNA sequences and to generate double-stranded breaks (DSBs). In this system, gene knockouts have been first carried out by nonhomologous end joining (NHEJ) for insertion or deletion of nucleotides. Homologous recombination (HR) has further caused gene replacements and insertions in targeted gene loci (Symington and Gautier, 2011). Gene mutation based on genome editing has been currently employed in wide variety of oilseed crops to improve the yield and quality of oil. Therefore, this genome editing has been introduced as modern breeding programs to facilitate rapid and particular oilseed crop varieties.

2.3 Conclusion

Plant biotechnology tools have proven an effective competent plant breeding program to develop genetically modified transgenic oilseed crops variety with healthy vegetable oil compositions. Recombinant biotechnology has modified genes associated with the biosynthesis and metabolism of fatty acids in oilseed crop to improve quality of edible oil. The pretreatment methodologies based on boiling, roasting, microwave, fermentation, and soaking have effectively reduced the concentration of antinutritional factors in oilseed crops. These technologies are currently used as potential candidates to improve yield and quality of vegetable edible oils. Oilseed crops are currently in great demand for production of value-added industrial products via molecular pharming and biofuel production. Mutant varieties of oil crops by molecular breeding may be proved as valid alternative to transgenic crops for the supply of fatty acids. We finally conclude that the molecular tools along with pretreatment technologies synergistically improve the quality of edible oils from oilseed crop.

Acknowledgement

The authors would like to thank RAMA University (Kanpur) and NIMS University (Rajasthan) for providing library and internet facilities for completion of this chapter.

References

Ahmadzadeh, A.R., Alizadeh, B., Shahryar, H.A. and Narimani-Rad, M., 2012. Path analysis of the relationships between grain yield and some morphological characters in spring safflower (Carthamus tinctorius) under normal irrigation and drought stress condition. *Journal of Medicinal Plants Research*, *6*(7), pp. 1268-1271.

Ajayi, I.A., Oderinde, R.A., Ogunkoya, B.O., Egunyomi, A. and Taiwo, V.O., 2007. Chemical analysis and preliminary toxicological evaluation of Garcinia mangostana seeds and seed oil. *Food Chemistry*, *101*(3), pp. 999-1004.

Alhusaini, S., McGee, K., Schisano, B., Harte, A., McTernan, P., Kumar, S. and Tripathi, G., 2010. Lipopolysaccharide, high glucose and saturated fatty acids induce endoplasmic reticulum stress in cultured primary human adipocytes: Salicylate alleviates this stress. *Biochemical and Biophysical Research Communications*, *397*(3), pp. 472-478.

Amini, F., Saeidi, G. and Arzani, A., 2008. Study of genetic diversity in safflower genotypes using agro-morphological traits and RAPD markers. *Euphytica*, *163*(1), pp. 21-30.

Anthea, M., Hopkins, J. and McLaughlin, C.W., 1993. The kidneys. In *Human Biology and Health* (Ed. M. Anthea). New Jersey, USA: Prentice Hall, pp. 1017-1055.

Anwar, F., Hussain, A.I., Iqbal, S. and Bhanger, M.I., 2007. Enhancement of the oxidative stability of some vegetable oils by blending with Moringa oleifera oil. *Food Chemistry*, *103*(4), pp.1181-1191.

Ariffin, A.A., Bakar, J., Tan, C.P., Rahman, R.A., Karim, R. and Loi, C.C., 2009. Essential fatty acids of pitaya (dragon fruit) seed oil. *Food Chemistry*, *114*(2), pp. 561-564.

Atole, N.A., Rathi, S.R., Ratnaparkhi, R.D. and Janjal, S.M., 2018. Research Article Genetic diversity analysis in selected germplasm lines of safflower (Carthamus tinctorius L.). *Electronic Journal of Plant Breeding*, *9*(1), pp. 213-219.

Bailey, A.E. and Hui, Y.H., 1996. Industrial oil and fat products *Bailey's Industrial Oil and Fat Products*.

Barazzoni, R., Zanetti, M., Cappellari, G.G., Semolic, A.N.N.A., Boschelle, M., Codarin, E., Pirulli, A., Cattin, L. and Guarnieri, G., 2012. Fatty acids acutely enhance insulin-induced oxidative stress and cause insulin resistance by increasing mitochondrial reactive oxygen species (ROS) generation and nuclear factor-κB inhibitor (IκB)–nuclear factor-κB (NFκB) activation in rat muscle, in the absence of mitochondrial dysfunction. *Diabetologia*, *55*(3), pp. 773-782.

Bisht, T.S., Sharma, S.K., Sati, R.C., Rao, V.K., Yadav, V.K., Dixit, A.K., Sharma, A.K. and Chopra, C.S., 2015. Improvement of efficiency of oil extraction from wild apricot kernels by using enzymes. *Journal of Food Science and Technology*, *52*(3), pp. 1543-1551.

Black, J.M., Nesheim, M.C. and Kinsella, J.E., 1994. Dietary level of maize oil affects growth and lipid composition of Walker 256 carcinosarcoma. *British Journal of Nutrition, 71*(2), pp. 283-294.

Bora, P., 2014. Anti-nutritional factors in foods and their effects. *Journal of Academia and Industrial Research, 3*(6), pp. 285-290.

Bruinsma, J., 2003. World agriculture: towards 2015/2030: an FAO perspective. Earthscan. Food and Agriculture Organization, London/Rome.

Bulbula, D.D. and Urga, K., 2018. Study on the effect of traditional processing methods on nutritional composition and antinutritional factors in chickpea (Cicer arietinum). *Cogent Food & Agriculture, 4*(1), p. 1422370.

Cao, S.S., Luo, K.L. and Shi, L., 2016. Endoplasmic reticulum stress interacts with inflammation in human diseases. *Journal of Cellular Physiology, 231*(2), pp. 288-294.

Carr, R., 1993. Oilseed harvesting, storage and transportation. In: *Proceedings of the World Conference on Oilseed Technology and Utilization*. Applewhite, Th. H. (Ed.). AOCS Press. Champaign, IL.

Carreiro, A.L. and Buhman, K.K., 2019. Absorption of dietary fat and its metabolism in enterocytes. In *The Molecular Nutrition of Fats* (pp. 33-48). Academic Press.

Chang, H.T., Han, H.X. and Tu, P.F., 1999. Chemical constituents and pharmacological effects of safflower. *Xiandai Yaowu Yu Linchuang, 14*, pp. 201-203.

Chaurasia, B. and Summers, S.A., 2015. Ceramides–lipotoxic inducers of metabolic disorders. *Trends in Endocrinology & Metabolism, 26*(10), pp. 538-550.

Chen K, Gao C. Targeted genome modification technologies and their applications in crop improvements. *Plant Cell Rep*. 2013;33:575-583.

Christian M, Cermak T, Doyle EL, Schmidt C, Zhang F, Hummel A, *et al.* Targeting DNA double-strand breaks with TAL effector nucleases. *Genetics*. 2010;186:757-761.

Cordain, L., Eaton, S.B., Sebastian, A., Mann, N., Lindeberg, S., Watkins, B.A., O'Keefe, J.H. and Brand-Miller, J., 2005. Origins and evolution of the Western diet: health implications for the 21st century. *American Journal of Clinical Nutrition, 81*(2), pp. 341-354.

Czyż, K., Bodkowski, R., Herbinger, G. and Librowski, T., 2016. Omega-3 fatty acids and their role in central nervous system-a review. *Current Medicinal Chemistry, 23*(8), pp. 816-831.

Dajue, L. and Mündel, H.H., 1996. *Safflower, Carthamus Tinctorius L.* Biodiversity International.

Danso-Baoteng, E., 2011. Effect of enzyme and heat pretreatment on sunflower oil recovery using aqueous and hexane extractions. *Engineering Technology, 80*, pp. 839-845.

Das, U.N., 2010. Essential fatty acids and their metabolites in the context of hypertension. *Hypertension Research, 33*(8), pp. 782-785.

Datta, N., 1976. Vegetable oils from some nonconventional sources. *Economic Botany, 30*(3), pp. 298-298.

De Mello, M.L., Bora, P.S. and Narain, N., 2001. Fatty and amino acids composition of melon (Cucumis melo var. saccharinus) seeds. *Journal of Food Composition and Analysis, 14*(1), pp. 69-74.

Diakogiannaki, E., Welters, H.J. and Morgan, N.G., 2008. Differential regulation of the endoplasmic reticulum stress response in pancreatic β-cells exposed to long-chain saturated and monounsaturated fatty acids. *Journal of Endocrinology, 197*(3), pp. 553-563.

Divine Bup Nde, and Anuanwen Claris Foncha, Optimization Methods for the Extraction of Vegetable Oils: A Review, 2020, *Processes.* 8 (2): 209.

Dyall, S.C. and Michael-Titus, A.T., 2008. Neurological benefits of omega-3 fatty acids. *Neuromolecular Medicine, 10*(4), pp. 219-235.

Elarbi, M.B., Khemiri, H., Jridi, T. and Hamida, J.B., 2009. Purification and characterization of α-amylase from safflower (Carthamus tinctorius L.) germinating seeds. *Comptesrendus Biologies, 332*(5), pp. 426-432.

Ennouri, M., Fetoui, H., Bourret, E., Zeghal, N. and Attia, H., 2006. Evaluation of some biological parameters of Opuntia ficus indica. 1. Influence of a seed oil supplemented diet on rats. *Bioresource Technology, 97*(12), pp. 1382-1386.

Esuoso, K., Lutz, H., Kutubuddin, M. and Bayer, E., 1998. Chemical composition and potential of some underutilized tropical biomass. I: fluted pumpkin (Telfairia occidentalis). *Food Chemistry, 61*(4), pp. 487-492.

Food and Agriculture Organization of the United Nations, 2010. Fats and fatty acids in human nutrition: Report of an expert consultation. *FAO Food Nutr Pap, 91*, pp. 1-166.

Francis, G., Kerem, Z., Makkar, H. P., & Becker, K. (2002). The biological action of saponins in animal systems: A review. *British Journal of Nutrition*, 88(6), 587-605.

Gao C. Genome editing in crops: from bench to field. *Natl Sci Rev*. 2015; 2:13-15.

Garcıa-López, P.M., Muzquiz, M., Ruiz-Lopez, M.A., Zamora-Natera, J.F., Burbano, C., Pedrosa, M.M., Cuadrado, C. and Garzón-De la Mora, P., 2001. Chemical composition and fatty acid profile of several Mexican wild lupins. *Journal of Food Composition and Analysis, 14*(6), pp. 645-651.

Gemede, H. F., & Ratta, N. (2014). Antinutritional factors in plant foods: Potential health benefits and adverse effects. *International Journal of Nutrition and Food Sciences*, 3(4), 284-289.

Gervajio, G.C., 2005. Fatty acids and derivatives from coconut oil. *Bailey's Industrial Oil and Fat Products, 6*(6).

Golkar, P. and Mokhtari, N., 2018. Molecular diversity assessment of a world collection of safflower genotypes by SRAP and SCoT molecular markers. *Physiology and Molecular Biology of Plants, 24*(6), pp. 1261-1271.

Griffin, B.A. and Cunnane, S.C., 2009. Nutrition and metabolism of lipids. *Introduction to Human Nutrition*, p.86.

Grosshagauer, S., Steinschaden, R., Pignitter, M., Strategies to increase the oxidative stability of cold pressed oils, *LWT*, 106 (2019), pp. 72-77.

Gulfraz, M., Kasuar, R., Arshad, G., Mehmood, S., Minhas, N., Asad, M.J., Ahmad, A. and Siddique, F., 2009. Isolation and characterization of edible oil from wild olive. *African Journal of Biotechnology*, *8*(16), pp. 3734-3738.

Hamamci, C., Saydut, A., Tonbul, Y., Kaya, C. and Kafadar, A.B., 2011. Biodiesel production via transesterification from safflower (Carthamus tinctorius L.) seed oil. *Energy Sources, Part A: Recovery, Utilization, and Environmental Effects*, *33*(6), pp. 512-520.

Handa, V., Kumar, V., Panghal, A., Suri, S., & Kaur, J. (2017). Effect of soaking and germination on physicochemical and functional attributes of horse gram flour. *Journal of Food Science and Technology*, 54(13), 4229-4239.

Harijith, A., Ebenezer, D.L. and Natarajan, V., 2014. Reactive oxygen species at the crossroads of inflammasome and inflammation. *Frontiers in Physiology*, *5*, p. 352.

Hesse, A., 1949. Industrial biosyntheses: part I. Fats. *Advances in Enzymology and Related Areas of Molecular Biology*, *9*, pp. 653-704.

Hwang, L.S., 2005. Sesame oil. *Bailey's Industrial Oil and Fat Products*, *2*, pp. 547-552.

Jakobsen, M.U., O'Reilly, E.J., Heitmann, B.L., Pereira, M.A., Bälter, K., Fraser, G.E., Goldbourt, U., Hallmans, G., Knekt, P., Liu, S. and Pietinen, P., 2009. Major types of dietary fat and risk of coronary heart disease: a pooled analysis of 11 cohort studies. *American Journal of Clinical Nutrition*, *89*(5), pp. 1425-1432.

Jaybhaye, R. V., & Srivastav, P. P. (2015). Development of barnyard millet ready-toeat snack food: Part II. *Food Science Research Journal*, 6(2), 285–291.

Jinek M, Chylinski K, Fonfara I, Hauer M, Doudna JA, Charpentier E. A programmable dual-RNA-guided DNA endonuclease in adaptive bacterial immunity. *Science*. 2012, 337, 816-821.

Jorge, N. and Luzia, D.M.M., 2012. Characterization of seed oil Pachira aquatica Aublet for food utilization. *Acta Amazonica*, *42*(1), pp. 149-156.

Kaur, R., Kaur, M., & Gill, B., Phenolic acid composition of flaxseed cultivars by ultra-performance liquid chromatography (UPLC) and their antioxidant activities: effect of sand roasting and microwave heating, *J. Food Process. Preserv.*, 41 (2017), p. e13181.

Khan, N.A, Booker, H. and Yu, P., 2015. Effect of heating method on alteration of protein molecular structure in flaxseed: relationship with changes in protein subfraction profile and digestion in dairy cows. *Journal of Agricultural and Food Chemistry*, *63*(4), pp. 1057-1066.

Khokhar, S. and Chauhan, B.M., 1986. Antinutritional factors in moth bean (Vignaaconitifolia): varietal differences and effects of methods of domestic processing and cooking. *Journal of Food Science*, *51*(3), pp. 591-594.

Khosla, P., 2006. Palm oil: a nutritional overview. *Agro Food Industry Hi Tech*, *17*(3), p. 21.

Kim YG, Cha J, Chandrasegaran S. Hybrid restriction enzymes: zinc finger fusions to Fok I cleavage domain. *Proc Natl Acad Sci USA*. 1996;93:1156-1160.

Kittiphoom, S. and Sutasinee, S., 2015. Effect of microwaves pretreatments on extraction yield and quality of mango seed kernel oil. *International Food Research Journal, 22*(3), p. 960.

Kyle, D.J. and Ratledge, C., 1992. *Industrial applications of single cell oils.* AOCS Publishing/

Lee, Y.C., Oh, S.W., Chang, J. and Kim, I.H., 2004. Chemical composition and oxidative stability of safflower oil prepared from safflower seed roasted with different temperatures. *Food Chemistry, 84*(1), pp. 1-6.

Li, F., He, Z.S. and Ye, Y., 2002. Flavonoids from Carthamus tinctorius. *Chinese Journal of Chemistry, 20*(7), pp. 699-702.

Liu, R., Lu, M., Zhang, T., Zhang, Z., Jin, Q., Chang, M., & Wang, X., Evaluation of the Antioxidant Properties of Micronutrients in Different Vegetable Oils, European *Journal of Lipid Science and Technology*, 2020, 122, (2).

Lonnerdal, B. (2002). Phytic acid–trace element (Zn, Cu, Mn) interactions. *International Journal of Food Science & Technology*, 37(7), 749-758.

Mailer, R., 2006. Chemistry and quality of olive oil. Primefacts. Profitable and Sustainable Primary Industries.

Masurkar, A.A.K., Datar, A.G. and Vakil, B.V., 2015. New microbial sources of PUFA from the Arabian Sea and fresh water habitats of Indian Territory. *International Journal of Pharmaceutical Sciences and Research, 6*(1), p. 349.

Mazaheri, Y., Torbati, M., Azadmard-Damirchi, S. and Savage, G.P., 2019. Effect of roasting and microwave pre-treatments of Nigella sativa L. seeds on lipase activity and the quality of the oil. *Food Chemistry, 274*, pp. 480-486.

McKevith, B., 2005. Nutritional aspects of oilseeds. *Nutrition Bulletin, 30*(1), pp. 13-26.

Min, D.B., Jeffrey, M.B., Lipid oxidation of edible oil, Food Sci. Tech. NY: Marcel Dekker (2002), pp. 335-364.

Mohamed Ahmed, I. A., Musa Ozcan, M., Uslu, N., Juhaimi, F. A. L., Osman, M. A., Alqah, H. A. S., Ghafoor, K., & Babiker, E. E. (2020). Effect of microwave roasting on color, total phenol, antioxidant activity, fatty acid composition, tocopherol, and chemical composition of sesame seed and oils obtained from different countries. *Journal of Food Processing and Preservation.* 44.

Morrison, W.H., Hamilton, R.J. and Kalu, C., 1995. Sunflowerseed oil. In *Developments in Oils and Fats* (pp. 132-152). Springer, Boston, MA.

Nagaraj, G., 1995. Quality and utility of oilseeds. Directorate of oilseeds research, India.

Nagatsu, A., Zhang, H.L., Watanabe, T., Taniguchi, N., Hatano, K., Mizukami, H. and Sakakibara, J., 1998. New steroid and matairesinol glycosides from safflower (Carthamus tinctorius L.) oil cake. *Chemical and Pharmaceutical Bulletin, 46*(6), pp. 1044-1047.

Nageli, C. and Loew, O., 1878. On the chemical composition of yeast. *Justus Liebig's Annalen der Chemie*, 193 (2), pp. 322-348.

Nagy, K. and Tiuca, I.D., 2017. Importance of fatty acids in physiopathology of human body. In *Fatty acids*. IntechOpen.

Nakamura, S., Takamura, T., Matsuzawa-Nagata, N., Takayama, H., Misu, H., Noda, H., Nabemoto, S., Kurita, S., Ota, T., Ando, H. and Miyamoto, K.I., 2009. Palmitate induces insulin resistance in H4IIEC3 hepatocytes through reactive oxygen species produced by mitochondria. *Journal of Biological Chemistry*, *284*(22), pp. 14809-14818.

Namgaladze, D. and Brune, B., 2016. Macrophage fatty acid oxidation and its roles in macrophage polarization and fatty acid-induced inflammation. *Biochimica et Biophysica Acta (BBA)-Molecular and Cell Biology of Lipids*, *1861*(11), pp. 1796-1807.

Nimbkar, N. 2002. Safflower rediscovered. *Times Agric. J.* 2, pp. 32-36.

Nivetha, K., Rao, A.S. and Nair, A., 2021. Microbial production of Omega-3 fatty acids: An overview. *Journal of Applied Microbiology*.

Norde, M.M., Oki, E. and Rogero, M.M., 2019. C-Reactive Protein and Fatty Acids: Public Health Concerns and Implications. In *The Molecular Nutrition of Fats* (pp. 117-133). Academic Press.

Nzikou, J.M., Mvoula-Tsieri, M., Matouba, E., Ouamba, J.M., Kapseu, C., Parmentier, M. and Desobry, S., 2006. A study on gumbo seed grown in Congo Brazzaville for its food and industrial applications. *African Journal of Biotechnology*, *5*(24).

Nzikou, J.M., Mvoula-Tsieri, M., Matos, L., Matouba, E., Ngakegni-Limbili, A.C., Linder, M. and Desobry, S., 2007. Solanum nigrum L. seeds as an alternative source of edible lipids and nutriment in Congo Brazzaville. *Journal of Applied Sciences*, *7*(8), pp. 1107-1115.

Oberoi, H.S., Babbar, N., Dhaliwal, S.S., Kaur, S., Vadlani, P.V., Bhargav, V.K. and Patil, R.T., 2012. Enhanced oil recovery by pre-treatment of mustard seeds using crude enzyme extract obtained from mixed-culture solid-state fermentation of kinnow (Citrus reticulata) waste and wheat bran. *Food and Bioprocess Technology*, *5*(2), pp. 759-767.

Oelke, E.A., Oplinger, E.S., Teynor, T.M., Putnam, D.H., Doll, J.D., Kelling, K.A., Durgan, B.R. and Noetzel, D.M., 1992. Safflower. *Alternative Field Crops Manual*.

Ojha, P., Adhikari, R., Karki, R., Mishra, A., Subedi, U. and Karki, T.B., 2018. Malting and fermentation effects on antinutritional components and functional characteristics of sorghum flour. *Food Science & Nutrition*, *6*(1), pp. 47-53.

Oracz, J., Nebesny, E., & Żyżelewicz, D. (2014). Effect of roasting conditions on the fat, tocopherol, and phytosterol content and antioxidant capacity of the lipid fraction from cocoa beans of different Theobroma cacao L. cultivars. *European Journal of Lipid Science and Technology*. 116, 1002-1014.

Ozdemir, M. and Devres, O., 2000. Analysis of color development during roasting of hazelnuts using response surface methodology. *Journal of Food Engineering*, *45*(1), pp. 17-24.

Pare, A., Nema, A., Singh, V.K. and Mandhyan, B.L., 2014. Combined effect of ohmic heating and enzyme assisted aqueous extraction process on soy oil recovery. *Journal of Food Science and Technology, 51*(8), pp. 1606-1611.

Pearl, S.A., Bowers, J.E., Reyes-Chin-Wo, S., Michelmore, R.W. and Burke, J.M., 2014. Genetic analysis of safflower domestication. *BMC Plant Biology, 14*(1), p. 43.

Piorreck, M., Baasch, K.H. and Pohl, P., 1984. Biomass production, total protein, chlorophylls, lipids and fatty acids of freshwater green and blue-green algae under different nitrogen regimes. *Phytochemistry, 23*(2), pp. 207-216.

Ramadan, M.F. and Mörsel, J.T., 2003a. Oil cactus pear (Opuntia ficus-indica L.). *Food Chemistry, 82*(3), pp. 339-345.

Ramadan, M.F. and Mörsel, J.T., 2003b. Recovered lipids from prickly pear [Opuntia ficus-indica (L.) Mill] peel: a good source of polyunsaturated fatty acids, natural antioxidant vitamins and sterols. *Food Chemistry, 83*(3), pp. 447-456.

Ramadan, M.F., Sharanabasappa, G., Parmjyothi, S., Seshagiri, M. and Moersel, J.T., 2006. Profile and levels of fatty acids and bioactive constituents in mahua butter from fruit-seeds of buttercup tree [Madhuca longifolia (Koenig)]. *European Food Research and Technology, 222*(5), pp. 710-718.

Rattray, J.B.M., 1988. Yeasts. *Microbial lipids, 1*, pp. 555-697.

Ren, R., Shi, C., Cao, J., Sun, Y., Zhao, X., Guo, Y., Wang, C., Lei, H., Jiang, H., Ablat, N. and Xu, J., 2016. Neuroprotective effects of a standardized flavonoid extract of safflower against neurotoxin-induced cellular and animal models of Parkinson's disease. *Scientific Reports, 6*, p. 22135.

Ren, X., Wang, L., Xu, B., Wei, B., Liu, Y., Zhou, C., Ma, H., & Wang, Z. (2019). Influence of microwave pretreatment on the flavor attributes and oxidative stability of cold-pressed rapeseed oil. *Drying Technology.* 37, 397-408.

Roh, J.S., Han, J.Y., Kim, J.H. and Hwang, J.K., 2004. Inhibitory effects of active compounds isolated from safflower (Carthamus tinctorius L.) seeds for melanogenesis. *Biological and Pharmaceutical Bulletin, 27*(12), pp. 1976-1978.

Rombaut, N, R Savoire, J Castello, Hecke E. Van, N Rombaut, B Thomasset, R Savoire, and J.-L Lanoiselle. "Optimization of Oil Yield and Oil Total Phenolic Content During Grape Seed Cold Screw Pressing." *Industrial Crops and Products*. 63 (2015): 26-33.

Rosenthal, A., Pyle, D.L., Niranjan, K., Aqueous and enzymatic processes for edible oil extraction, *Enzym. Microb. Technol.*, 19 (1996), pp. 402-420.

Sabikhi, L. and Kumar, M.S., 2012. Fatty acid profile of unconventional oilseeds. *Advances in Food and Nutrition Research, 67*, pp. 141-184.

Sato, H., Kawagishi, H., Nishimura, T., Yoneyama, S., Yoshimoto, Y., Sakamura, S., Furusaki, A., Katsuragi, S. and Matsumoto, T., 1985. Serotobenine, a novel phenolic amide from safflower seeds (Carthamus tinctorius L.). *Agricultural and Biological Chemistry, 49*(10), pp. 2969-2974.

Shahidi, F., Zhong, Y., Lipid oxidation and improving the oxidative stability, *Chem. Soc. Rev.*, 39 (2010), pp. 4067-4079.

Shankar, D., Agrawal, Y.C., Sarkar, B.C., and Singh, B.P.N., Enzymatic hydrolysis in conjunction with conventional pretreatments to soybean for enhanced oil availability and recovery, *Journal of the American Oil Chemists' Society, 1997,* 74, 1543-1547.

Shrestha, K., & De, M. B., Effect of Seed Roasting on Canolol, Tocopherol, and Phospholipid Contents, Maillard Type Reactions, and Oxidative Stability of Mustard and Rapeseed Oils. *Journal of Agricultural and Food Chemistry, 2014, 62* (24), 5412-5419.

Shukla, B. D., Srivastava, P. K. and Gupta, R. K., 1992. *Oilseeds processing technology*. India.

Smith, J.R., 1996. *Safflower.* AOCS Press, Champaign, IL, p.624.

Sorho, S., Yaya, S., Augustin, A.A. and Lemée, L., 2006. Fatty acids and unsaponifiable composition of Cucumis amaris seeds oil. *Journal of Applied Sciences, 6*(15), pp. 3167-3169.

Spector, A.A. and Kim, H.Y., 2015. Discovery of essential fatty acids. *Journal of Lipid Research, 56*(1), pp. 11-21.

Strayer, D., Belcher, M., Dawson, T., Delaney, B., Fine, J., Flickinger, B., Friedman, P., Heckel, C., Hughes, J., Kincs, F. and Liu, L., 2006. Food Fats and Oils, Institute of Shortening and Edible Oils. Washington, DC.

Sujatha, M., Geetha, A.P., Sivakumar, P. and Palanisamy, N., 2008. Biotechnological interventions for genetic improvement of safflower. In *Proceedings of the 7th International Safflower Conference,* pp. 3-6.

Symington LS, Gautier J. Double-strand break end resection and repair pathway choice. *Annu Rev Genet.* 2011; 45:247-271.

Taha, E. and Matthäus, B., 2018. Effect of Roasting Temperature on Safflower Seeds and Oil. *Journal of Food and Dairy Sciences, 9*(3), pp. 103-109.

Temelli, F. Perspectives on supercritical fluid processing of fats and oils. *Journal of Supercritical Fluids.,* 2009, 47, 583-590.

Thorpe, R.F. and Ratledge, C., 1972. Fatty acid distribution in triglycerides of yeasts grown on glucose or n-alkanes. *Microbiology, 72*(1), pp. 151-163.

Tripathy, D. B., & Mishra, A. (2016). Renewable plant-based raw materials for industry. *Sustainable Inorganic Chemistry*, 353.

TV, A.K., Mani, I., Pramod, A., Samuel, D.V.K., Jha, S.K., Sahoo, P.K., Sinha, J.P. and Kar, A., 2018. Effect of extrusion technique on antinutritional factors of sorghum-soya blends. *Indian Journal of Agricultural Sciences, 88*(3), pp. 420-428.

Van Bruggen, J., Veth, P. and Sebastiaan, N.L., 1993. Method and device for reducing the amount of anti-nutritional factors in a mixture of raw material for animal feed. World Intellectual Property Organization (WO/1993/005664).

Velasco, L. and Fernandez-Martinez, J.M., 2001. Breeding for oil quality in safflower. In *Proceedings of the 5th International Safflower Conference,* Williston, North Dakota and Sidney, Montana, USA. Safflower: a multipurpose species with unexploited potential and world adaptability, pp. 133-137.

Vieira, S.A., Zhang, G., Decker, E.A., Biological implications of lipid oxidation products, *J. Am. Oil Chem. Soc.*, 94 (2017), pp. 339-351.

Wakelyn, P.J. and Wan, P.J., 2005. Solvent extraction to obtain edible oil products. *Handbook of Functional Lipids*, CRC Press, US, pp. 89-131.

Wannes, W.A., Mhamdi, B., Sriti, J. and Marzouk, B., 2010. Glycerolipid and fatty acid distribution in pericarp, seed and whole fruit oils of Myrtus communis var. italica. *Industrial Crops and Products*, *31*(1), pp. 77-83.

Waszkowiak, K., Mikołajczak, B., Kmiecik, D., Changes in oxidative stability and protein profile of flaxseeds resulting from thermal pre-treatment, *J. Sci. Food Agric.*, 2018, 98, pp. 5459-5469.

Weiss, E.A., 2000. *Oilseed crops*, Blackwell Science, London, p. 364.

Xiang, M.S., Tan, J.K. and Macia, L., 2019. Fatty acids, gut bacteria, and immune cell function. *Molecular Nutrition of Fats*, pp. 151-164.

Yang, Kai-Min, Fu-Lan Hsu, Chih-Wei Chen, Chin-Lin Hsu, and Ming-Ching Cheng. 2018. "Quality Characterization and Oxidative Stability of Camellia Seed Oils Produced with Different Roasting Temperatures". *Journal of Oleo Science*. 67 (4): 389-396.

Yoo, H.H., Park, J.H. and Kwon, S.W., 2006. An anti-estrogenic lignan glycoside, tracheloside, from seeds of Carthamus tinctorius. *Bioscience, Biotechnology, and Biochemistry*, *70*(11), pp. 2783-2785.

Young, G. and Conquer, J., 2005. Omega-3 fatty acids and neuropsychiatric disorders. *Reproduction Nutrition Development*, *45*(1), pp. 1-28.

Zhang, Yi, Karen Massel, Ian D. Godwin, and Caixia Gao. 2018. *Applications and potential of genome editing in crop improvement. Genome Biology*. https://doi.org/10.1186/s13059-018-1586-y.

Zhang, W., Cao, X., Liu, S.Q., Aroma modulation of vegetable oils- a review, *Crit. Rev. Food Sci. Nutr.*, 2020, 60, pp. 1538-1551.

3

Sustainability of Oils and Fats Over Petrochemicals

Swati Chaudhary

Department of Applied Science & Humanities, Lloyd Institute of Engineering and Technology, Greater Noida, Uttar Pradesh, India

Abstract

An alarming increase in industrialization and urbanization in combination with the increasing rate of global concerns, i.e., global warming and greenhouse gases over the past few years, compelled us to focus on the concept of "sustainable development". In this regard, the shift from "fossil feedstock" to "renewable resources" for daily uses can considerably contribute to sustainable development. Renewable resources mainly consist of plant- and animal-derived "Oils and fats". There are a number of advantages offered by fats and oils–based products in comparison to petrochemicals-based raw materials; they are low cost, with an environment friendly nature, a simple synthesis process, and lower toxicity, etc. In addition, fats and oils–based products covered a wide span of applications which includes pharmaceuticals, dyes, coatings, packaging, cosmetics, lubricants etc. Further, the maximum utilization of renewable resources based on fats and oils wherever and whenever possible is an important step towards sustainable development. Thus, the need of the hour is to focus on the development and mass production of oils and fats–based products in spite of petrochemical-based products. This chapter mainly focuses on outlining the sustainable characteristics of oils and fats over the petrochemicals. Herein, we have broadly covered the various properties of oils and fats in comparison with petrochemicals. Recent trends of applications have been also discussed.

***Keywords*:** Sustainable development, fossil fuels, petrochemicals, renewable resources

Email: swatichoudhary819@gmail.com

Divya Bajpai Tripathy, Anjali Gupta, Pooja Agarwal, Anuradha Mishra, and Arvind Kumar Jain (eds.) Oils and Fats as Raw Materials for Industry, (65–84) © 2024 Scrivener Publishing LLC

3.1 Oils and Fats as Renewable Feedstock

The alarming increase in industrialization and modernization resulted in exaggerated demands for energy. The major energy sources include fossil fuels which consist of coal, petroleum and natural gas. However, the limited and depleted stocks of these resources resulted in the need to develop alternative sustainable and environmentally friendly energy sources (Akizu-Gardoki *et al.*, 2022). Thus, the present chapter majorly discusses and compares the two types of energy sources, which are Oleochemicals and Petrochemicals.

3.1.1 What are Oleochemicals?

Oils and fats derived from plant and animal sources used to be the most essential renewable feedstock of various types of industries. This can be ascribed to the conversion of conventional oils and fats to Oleochemicals. Oleochemicals belong to a class of chemicals which are procured from natural oils and fats to be used as the raw material for a number of industries. The major reasons behind the increasing consumer demands for oleochemicals are their reversibility, sustainability, environment friendly nature and most importantly, the increasing prices of oils (Vennestrøm *et al.*, 2011; Biermann *et al.*, 2011).

In addition, oleochemicals can also be used as an alternative to petroleum-based products (or petrochemicals) which leads to the shifting of manufacturers from petrochemicals to oleochemicals. They can be produced via different chemical reactions such as hydrolysis of water, hydrolysis of alcohols, hydrogenation etc. Some commonly used oleochemicals are palm oil, soya oil, coconut oil, etc. (Gunstone, 2009).

3.1.2 Classes of Oleochemicals

The four major classes of Oleochemicals are briefly discussed below.

3.1.2.1 Acids

Oils and fats on hydrolysis with water resulted in the formation of free acids. These reactions are usually carried out at high pressure (20-60 bar) and temperature (250°C). Lower temperature is preferred for highly unsaturated oils as at high temperature discoloration takes place and both fatty acids and glycerol will get distilled. The synthesized acids are utilized in bulk quantities to synthesize esters, soaps and amines (Katagi *et al.*, 2020; Metzger, 2009).

3.1.2.2 Esters

Different methods are available to synthesize esters using fats and oils using acidic, alkaline or enzymatic catalysts. The most preferred industrial method to synthesize esters is the reaction of triacylglycerols (natural oils and fats) with methanol in the presence of suitable catalysts. This reaction is also known as Methanolysis. This reaction is carried out on a wide scale for preparing methyl ester to be used as biodiesel, solvent and in the production of alcohols (Round *et al.*, 2019).

3.1.2.3 Alcohols

The synthesis route of alcohols from natural fats and oils includes the hydrogenolysis of free acid, methyl ester or glycerol ester. Out of all the above-mentioned methods, the second method is most feasible. Catalytic reductions are usually performed at pressure (250-300 bar) and temperature (~200°C) (Noweck and Grafahrend, 2006).

3.1.2.4 Fatty Amines

Fatty amines are basically the intermediates for synthesizing different types of nitrogen containing compounds (majorly used as the intermediates). The synthesis route that is typically followed for preparing fatty amines is acids are converted to nitriles via the formation of corresponding amines by reacting with ammonia in the presence of catalysts such as manganese acetate, bauxite, zinc oxide or cobalt salts at a temperature of 280-360°C. Further, nitriles are converted to primary or secondary amines via hydrogenation in presence of ammonia and nickel or cobalt catalyst at 120-180°C and 20-40 bar. The obtained products can further be converted into tertiary amines or quaternary ammonium compounds on reaction with formaldehyde or alkylating agents respectively. All these synthesized nitrogen-containing compounds can be further used as cationic surfactants (Cha *et al.*, 2020).

3.1.3 Applications of Oleochemicals

Oleochemical-based products are used in a wide variety of applications. Some of them are discussed below.

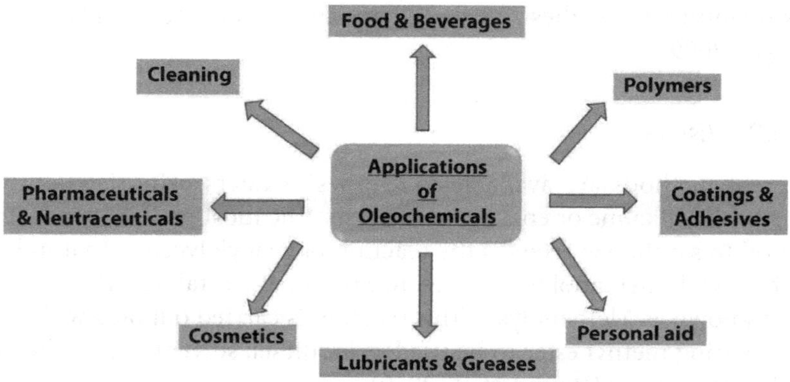

Figure 3.1 Applications of oleochemicals.

3.1.3.1 Pharmaceuticals & Nutraceuticals

Oleochemicals can be used as preservatives, ointments and viscosifiers. For instance, Isopropyl myristate is used as moisturizer in the medicinal products. Further, Glycerine (99.7%) is used in the composition of cough syrup as the emollient (ointment) (Karlheinz Hill, 2007).

3.1.3.2 Cosmetics

The major advantage of oleochemicals lies in the production of customer safety products. Thus, it can also be used for the production of cosmetic products such as Octyl stearate present in lipstick and eyeshadow as a moisturizer is a palm plant–based oleochemical. Similarly, polyhexanide is an aquatic preservative present in many makeup products (Xu *et al.*, 2007).

3.1.3.3 Lubricants & Greases

Oleochemical-based lubricants and greases are mainly utilized in automobile industries as these materials possess a property to reduce the friction between the moving parts. Rubber-based Stearic acid is primarily used in the synthesis of lubricants and greases. Further, cold and heat resistant properties of Dioctyl sebacate make it an ideal alternative for engine lubricating oils as a base (Rupilius and Ahmad, 2007).

3.1.3.4 *Cleaning*

Oleochemical-derived products have been successfully utilized for industrial and household cleaning as wetting agents, additives, foaming agents and sanitizing agents. For example, coconut fatty acid can be used as a conditioner for leather-based products and as a wax for polishing wooden furniture (Ghazali *et al.*, 2021).

3.1.3.5 *Food & Beverages*

Oleochemicals can also be used as food packaging products and sanitizers for food contact surfaces; stearic acid, for example, is used as a hardening agent in the food and beverages industry. In addition, it is also used as the molding agent while preparing candies. Calcium stearate is used as the food humectant also (Jeya Henry, 2009 and Rios *et al.*, 2014).

3.1.3.6 *Coatings & Adhesives*

Eco-friendly behavior of oleochemicals-based products make it likely that they will be used in the form of coatings and adhesives in the engineering and building industries. Zinc stearate is commonly used as the lubricating agent in paints and coatings. Castor oil is also used as the varnishing agent for industrial sealing purposes (Patel *et al.*, 2012).

3.1.3.7 *Personal Aid*

The antigenic and non-carcinogenic properties of oleochemicals are utilized in the production of personal care products. Thus, they play an important role in self-aid as the ingredient of many hair and skin care products. For instance, Glycerine derived from vegetable oil is commonly known and used for its moisturizing properties (Mbaraka and Shanks, 2006).

3.1.3.8 *Polymers*

Oleochemicals have been also utilized in the synthesis of monomers and polymers. Polymer materials such as polymerized soybean oil, castor oil and linseed oil are used as the drying oils. In addition, fatty acid eaters, fatty acid amides and waxes can be used as lubricants. Further, epoxides and soaps can also be used as stabilizers and plasticizers (Meier *et al.*, 2007).

3.2 Petrochemicals as Non-Renewable Feedstock

3.2.1 What are Petrochemicals?

Despite knowledge of the limited availability of fossil fuels, the number of fossil fuels–related applications are enormous. Out of all these applications, a major one is related to petroleum. Petroleum is basically a complicated mixture of hydrocarbons formed by the degradation of fossil leftovers. Further, petroleum is processed to convert into some fruitful parts known as fuels (CNG, LPG, petrol, etc.) or lubricants. In addition, 10% of petroleum is utilized to synthesize a range of chemicals called petrochemicals. Petrochemicals are a class of chemicals procured from hydrocarbons of petroleum or natural gas directly or indirectly (Kubičková and Kubička, 2010).

Further, a number of gaseous hydrocarbons have been also obtained during the refining process of petroleum. These gaseous hydrocarbons basically are the hydrocarbons up to five carbon atoms such as methane, ethane, propane, butane and pentane. They have emerged as a potent solution for a number of applications over the passage of time; the most important is utilizing gaseous hydrocarbons as starting material for the synthesis of petrochemicals. This huge dependency on petrochemicals paved the way for converting higher hydrocarbons to smaller hydrocarbons via the process of cracking (Tripathy *et al.*, 2017).

3.2.2 Classification of Petrochemicals

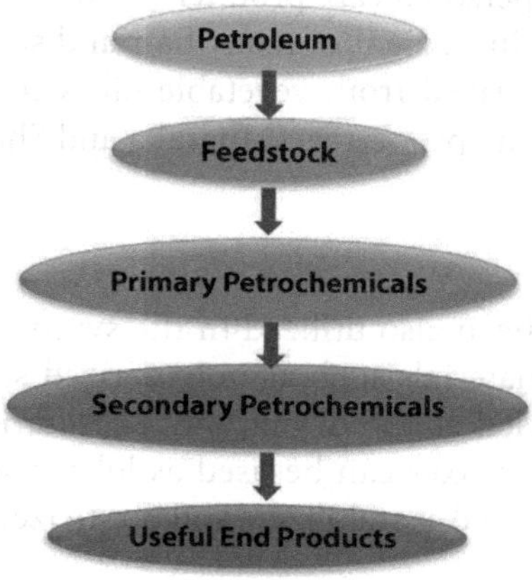

Figure 3.2 Classification of petrochemicals.

3.2.2.1 *On the Basis of Processing*

On the basis of complex processing of petrochemicals starting from petroleum to production of useful products, petrochemicals are broadly classified into four categories:

1. Feedstock
2. Primary Petrochemicals
3. Intermediate Petrochemicals
4. Downstream Petrochemicals

3.2.2.1.1 Feedstock

Feedstock basically is the first material which has been used for the synthesis of petrochemicals. Two most primarily used feedstocks for the production of petrochemicals are (1) Natural gas, (2) Naphtha and reformed naphtha. Natural gas is present in the natural environment in association with petroleum while naphtha is a fraction which has been obtained while refining petroleum takes place. Many industries opted for natural gas as starting material while some others preferred naphtha over natural gas as starting material. Further, the primary criteria for selecting feedstock are availability of particular feedstock and availability of technology (Palankoev *et al.*, 2019).

3.2.2.1.2 Primary Petrochemicals

The molecules which have been obtained from feedstock can further be used as petrochemical intermediates under the category of Primary petrochemicals. Different types of feedstock such as natural gas, naphtha and reformed naphtha resulted in the formation of different types of Primary petrochemicals (Li *et al.*, 2016).

3.2.2.1.3 Secondary Petrochemicals

Secondary petrochemicals are those petrochemicals which have been obtained from primary petrochemicals via chemical reaction. These are also known as Intermediate petrochemicals. These intermediate petrochemicals can be used directly or will be processed up to produce derivatives of petrochemicals via chemical reaction or to produce useful products via a series of reactions (Minakov *et al.*, 2019).

3.2.2.1.4 Downstream Petrochemicals

Downstream petrochemicals are those petrochemicals that have been obtained at a later stage via a series of chemical reactions (Lima *et al.*, 2016).

3.2.2.2 On the Basis of Primary Petrochemicals

It is not possible to list and discuss all the types of petrochemicals here; thus broadly on the basis of the number of primary petrochemical intermediate, petrochemicals are classified according to the following types.

3.2.2.2.1 Petrochemicals from Methane

Methane is the major hydrocarbon component of natural gas. In addition, it is also obtained as the by-product during the petroleum-refining process. The major petrochemicals obtained from methane are its chlorinated products, unsaturated hydrocarbons (namely, ethylene, propylene and acetylene), carbon black, hydrogen and methanol (Wang *et al.*, 2018).

3.2.2.2.2 Petrochemicals from Ethylene

As we have discussed in the previous section, ethylene is obtained from pyrolysis of natural gas or from naphtha by cracking. Ethylene consists of carbon-carbon double bond, which makes it highly reactive as it can be converted into a number of different petrochemicals and useful end products. The major petrochemicals obtained from ethylene are ethylene oxide, ethanol, polyethylene, ethylene glycol, ethyl benzene, vinyl chloride and dichloroethane (Elreedy and Tawfik, 2015).

3.2.2.2.3 Petrochemicals from Propylene

Pyrolysis of natural gas or cracking of naphtha resulted in the formation of propylene which is also an unsaturated hydrocarbon. The major petrochemicals obtained from propylene are glycerol, polypropylene, isopropanol and isopropyl benzene (cumene) (Bayoumi and McCaslin, 2017).

3.2.2.2.4 Petrochemicals from Acetylene

Acetylene, also called ethyne, is also obtained from natural gas via pyrolysis. Further, presence of triple bond also makes it highly reactive. The major petrochemicals obtained from propylene are acetaldehyde, vinyl chloride, vinyl acetate and acrylonitrile (Chen *et al.*, 2021).

3.2.2.2.5 Petrochemicals from Butadiene

Butadiene being a diene consisting of two alternating double bonds is mainly obtained via cracking of naphtha. It is a monomer for synthesis of polybutadiene which acts as a substitute for natural rubber. Butadiene in combination with styrene undergoes copolymerization to form Buna-S (Sadeghi-Yarandi *et al.*, 2020).

3.2.2.2.6 Petrochemicals from Benzene

Naphtha undergoes catalytic reforming (aromatization) to convert aliphatic hydrocarbons to aromatic hydrocarbons. Major petrochemicals obtained from aromatization are nitrobenzene, cyclohexane, chlorobenzene, ethyl benzene, cumene, linear and branched chain alkyl benzene (Wang *et al.*, 2019).

3.2.3 Applications of Petrochemicals

Similar to Oleochemicals, Petrochemicals cover a wide range of applications starting from daily life for example, plastic carry bags, to explosives. Some of the major classification of petrochemicals applications are discussed below.

3.2.3.1 Preservatives

Petrochemicals are majorly used as preservatives for cosmetics and food additives to retain their durability. In this regard, a biological synthesis of vitamins was reported making use of petrochemical waste from corn syrup and meat industry using microbe, B. megaterium. In addition, it is also

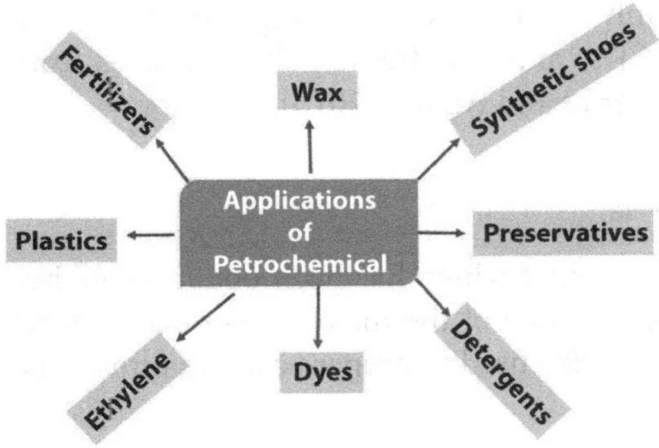

Figure 3.3 Applications of petrochemicals.

used in the production of vitamins such as Acetylsalicylic acid (ASA). This microbe is supposed to have the high ability to utilize a large number of carbon sources (Berlin and Balakshin, 2014).

3.2.3.2 Wax

Petrochemicals can also be used as wax which is further used to prepare candles. Wax is also used to cast statues, showpieces and polishes. A high wax content emulsion was reported in which cream-coloured paraffin wax emulsion was synthesized with high stability, dispersibility and volatility (Kanaujia, 2018).

3.2.3.3 Detergent

Detergents have been also synthesized via a variety of petrochemical products. The major ingredients of detergents are oils and petrochemicals. Detergents are broadly classified into two types: (i) soap detergents, and (ii) soapless detergents (Zhao *et al.*, 2019).

3.2.3.4 Dyes

Dyes are also one of the familiar petroleum products. These products consist of various colors, mainly the ink used in pens. One method of synthesizing printing dye composed of polypropylene acid sodium alginate was proposed. This printing gum has the characteristics of structural viscosity, permittivity and stability (Tripathy *et al.*, 2017).

3.2.3.5 Plastics

Polyester is mainly used for synthesizing plastic essentials. In addition, compact discs (CDs) and cassettes are also prepared from petrochemical oil or polyester (Mrad and Alfares, 2016).

3.2.3.6 Ethylene

Ethylene is a by-product which has been obtained via heating and distillation of oil. It is also one of the major constituents of petrochemical oil. Further, ethylene is also used to prepare camera films, trash bags, containers etc.

3.2.3.7 Fertilizers

Petrochemical products can also be used as fertilizers to protect crops from damage. Fertilizers are also used for increasing crop production.

3.2.3.8 Synthetic Shoes

Petroleum products are also utilized in the synthesis of synthetic shoes also known as sneakers. They are so designed that their sole retains the same elasticity in all types of weather, unlike the sole prepared from natural rubber, which expands when cooled and contracts on heating.

3.3 Oils and Fats vs. Petrochemicals

The overall performance of Oleochemicals (oils & fats) has been compared with petrochemicals on the basis of the following parameters which are discussed in the sections below.

3.3.1 Comparison of Production

The petroleum-based resources mainly consist of long alkyl chains with substituted aromatic compounds. Major fraction (approx. 80%) of these resources are converted into fuels and the rest is converted into petrochemicals. On the other hand, vegetable oils and fats are mainly used for domestic (food & feed) purposes. Only a minor part is further processed to produce oleochemicals and biofuels (Nachtergaele *et al.*, 2019).

3.3.2 Comparison of Processing

The petrochemical industry produces fuels via the processes of catalytic cracking, hydrocracking and extraction in a very systematic way. Further, for converting crude petroleum into petrochemicals, there is a need for additional reactions such as dehydrogenation, epoxidation, chlorination, oxidation, etc., to introduce functional groups which are not present in crude petroleum. While functional groups such as unsaturated ester are already present in fats and oils, these groups can be converted into fatty alcohols via transesterification and hydrogenation. Thus, there is a kind of competition between similar fatty alcohols from products derived from petroleum resources and renewable oils and fats (Chin *et al.*, 2019).

3.3.3 Comparison of Fuel Efficiency

Petrochemistry produces efficient fuels but there is a requirement of additional reaction to produce fuels via petrochemicals. On the other hand, when vegetable oils are used as biodiesel, the presence of functional groups such as unsaturated esters resulted in the production of fuel with low energy content and stability (Shah *et al.*, 2016).

3.3.4 Comparison of Environmental Adaptability

Most of the processes used in the production of petrochemical-based products are not environmentally friendly which resulted in environmental degradation, while on the other hand, the production of "green diesel" by hydrotreatment resulted in diesel oil with the same efficiency as produced from petrochemicals. The term "green" here indicates the environmentally friendly techniques used to produce fats and oils–based fuels as compared to petrochemicals-based fuels (Hosney *et al.*, 2018).

3.3.5 Comparison of Sustainability

Sustainable development is the need of the hour owing to increasing urbanization and industrialization. Thus, in terms of sustainability, oils and fats derived from plant and animal sources are renewable sources with limitless stock. On the other hand, fossil fuel–derived petrochemicals are non-renewable sources with limited stock which goes on decreasing every day due to the higher demands (Tripathy *et al.*, 2017; Zhou *et al.*, 2016).

3.4 Trends in the Oleochemical Industry

The regular depletion and environment hazardous characteristic of non-renewable sources paved the way for dependency on the renewable factors. Apart from that, there are several advantages such as availability, feasibility, easy synthesis, renewability, cost efficiency, durability, eco-friendliness, etc., offered by oleochemicals-based products.

In addition, the rising demand for oleochemicals-based substitutes has been credited to the following reasons also:

- Easy availability of raw materials: owing to the renewable behaviour of oleochemicals-based raw materials in comparison to petrochemical-based ones.

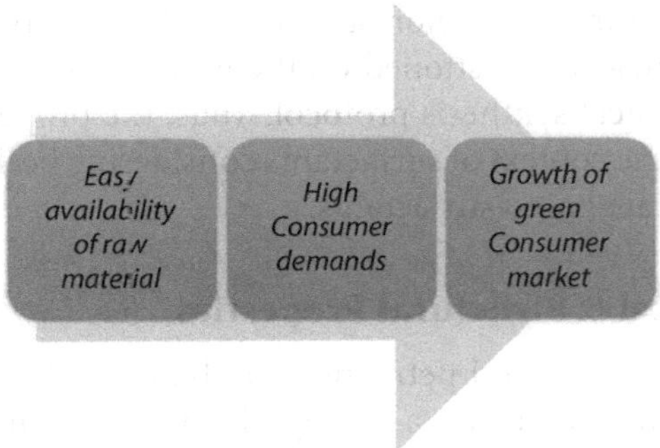

Figure 3.4 Trends in the oleochemicals industry.

- High demand from consumers: owing to the oleochemicals-based products durability, feasibility and utility as compared to petrochemicals-based products.

Growth of the green consumer market: owing to the global expansion of oleochemicals-based industry due to their characteristic advantages, as already mentioned (Tesser *et al.*, 2020).

3.5 Oleochemicals & Petrochemicals Surfactants

Surfactants, basically, are also known as Surface active agents used for removing dirt or grease from dishes, clothes, skin, etc., by holding them as solution/suspension in solution. Further, regarding oleochemicals- and petrochemicals-based surfactants, both have some pros and cons as follows. Oleochemicals-based surfactants are supposed to be better as they are environmentally friendly, but on the other hand, their persistent usage

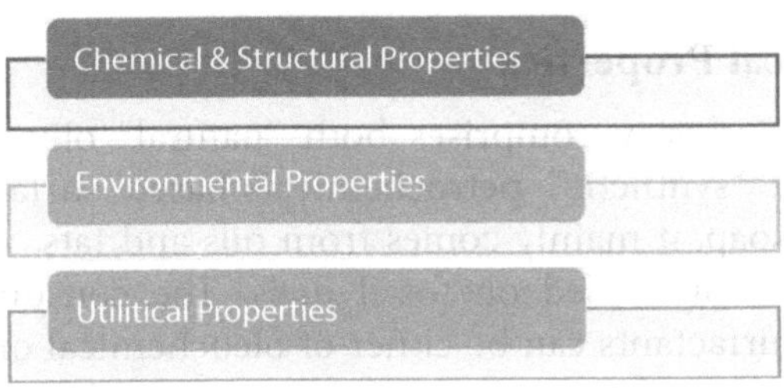

Figure 3.5 Oleochemicals & petrochemicals surfactants.

led to the global problem of deforestation. Similarly, petrochemicals-based surfactants are majorly questioned on the basis of their limited feedstock and tedious chemical synthesis protocol. Thus, the properties will be discussed below in the context of surfactants considering both oleochemicals- and petrochemicals-based surfactants.

3.5.1 Chemical & Structural Properties

Both the oleochemical- and petrochemical-based surfactants possess the same basic organization such as a hydrophilic (water-loving) tail and hydrophobic (water-repelling) head. The role of the tail is to bind the grease/dirt particles while the head pushes the grease/dirt particle-surfactant couple to the water for further removal of grease or dirt particles. However, there are some minor differences between both the reactants such as in the case of oleochemical surfactants, hydrocarbon chains are always even numbered and linear while in petrochemical surfactants, hydrocarbon chains are slightly branched and contain both odd and even numbers of carbon atoms. These differences may lead to significant changes in the cleaning performance of respective surfactants. Sometimes, it has been also observed that the structure of hydrocarbon chains in petrochemical surfactant is more complex than in oleochemical surfactant.

3.5.2 Environmental Properties

The eco-friendly behaviour of oleochemical-based surfactants proved to be more advantageous than petrochemical-based ones. The raw materials of oleochemical surfactant are oils and fats derived from plants and animals; these are renewable, biological and environmentally friendly while the raw materials (fossil fuels) such as coal, petroleum and natural gas of petrochemical surfactants are limited in stock, costly and environmentally hazardous. Thus, we should always go for oleochemical surfactants in comparison to petrochemical surfactants, if environmental feasibility is considered.

3.5.3 Utilitical Properties

The surfactant industry comprises both “natural” oleochemicals-based surfactants and “synthetic” petrochemicals-based surfactants. Further, coming to the soap, it mainly comes from oils and fats, while alkyl benzene sulphonates are based on fossil fuels. The consumption of Fatty alcohol-based surfactants can be either of oleochemical or petrochemical origin based on the relative merits of the two sourcing groups (Salimon *et al.*, 2012).

3.6 Oleochemicals-Based Products

The list of oleochemicals-based products available primarily are as follows:

- Soap Noodles
- Soap Noodles 80:20 Snow White
- Soap Noodles 90:10 Snow White
- Soap Noodles TFM 75% MinSoap Noodles TFM 72% MinSoap Noodles TFM 65% Min
- Fatty Acid & Fatty Alcohol
- Stearic Acid Triple Pressed
- Cetyl Alcohol
- Cetyl Stearyl Alcohol
- Stearic Acid
- Glycerine
- Refined Glycerine
- Refined Glycerine 99.7% Min USP
- Refined Glycerine 99.7% Min USP
- Lauryl Mystryl Alcohol C12-14
- Oleic Acid 75% Min
- Distilled Coconut Fatty Acid
- Palm Fatty Acid Distillate
- Care Chemicals
- Sodium Lauryl Ether Sulfate SLES TN 70%
- Coconut Fatty Acid Diethanolamide-CDE
- Glycerol Mono Stearate (GMS)
- Medium Chain Triglycerides (MCT)

3.7 Conclusion

The increasing energy demands with the rising population paved the way for the large-scale production of energy resources. Thus, the present chapter focused initially on the introduction, types and applications of oleochemicals and petrochemicals. Further, different aspects such as production, processing, fuel efficiency, environmental adaptability and sustainability have also been discussed to compare the efficiency of oleochemicals with petrochemicals. Different aspects have different conclusions such that in terms of environmental adaptability and sustainability, oleochemicals are more efficient than petrochemicals. On the other hand, fuel efficiency of petrochemical-based resources are higher than oleochemical-based fuels.

References

Akizu-Gardoki, O., de Ulibarri, B., Iturrondobeitia, M., Minguez, R., & Lizundia, E. (2022). Ecodesign coupled with Life Cycle Assessment to reduce the environmental impacts of an industrial enzymatic cleaner. *Sustainable Production and Consumption*, 29, 718-729.

Bayoumi, A., & McCaslin, R. (2017). Internet of things – a predictive maintenance tool for general machinery, petrochemicals and water treatment. In *Advanced Technologies for Sustainable Systems* (pp. 137-146). Springer, Cham.

Berlin, A., & Balakshin, M. (2014). Industrial lignins: analysis, properties, and applications. *Bioenergy Research: Advances and Applications*, 315-336.

Biermann, U., Bornscheuer, U., Meier, M. A., Metzger, J. O., & Schäfer, H. J. (2011). Oils and fats as renewable raw materials in chemistry. *Angewandte Chemie International Edition*, 50(17), 3854-3871.

Cha, H. J., Hwang, S. Y., Lee, D. S., Kumar, A. R., Kwon, Y. U., Voß, M., ... & Park, J. B. (2020). Whole-cell photoenzymatic cascades to synthesize long-chain aliphatic amines and esters from renewable fatty acids. *Angewandte Chemie*, 132(18), 7090-7094.

Chen, Y., Du, Y., Wang, Y., Krishna, R., Li, L., Yang, J., & Mu, B. (2021). A stable metal–organic framework with well-matched pore cavity for efficient acetylene separation. *AIChE Journal*, 67(5), e17152.

Chin, S. Y., Azizan, N., Ahmad, M. A. A., & Kamaruzaman, M. R. (2019). Ion Exchange resins catalysed esterification for the production of value added petrochemicals and oleochemicals. In *Applications of Ion Exchange Materials in Chemical and Food Industries* (pp. 75-98). Springer, Cham.

Elreedy, A., & Tawfik, A. (2015). Effect of hydraulic retention time on hydrogen production from the dark fermentation of petrochemical effluents contaminated with *Ethylene Glycol. Energy Procedia*, 74, 1071-1078.

Ghazali, R., Ishak, S. A., & Zolkarnain, N. Palm-based Oleochemicals in Non-Food Applications: Meeting the Environmental Regulatory Challenges.

Gunstone, F. (2009). *Oils and fats in the food industry*. John Wiley & Sons.

Henry, J. (2009). Processing, manufacturing, uses and labelling of fats in the food supply. *Annals of Nutrition & Metabolism*, 55(1/3), 273-300.

Hill, K. (2007). Industrial development and application of biobased oleochemicals. *Pure and Applied Chemistry*, 79(11), 1999-2011.

Hosney, H., Nadiem, B., Ashour, I., Mustafa, I., & El-Shibiny, A. (2018). Epoxidized vegetable oil and bio-based materials as PVC plasticizer. *Journal of Applied Polymer Science*, 135(20), 46270.

Kanaujia, P. K. (2019). Gas Chromatography| Petroleum and Petrochemical Applications. *Encyclopedia of Analytical Science*, 217-231.

Katagi, K. S., Kadam, N. S., Mane, S. G., & Akki, M. (2020). Unique oleochemical data source: Deployed for the optimization of fatty acids of seed oils for production of liquid fuel. *Chemical Data Collections*, 25, 100324.

Keim, W. (2010). Petrochemicals: Raw material change from fossil to biomass. *Petroleum Chemistry*, 50(4), 298-304.

Kubičková, I., & Kubička, D. (2010). Utilization of triglycerides and related feedstocks for production of clean hydrocarbon fuels and petrochemicals: a review. *Waste and Biomass Valorization*, 1(3), 293-308.

Li, X., Li, G., Li, J., Yu, Y., Feng, Y., Chen, Q., & Wang, Y. (2016). Producing petrochemicals from catalytic fast pyrolysis of corn fermentation residual by-products generated from citric acid production. *Renewable Energy*, 89, 331-338.

Lima, C., Relvas, S., & Barbosa-Póvoa, A. P. F. (2016). Downstream oil supply chain management: A critical review and future directions. *Computers & Chemical Engineering*, 92, 78-92.

Mbaraka, I. K., & Shanks, B. H. (2006). Conversion of oils and fats using advanced mesoporous heterogeneous catalysts. *Journal of the American Oil Chemists' Society*, 83(2), 79-91.

Meier, M. A., Metzger, J. O., & Schubert, U. S. (2007). Plant oil renewable resources as green alternatives in polymer science. *Chemical Society Reviews*, 36(11), 1788-1802.

Metzger, J. O. (2009). Fats and oils as renewable feedstock for chemistry. *European Journal of Lipid Science and Technology*, 111(9), 865-876.

Minakov, A. V., Shebeleva, A. A., Strizhak, P. A., Chernetskiy, M. Y., & Volkov, R. S. (2019). Study of the Weber number impact on secondary breakup of droplets of coal water slurries containing petrochemicals. *Fuel*, 254, 115606.

Mrad, M., & Alfares, H. (2016). Optimum multi-period, multi-plant, and multi-supplier production planning for multi-grade petrochemicals. *International Journal of Industrial Engineering-Theory, Applications and Practice*, 23(6), 399-411.

Nachtergaele, P., De Meester, S., & Dewulf, J. (2019). Environmental sustainability assessment of renewables-based propylene glycol at full industrial scale production. *Journal of Chemical Technology & Biotechnology*, 94(6), 1808-1815.

Noweck, K., & Grafahrend, W. (2006). Fatty alcohols. *Ullmann's Encyclopedia of Industrial Chemistry*, 14, 14.

Palankoev, T. A., Dementiev, K. I., & Khadzhiev, S. N. (2019). Promising processes for producing drop-in biofuels and petrochemicals from renewable feedstock. *Petroleum Chemistry*, 59(4), 438-446.

Patel, M. M., Patel, C. J., & Patel, N. K. (2012). Study of Thermal Properties on UV-Curable Coatings Derived from Oleochemical Polyols. *Chemical Science Transactions*, 1(2), 289-296.

Rios, R. V., Pessanha, M. D. F., Almeida, P. F. D., Viana, C. L., & Lannes, S. C. D. S. (2014). Application of fats in some food products. *Food Science and Technology*, 34(1), 3-15.

Round, J. W., Roccor, R., & Eltis, L. D. (2019). A biocatalyst for sustainable wax ester production: re-wiring lipid accumulation in Rhodococcus to yield high-value oleochemicals. *Green Chemistry*, 21(23), 6468-6482.

Rupilius, W., & Ahmad, S. (2007). Palm oil and palm kernel oil as raw materials for basic oleochemicals and biodiesel. *European Journal of Lipid Science and Technology*, 109(4), 433-439.

Sadeghi-Yarandi, M., Golbabaei, F., & Karimi, A. (2020). Evaluation of pulmonary function and respiratory symptoms among workers exposed to 1, 3-Butadiene in a petrochemical industry in Iran. *Archives of Environmental & Occupational Health*, 75(8), 483-490.

Salimon, J., Salih, N., & Yousif, E. (2012). Industrial development and applications of plant oils and their biobased oleochemicals. *Arabian Journal of Chemistry*, 5(2), 135-145.

Shah, J., Arslan, E., Cirucci, J., O'Brien, J., & Moss, D. (2016). Comparison of oleo-vs petro-sourcing of fatty alcohols via cradle-to-gate life cycle assessment. *Journal of Surfactants and Detergents*, 19(6), 1333-1351.

Tesser, R., Vitiello, R., Russo, V., Turco, R., Serio, M. D., Lin, L., & Li, C. (2020). Oleochemistry products. In *Industrial Oil Plant* (pp. 201-268). Springer, Singapore.

Tripathy, D. B., & Mishra, A. (2016). Renewable plant-based raw materials for industry. *Sustainable Inorganic Chemistry*, 353.

Tripathy, D. B., Yadav, A., & Mishra, A. (2017). Applications of petrochemicals: A mini review. *Recent Adv. Petrochem.* Sci, 2, 67-69.

Vennestrøm, P. N. R., Osmundsen, C. M., Christensen, C. H., & Taarning, E. (2011). Beyond petrochemicals: the renewable chemicals industry. *Angewandte Chemie International Edition*, 50(45), 10502-10509.

Wang, A., Austin, D., He, P., Mao, X., Zeng, H., & Song, H. (2018). Direct catalytic co-conversion of cellulose and methane to renewable petrochemicals. *Catalysis Science & Technology*, 8(21), 5632-5645.

Wang, Z., Liu, H., & Wang, J. (2019). Optimization of the separation unit of methanol to propylene (MTP) process and its application. *Chinese Journal of Chemical Engineering*, 27(5), 1089-1093.

Xu, Y. X., Hanna, M. A., & Josiah, S. J. (2007). Hybrid hazelnut oil characteristics and its potential oleochemical application. *Industrial Crops and Products*, 26(1), 69-76.

Zhao, D., Thallapally, P. K., Petit, C., & Gascon, J. (2019). Advanced Porous Materials: Design, Synthesis, and Applications in Sustainability. *ACS Sustainable Chemistry & Engineering*, 7(9), 7997-7998.

Zhou, Y. J., Buijs, N. A., Zhu, Z., Qin, J., Siewers, V., & Nielsen, J. (2016). Production of fatty acid-derived oleochemicals and biofuels by synthetic yeast cell factories. *Nature Communications*, 7(1), 1-9.

4

Oils and Fats in the Food Industry

Garima Gupta* **and Priyanka Singh**

Department of Biotechnology, IILM College of Engineering and Technology, Greater Noida, India

Abstract

The global market for fats and oils was estimated at USD 236.7 billion in 2021 and is expected to grow exponentially in the period 2021-26 to reach USD 285.2 billion. Fats and oil play a major part in a healthy diet and the entire food industry aims to provide these either as a primary product or by incorporating them in various packaged food products. The major functions of fats are energy molecules, absorption of fat-soluble vitamins, providing essential fatty acids, preventing friction between internal organs and helping maintain body temperature. The choice of fats or oils in various food groups is based upon its functionality, like the amount of solid fat present at room temperature. Healthy diet focuses on the quantity and quality of fat consumed. The demand for desired fats and oils by nutritionists to maintain a balance between production and supply is very difficult. The farmers along with food processors have to work towards the current needs like reduced demand of trans acid and increased requirements of omega-3 fatty acids due to increase in obesity among people. Presently almost all of the fats used in the food industry are sourced from plants and only one-sixth of the supply is of animal origin. Fats are extracted, refined, clarified and modified to optimize their chemical, physical and nutritional characteristics to satisfy the demand of health-conscious consumers.

Keywords: Fatty acids, extraction, refining, modification process, blending, hydrogenation, nutritional properties

**Corresponding author*: garima.gupta@iilm.edu

Divya Bajpai Tripathy, Anjali Gupta Pooja Agarwal, Anuradha Mishra, and Arvind Kumar Jain (eds.) Oils and Fats as Raw Materials for Industry, (85–116) © 2024 Scrivener Publishing LLC

4.1 Introduction

Formulated food products contain several components which contribute to the structure, texture, sensory impression or flavour of the product. As the rate of obesity and hypertension is increasing, the demand for nutritious food and for less-fat-containing food is increasing and influencing market trends. Companies are trying to develop products with low fat or fat substitutes, but as it determines the physicochemical and sensory characteristics of products it becomes challenging to develop a product with similar texture, flavour and satiating effect. Oils and fats are a type of lipid that differ from each other in terms of their physical state at room temperature. The presence of a number of fatty acids determines these physical characteristics [1]. Mostly, animal fats are found in solid form, vegetable fats in liquid form and the presence of some tropical vegetable fats that are commonly referred to as "butters" (e.g., cocoa butter).

4.1.1 Fatty Acids

Many natural fatty acids are found in our bodies and the food we consume. Fatty acids are the building blocks of the fat. Most food scientists only need to be conversant with about 20 of them. The commonly found fatty acids consist of a single linear chain of 8–22 carbon atoms which typically feature unsaturated centres present at some places in the linear chain, though there are exceptions [2].

4.1.2 Triacylglycerols

Glycerol esters are the most common form of fatty acids. Each of the three carbon atoms in glycerol consists of a hydroxyl group. This produces a number of different esters [3]. Classification of glycerol esters is in three types, namely Monoacylglycerols, Di-acylglycerols, or Triacylglycerols.

4.1.3 Waxes of Ester

Wax is a material which has specific properties characteristics and often consists of aldehydes, alcohols and hydrocarbons among other medium- and long-chain molecules. As a result, RCOOR stands for the ester of the acid RCOOH and the alcohol ROH [4]. Both vegetable (carnauba, jojoba) and animal (beeswax, wool wax) waxes are used. Some solvent-extracted oils may need to be dewaxed.

4.1.4 The Phospholipids

(Also called phosphatides) are by far the most important biological compounds. Every living plant or animal cell is a sac that houses vital life-sustaining activities. A porous membrane is composed of phospholipids [5]. The membrane allows molecules to go in and out of the cell in a regulated manner. The nature of the cell membranes determines these movements. The lipid bilayer has crucial qualities as it is Amphiphiles, making it useful in medications and food [6].

4.1.5 Sterols

Sterols are present in oils and fats and they share a portion of its properties. Cholesterol is related with coronary illness (CHD) and numerous proposals concerning lipid admission are connected with its effect on the cardiovascular framework. Many vegetable oils consist of 1000-5000 mg/kg of phytosterols [7]. In refined oils it is available at low level; however, some has been taken out in the aeration interaction. Sterols can be recuperated from deodorizer distillate (alongside different mixtures) and these give a significant wellspring of forerunners of numerous drug steroids. Spreads and other food varieties consist of added Phytosterols in order to decrease cholesterol retention [8]. Cholesterol is a zoosterol essentially of creature beginning and is absent in plant frameworks at any critical level.

4.1.6 Tocols

Tocols are heterocyclic compounds comprising a side chain of C16 isoprenoid which is fat soluble. The tocols have two valuable but not identical properties: they exhibit vitamin E activity and they are powerful, natural, fat-soluble antioxidants [9]. The total tocol content in crude vegetable oils can range from 10 ppm (equivalent to mg/kg) in coconut oil to 1370 ppm in soybean oil. Natural blends of tocopherols are used as antioxidants at levels up to 500ppm, often in conjunction with ascorbyl palmitate, which prolongs tocols' antioxidant activity [10].

4.1.7 Hydrocarbons

Hydrocarbons are present in small quantities in oils and fats. These include alkanes, squalene and carotenes. The biological precursor of vitamin A is carotene [11].

4.2 Sources of Oils and Fats

Fats including butter, lard, tallow, chicken fat, and fish oils fat are a major agricultural product that is consumed in the form of milk, cream, butter, and cheese. They are classified as saturated, monounsaturated, or polyunsaturated based on their structural placement from the fatty acid group [12]. Plant sources are primarily polyunsaturated, with saturated amounts coming primarily from animal sources. Palmatic, oleic, stearic, lauric, and linoleic acid are the major fatty acids. Fats and oils affect humans during fatty acid metabolism and are involved in a number of organs and systems. The major ones include endocrine system, sensory taste buds, transport of fat-soluble vitamins, regulating cholesterol density inside the cells, organ protection and metabolism regulation.

Vegetable, animal, and marine fats and oils are all possible. Corn oil, sunflower oil, soybean oil, cottonseed oil, peanut oil, olive oil, canola oil, pumpkin seed oil, safflower oil, grape seed oil, sesame oil bran oil, argan oil, palm oil, linseed oil, and coconut oil are examples of vegetable fats. Soybean is one of the most frequent vegetable oils. Vegetable oils from sunflower, olive, sesame, mustard, etc., contain a lower amount of saturated fats. On the other hand, oils extracted from sources like coconut, palm, tallow, etc., contain a good amount of saturated fats [13].

Lard, tallow, and butterfat are animal fats, while fish oils include salmon oil, cod liver oil, etc. and tabulated in Table 4.1 are the different applications of edible oils from various sources.

Table 4.1 Applications of edible oils from various sources.

S. no	Sources	Applications
1.	Butterfat	Cooking with butterfat and including it in specialty recipes is common. Anhydrous butterfat is known as ghee in India, where it is manufactured in enormous amounts and frequently combined with buffalo milk fat.
2.	Lard	Lard works incredibly well as a shortening in baking because of its comparatively large fat crystals. Lard-based pie crusts have a tendency to be flakier than butter-based ones. To combine the shortening qualities of lard with the flavour of butter, many cooks use both types of fat in their pastries.

(Continued)

Table 4.1 Applications of edible oils from various sources. (*Continued*)

S. no	Sources	Applications
3.	Tallow	Tallow has been a common ingredient and cooking oil in traditional meal preparation.
5.	Soybean oil	Shortenings and margarines made of soybean oil stop the development of a gluten matrix in baked goods, resulting in a delicate, moist product.
6.	Mustard oil	The olive oil of Eastern India is mustard oil. It can also be used as salad oil in addition to being a cooking medium.
7.	Sunflower oil	It has been used to prepare meals such as frozen pre-fried foods and French fries at home, in fast-food establishments, and in the industry. Cooking, salad dressings, and the production of fat spreads all need sunflower oil.
8.	Coconut oil	The use of coconut oil in cooking is likely its most well-known application, although it is also widely employed in baked goods, processed meals, and infant formula by the food industry.
9.	Groundnut oil	It is used for deep frying, sautéing, and dressings because of its stability. All culinary methods, including frying, grilling, sautéing, seasoning, and tempering, are suited for this oil.

4.2.1 Animal Sources

4.2.1.1 Butterfat

Butterfat is commonly derived from cow's milk. Butterfat, water, and salt are mixed together. Vitamin A and, to a lesser level, vitamin D are abundant in the butter list. It contained 29-32% monounsaturated fatty acids, 2-4% polyunsaturated fatty acids, and saturated fatty acids comprise 12-32% [14]. Its unusual flavour and bright colour contribute to its widespread popularity. Because of cheaper prices, improved and consistent quality, and health factors, its popularity as a table spread has been reduced and on the other hand the demand for margarine has increased. Butterfat is also incorporated into dairy products like cheese, whipping cream, ice cream, etc.

4.2.1.2 Tallow

Beef cattle provide the majority of edible tallow. It is tougher and firmer than lard at room temperature. It comprises 54.9% saturated fatty acids and at room temperature it is more firm and tough as compared to lard. Unsaturated fatty acids account for 40.9%. Polyunsaturated fatty acids account for 4.2%.

4.2.2 Oils from Fish

Fish oils are found in the bodies or livers of a variety of fish, including cod, whales, salmon, and so on. Fatty acid content changes not just from one species to the next, but also within the species.

4.2.2.1 Cod Liver Oil

It is procured from cod fish and contains a good amount of omega 3 fatty acids, vitamin A and D. The main omega 3 fatty acids present in cod liver oil are Eicosapentaenoic acid and docosahexaenoic acid.

4.2.3 Vegetable-Oil

4.2.3.1 Soybean Oil

This is the major vegetable oil produced and consumed globally derived from soybean plant seeds, it contains mainly polyunsaturated fatty acids estimated at 61%, followed by mono unsaturated fatty acids and saturated fatty acids at 25% and 1% respectively [15]. Soybean oil has 89% of linolenic acid and 11% of alpha linolenic acid. The major application of soybean oil is in mayonnaise, shortenings, cooking oil, margarine, etc.

4.2.3.2 Palm Oil

Extracted from palm fruits, it is the second most produced oil globally because of its high yield and efficient processing. It consists of a well-balanced fatty acid makeup. Primary acid contents were palmitic acid (5%) and oleic acid (39%), with linoleic acid (UIl%) and barely a trace of linoleic acid [16].

The oil is relatively resilient to oxidative degradation due to low quantities of linolenic acid and nearly no linolenic acid. Some palm oils are better than others, depending on the type. They are used in cooking/frying oils, margarine, shortening, specialty cans and spray-dried products.

4.2.3.3 Canola Oil

Derived from a new species of rapeseed plant, it is considered as a healthy oil due to low content of saturated fatty acids. It occupies third place in the order of oil and grease production. Its lower stability is due to chlorophyll, linolenic acid and their metabolic intermediates. It is rich in tocopherol. It is mainly used for shortening, cooking margarine and frying.

4.2.3.4 Sunflower Oil

Grown in defined geographical areas, it still holds the fourth position globally in terms of popularity. Based upon its low oxidative stability, its applications in frying are limited, but it is good in manufacturing salad dressings and in baking. It contains saturated fat, oleic acid and linoleic acid. It has a good stable taste.

4.2.3.5 Coconut Oil

It is extracted from copra obtained from coconut palm tree. This oil is liquid at 25° Celsius and above due to the presence of a large number of liquid acids, but is solid at room temperature. It contains a maximum amount of lauric acid which is estimated to be around 50%. It forms foam when heated, thus it is not mixed with other oils. It has numerous applications due to its higher oxidative stability which includes frying oil, infant foods, margarine production, sports food, etc.

4.2.3.6 Palm Kernel Oil

In the international market, lauric oil is second only to coconut oil. It has a comparable fatty acid content and qualities to coconut oil, although it contains slightly less short-chain fatty acids and has a higher permeability. Palm kernel oil contains 8% lauric acid as the primary fatty acid. 15% oleic acid and 16% myristic acid. No other fatty acids account for more than 10% of the total. It is used in the manufacturing of various products like filling cream of biscuits, margarine, non-dairy whipping and ice creams.

4.2.3.7 Cottonseed Oil

The cotton plant's seeds are used to make this oil. Oil is a by-product of the textile industry, and it is dependent on the use of cotton. As a result, only a little amount of oil is traded. Cotton seed oil is a dark reddish-brown liquid with a strong flavour and odour. Cotton seed oil, on the other hand, cannot be treated to have the same oxidative and flavour stability. Its applications involve manufacturing of salad oils, margarine, etc., and in deep frying of various products.

4.2.3.8 Groundnut (Peanut) Oil

The seed of Arachis hypogaea, often known as peanut, earthnut, or groundnut, is used to extract groundnut oil. Groundnut oil is high in digestible protein, unsaturated, and has a roasted nutty flavour. In the United States, light yellow crude oil is mostly used for frying and cooking. Groundnut oil is used in the preparation of short dishes, and margarine. It can be used for deep frying, seasoning and also to make dressings.

4.3 Methods of Extraction

The oil is extracted from the source before it can be incorporated into any food product. The main step involved isolation of crude oil by various extraction methods; this crude oil is refined and finally modified for human consumption (Figure 4.1).

4.3.1 Extraction

Pressing of oily fruits like olive is the dominant extraction method, whereas in case of oil seeds pressing is, whereas in case of oilseeds Bing is followed by solvent extraction in order to achieve maximum extraction. The protein-rich residue left after extraction is used as animal feed or as a health supplement by humans. Extraction of animal fats is achieved by heating.

4.3.2 Refining

The crude oil extracted from various sources requires refining prior to human consumption. The objective of refining is to remove non-desirous pigments, metals, peroxides, add other oxidation products; on the other

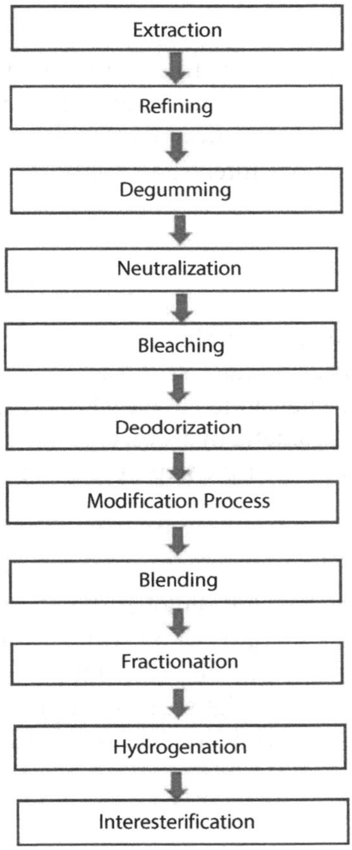

Figure 4.1 Unit operations involved in extraction and processing of edible oils.

hand, we need to retain some products. The various process involved in refining are described below.

4.3.3 Degumming

This step involves removal of phospholipids by treating crude oils with water or mild acid. This forms a gum which is separated by centrifugation. EDTA or enzymes can also be used for degumming which helps in the removal of trace metals.

4.3.4 Neutralization

This step involves removal of free acids by using aqueous alkali. Saponification leads to loss of some oil and this soap requires treatment before it is released into the environment. Steam distillation for removal of free acids is gaining more interest as it is an environmentally friendly method.

4.3.5 Bleaching

This step removes the undesired pigments present in oil by heating oil at 90 to 120 degrees Celsius along with bleaching earth preactivated with acid. Activated carbon, when added to bleaching earth, helps in removal of non-volatile polycyclic aromatic hydrocarbons. The higher cost of bleaching and its disposal is an expensive step in refining.

4.3.6 Deodorisation

This is the final refining step and removes the off flavor producing oxidative products. It involves sparging of oil with steam at high temperatures (170-250^0 C) and low pressure. The temperature should be controlled in case of polyunsaturated oils as higher temperatures will lead to isomerization of cis to trans pause which will cause loss of nutrient properties.

4.3.7 Modification Process

The properties of fats and oils need to be modified based upon their end use. Modification methods involve blending of oils, Bama fractionation of fat hydrogenation and interesterification. Esterification can be performed by enzymatic or chemical catalyst. The cost and effectiveness off these processes helps in deciding the various unit operations.

4.3.8 Blending

This involves mixing various oils and fats to obtain and improve products with the desired physical and nutritional properties. Various vegetable oils are mixed to make a product with the desired fatty acid composition. Addition of certain active compounds like antioxidants will increase their oxidative stability.

4.3.9 Fractionation

It separates high and low melting triacylglycerol into STRN and Pauline fraction. This step is performed under controlled conditions of low temperature. The most commonly applied procedure is dry fractionation. Winterisation refers to oil storage during winter which leads to crystallization of saturated triacylglycerol. B vaccine is a different phenomenon of removing wax from solvent extracted oils by storing at defined temperature followed by filtration.

4.3.10 Hydrogenation

It is a chemical process which involves incorporation of hydrogen on unsaturated fats and oils to convert them into saturated forms. This process enhances the shelf life of the oils by providing oxidative stability. It also increases the melting point of the facts which accounts for their improved applications as it spreads. On the other hand, this process generates trans-fat which is not recommended because it leads to heart disease and increased risk of stroke and diabetes.

4.3.11 Interesterification

It involves Matty rearrangement of fatty acids by breakdown and formation of Ester bonds in order to make them more desirable in the food industry. Interesterification process can be performed by either chemical or enzymatic method.

4.3.12 Chemical Method

It involves the application of alkaline catalyst like dash or any which randomly distributes the fatty acids in the product. This increases the melting point of fats but also leads to some loss of oil do two-eighths conversion two easters and assets. The major application of this process is in the production of spreads without applying hydrogenation.

4.3.13 Enzymatic Method

Interesterification with lipases has advantages over chemical processes as it is more specific and generates less by-products. This leads to less unit operations required to purify and thus reduces cost. Lipases are specific for the site of action on fatty acid and to a defined acidic or ester groups. Although there are many advantages of using lipases for interesterification, their high cost as compared to chemical catalysts makes their application restricted.

4.4 Constituents of Fat and Oil

4.4.1 Triglycerides

Three fatty acids are joined to one glycerol molecule to form a triglyceride. It's a simple triglyceride if all three fatty acids are the same. Chemists call the chemical components found in fat before it is divided into triglycerides. Because natural fats contain a variety of fatty acids, a large number of distinct triglycerides can be found in nature [17]. The fatty acid or acids they contain are used to name them. Triolein is the triglyceride of oleic acid, tripalmitin is the triglyceride of palmitic acid, tristearin is the triglyceride of stearic acid, and monopalmitin-distearin is the triglyceride of stearic acid.

Despite the large variety of triglycerides found in nature, the triglycerides of these seven acids account for the vast majority of natural fats and oils. Fats and oils are almost always made up of various quantities of triglycerides. One triglyceride predominates in some fats, another in others, and others are present in significant levels in yet others. There appears to be no single triglyceride in any natural fat or oil. The features of the triglyceride from which they are made, as well as the proportions of this triglyceride to one another, determine the attributes of different fats and oils.

4.4.2 Mono- and Diglycerides are Two Types of Fatty Acids

Glycerides are commercially produced by reacting glycerol with triglycerides or esterifying glycerol with fatty acids. The regular digestion of triglycerides results in the formation of mono- and diglycerides in the digestive tract [18]. Both animal fats and vegetable oils contain trace levels of them naturally. Oil primarily constituted of diglycerides has also been utilised to replace triglyceride-based oil. Below are illustrations of mono- and triglyceride molecular structures.

Free fatty acids are found in fats, as the name implies. Several percent of free fatty acids can be found in some unprocessed oils. The refining procedure lowers the quantities of free fatty acids. Free fatty acid content in fully refined fats and oils is typically less than 0.1%.

4.4.3 Saturated and Unsaturated Fatty Acids

The term "saturated fatty acid" refers to a fatty acid that includes the maximum amount of hydrogen possible. In terms of hydrogen, it is saturated. Saturated acids include myristic, lauric, palmitic, and stearic acids. At normal temperatures, they are solids [19]. Many fatty acids are called "naturally saturated," meaning they cannot be produced "harder" than their original form. As previously stated, the levels of saturates are often higher in fats that are solid at room temperature. Fatty acids, saturated ones are extremely stable, meaning they don't quickly go rancid, which means they keep well (shelf life).

Government recommendations, on the other hand, advise people to limit their consumption of saturated fats because they can raise blood cholesterol levels, which is one of the leading causes of heart disease. Most animal fats, such as meat, butter, cheese, and cream, are high in saturated fat and should thus be consumed in moderation.

4.4.3.1 Unsaturated Fatty Acids

When the maximum amount of hydrogen is not present in the fatty acid molecule it is classified as unsaturated. In terms of hydrogen, it is unsaturated. Oleic, linoleic, and linolenic acids are examples of unsaturated acids. At room temperature, they are liquids. Unsaturated fatty acids are classified into three types:

Monounsaturated Fatty Acids (MUFA) are fatty acids that contain only single double bond in their structure. They are relatively resistant to oxidation and the development of rancidity, and are now regarded as the healthiest type of fat to consume. Monounsaturated Olive and Rapeseed oils are the most prevalent sources.

Polyunsaturated Fatty Acids (PUFA) are fatty acids that have two or more double bonds in their chemical structure. Because they are the least resistant to oxidation, they are best employed in cold applications. Sunflower seed oil is the most prevalent source of fatty acids.

Trans Fatty Acids (TFA) are made up of two types of fatty acids: hydrogenated vegetable oils and animal fats.

4.4.3.2 Phosphatides

These are commonly known as phospholipids, and are made up of an alcohol (typically glycerol) and fatty acids, as well as a phosphate ester.

During the degumming and refining processes, the majority of the phosphatides are removed from the oil. Phosphatides are a key source of lecithin, a natural emulsifier.

4.4.3.3 *Sterols*

Oils from animals and vegetables both include sterols. Cholesterol is the major sterol found in animal fats and only in trace amounts in vegetable oils. Phytosterols are a group of sterols found in vegetable oils and plants. The most well-known vegetable oil sterols are stigmasterol and sitosterol. When it is included in food products, sitosterol has been demonstrated to lower blood and LDL cholesterol.

4.4.3.4 *Tocopherols and Tocotrienols*

These are essential minor components of most vegetable lipids. They act as antioxidants and sources of vitamin E, which helps to prevent rancidity. Alpha-tocopherol is known to have the highest vitamin E activity and the lowest antioxidant activity among the tocopherols. The antioxidant activity of delta tocopherol is the highest. Most vegetable oils include tocopherols, which are largely eliminated during processing. The largest concentrations are seen in corn and soybean oils. Animal fats do not contain significant levels of tocopherols. Palm oil is the most abundant source of tocotrienols.

4.4.3.5 *Pigments*

Carotenoids are naturally occurring yellow to deep red pigments found in fats and oils. They mostly contain carotenes like lycopene and xanthophylls like lutein. The highest concentration of carotene is found in palm oil. Chlorophyll is the green pigment that gives plants their colour and is required for photosynthesis. Among typical vegetable oils, canola oil has the highest chlorophyll content. The naturally occurring quantity of chlorophyll in oils can give them a green tint at times. Cottonseed oil contains the colour gossypol [20].

4.4.3.6 *Alcoholic Fatty Acids*

Fatty alcohols are found in most edible lipids; long-chain alcohols play a minor role. Waxes found in some vegetable oils contain a modest quantity of esterified fatty acids. Some marine oils have higher concentrations.

4.5 Physical Properties

Melting point, polymorphism, and crystal structure Long-chain compounds frequently exist in multiple crystalline forms in the solid state, and so have multiple melting points. Polymorphism is a scientific and technical trait of importance. This is critical for successful mixing of materials consisting of fat, for example, spreads that should achieve and keep a specific appearance during preparation and storage. Chocolate bloom and Graininess in margarine are caused by polymorphism alterations. Nuclear Magnetic Resonance spectroscopy, IR spectroscopy, differential scanning calorimetry and X-ray diffraction techniques are the most common experimental approaches used to investigate melting and crystallization phenomenon [21].

However, the length of the most stable form might sometimes imply a dimeric or trimeric unit. The molecules adopt the tilt angle at this orientation; they are tightly packed, and this results in the highest physical stability and melting point.

4.5.1 Acids Alkanoic and Alkanoic

The melting temperatures of acids when Acids which contain one less carbon atom melt slower than odd acids. At 120–125°C, the two saturated acid curves merge. Unsaturated acid melting points are determined by the length of the chain.

4.5.2 Fat Analysis

The nature of analytical processes were mostly chemical. Many chemical reagents along with solvents were used. Many of these processes are being replaced by techniques such as chromatography and spectroscopy, which are rapidly replacing them. These are often faster, take less labour, are more accurate, and use less material. However, the equipment is more advanced and pricier. Burettes and older glass pipettes have mainly been supplanted by spectrometers and chromatography systems.

Before every test, a representative sample of material must be obtained, maybe transported, and stored before any measurements can be taken. All of these stages have defined procedures. Even the most meticulous analysis will be useless if care is not done in all of these areas.

4.5.3 Content of Oil

Many methods are used for quantitatively extracting lipid from a sample, each of which is dependent on the nature of the matrix. The biological materials are extracted using chloroform–methanol according to well-established techniques [22]. The Folch extraction involves shaking pulverised or homogenised tissue with a 2:1 mixture of chloroform and methanol, then partitioning the organic extract with an aqueous potassium chloride solution. The volume of the mixed layers consisting of chloroform/methanol/water should be in the ratio 8:4:3.

Saturated and unsaturated acids are found in oils and fats. The value of iodine depends on the reaction with iodine monochloride or any other halogen compound kept under controlled conditions has traditionally been used to determine average unsaturation. Many specifications relating to oils and fats still mention the value. It has many drawbacks. Therefore, this procedure is very time-consuming, labour-intensive, and involves the use of hazardous chemicals and solvents [23]. As a result, iodine value is currently frequently estimated from the composition of fatty acid measured by Gas Chromatography using iodine values for each and every component.

4.5.4 Saponification

In most specifications for crude and refined oils, the amount of free acid is mentioned. It can be represented as an acid value or also as free fatty acid percentage by titration using a standard sodium hydroxide solution. Free fatty acid is stated in most oils as oleic acid. The quantity of alkali required to hydrolyze (saponify) a fat is a measure of the acyl chains' average chain length, albeit this number is influenced by other factors [24].

4.5.5 Melting Characteristics, Solid Fat Content, and Qualities at Low Temperatures

Fats are complex combinations of numerous triacylglycerols that might be in solid or liquid form at room temperature. Because fats are a mixture of solid and liquid components, most of the spreads are found in plastic solid form which starts deforming when external pressure is applied during the spreading action [25]. The proportion of these two phases changes with temperature, hence knowing the ratio of solid and liquid at a range of temperatures is commonly required. This is crucial when evaluating the quality of spreadable and confectionery fats. The temperature at which solids first form after cooling is also crucial.

4.5.6 Value of Peroxide, Anisidine, Stability and Shelf Life

Autoxidation is the most prevalent oxidative reaction and occurs during an induction period during which the degeneration is not severe, and knowing the length of this period is useful for food producers. There are several phases of oxidation that can be identified, and tests for each stage are available. The primary oxidation products are allylic hydroperoxides, which are measured as peroxide value or as conjugated dienes formed during PUFA oxidation [26]. Unsaturated aldehydes form secondary oxidation products, which are quantified by the anisidine value. Short-chain acids form tertiary-oxidation products, which are measured by the term called as oil stability index (OSI).

Hydroperoxide measurement is the most common way of determining oxidation state. Potassium iodide reacts with these molecules to form iodine. It may then be quantified volumetrically using sodium thiosulphate. The peroxide value of freshly refined material should be less than one. When a fat's peroxide value exceeds 10, it's called rancid.

The anisidine value is determined by measuring the intensity of the chromophore at 350 nm emitted by molecules of the type ArNCHCHCHR, which are formed by reacting anisidine (4-MeOC6H4NH2) along with carbonyl compounds, primarily 2-enals (RCHCHCHO). Because this value fluctuates based on the enals present, it can only be compared with results for a particular type of oil.

4.5.7 Gas Chromatography

It is a technique for analysing gases. This is basically determined by Gas Chromatography of the esters generated from the triacylglycerols and determines the concentration of fatty acids [27]. GC methods can also be used to determine minor components including sterols, sterol esters, and tocopherols. The techniques for creating the methyl esters, the GC settings, and the ways in which the data are presented must all be considered when examining fatty acid composition. Component acids which are present in the form of methyl esters are separated and quantified using GC.

4.5.8 Infrared Spectroscopy and Fourier Transform Infrared

In recent years, Fourier transform infrared (FTIR) spectroscopy has advanced significantly, and it is currently employed as an alternate method of assessing various parameters relevant to lipid analysts. In one second, a Fourier Transform Infrared spectrometer can record the whole spectrum, which may then be combined with other scans using this technique in order to produce a traditional infrared spectrum [28].

4.6 Chemical Characteristics

4.6.1 Hydrogenation

In its most basic form, hydrogenation entails a reaction between the unsaturated centres of an oil or fat and a metallic catalyst. A reaction comprising solids, liquids, and gases occurs on the surface of a solid at the right pressure along with temperature. Due to hydrogenation the physical, chemical, and nutritional qualities of the product have changed. Between these, sometimes concessions must be made. Every year, millions of tonnes of soybean and other unsaturated vegetable oils containing oleic, linoleic, and linolenic acids are hydrogenated, as are decreasing levels of fish oils with more complicated unsaturation patterns [29]. This response can be carried out in three different ways:

A. Brush hydrogenation is a brief procedure used to lower the quantity of linolenic esters (18:3) in soybean and rapeseed/canola oils to roughly 4%, extending shelf life. This isn't a particularly lengthy hydrogenation procedure. Aside from the reduction in linolenate, the fatty acids are largely unchanged, and trans isomer formation will be negligible. Several products are created by hydrogenating soybean oil with an iodine value (IV) of 132 to oils with IVs of 110 (brush hydrogenation), 97, 81, and 65.

B. A common method of extending the usefulness of oils and fats is Partial hydrogenation. This is a more comprehensive reaction than brush hydrogenation, in which the quantity of polyunsaturated fatty acids is greatly reduced and a significant fraction of trans 18:1 is generated. As trans esters have higher melting point than their cis isomers, this causes a rise in melting point, which influences spread ability, oral reaction, and baking performance. There are two more changes to consider. The elimination of polyunsaturated fatty acids completely or partially improves oxidative stability.

C. Complete hydrogenation is a more involved procedure that converts practically all unsaturated acids to their saturated equivalents. This yield is then interesterified by blending it with a liquid (unsaturated) oil. Stearic acid is abundant in hard stock, which is generated from the unsaturated fatty acids that were previously present. It includes relatively little trans acids and just a small amount of unsaturated acid. On the other hand, the word "hydrogenated" may be required on the label.

Since it is an exothermic reaction, it requires cooling and churning in order to distribute the heat. The nature of the oil being treated is one of the most important aspects to consider; also the degree to which hydrogenation is sought.

4.6.2 Oxidation of the Atmosphere

Unsaturated fats, like other unsaturated products like rubber and paints, decay as a result of reactions with oxygen (air), which causes oxidative rancidity in fat-containing foods (another form of rancidity results from hydrolysis). There are two types of oxidative degradation (autoxidation and photo-oxidation). These processes produce comparable, but not identical, unstable allylic hydroperoxides, which degrade into volatile short-chain molecules (mostly aldehydes), which are responsible for the oxidative rancidity smells and flavours [30].

These chemicals have low but varying threshold levels, which means that only small amounts are required to elicit their negative effects. Heat, light, the presence of pro-oxidants (copper and iron) and antioxidants (natural or synthetic), the presence of already-oxidized material, and air all influence oxidation processes, so all of these factors must be considered when handling, transporting, and storing oils and fats and foods containing them. The amount and kind of unsaturation has a significant impact on the rate of oxidation [31].

This is why (highly unsaturated) fish oils are unstable, and why oils containing linolenic acid (soybean and rapeseed/canola oils) have a short shelf life, and why people prefer oils with lower quantities of this acid, which can be obtained through breeding or brush hydrogenation. Polyunsaturated acids (linoleic, linolenic, and other polyunsaturated acids) oxidise significantly faster than monounsaturated oleic acid. The quantity of doubly activated allylic groups contained in PUFA is highly linked to oxidative degradation.

4.6.3 Changes in Temperature

When heated, unsaturated centres may suffer unfavourable modifications. When the temperature approaches 180^{0}C and the fatty acyl chains include three or more methylene interrupted olefinic centres, this is especially essential. This problem can occur during the deodorization of soybean. Although the likelihood of oxidative change is increased at higher temperatures, unfavourable alterations can occur even when there is no air present [32]. Stereoisomerism of linolenic acid with three double bonds occurs from 220°C upwards and is faster than stereoisomerism of linoleate with just two double bonds, according to studies on the refining of vegetable oils: geometrical isomerism; polymerisation. These alterations are nutritionally undesirable since the nutritional value of polyunsaturated fatty acids is dependent on them maintaining their all-cis unsaturation pattern. It's possible that detecting these alterations will necessitate the use of specialised analytical methods [33].

4.6.4 The Carboxyl/Ester Function's Reactions

Glycerol esters are oils and fats, and the ester functionalities are reactive centres that can be altered in a variety of ways. They can be hydrolyzed, for example. The suffix -lysis denotes splitting, hence hydrolysis denotes splitting with water, with fatty acids and glycerol as the final products. A catalyst is usually required for this reaction. Lipolysis is the term for a reaction that is catalysed by a lipase. Although hydrolysis is significant in the oleochemical sector, the conversion of glycerol esters is vital in the food industry. Methanolysis and glycerolysis are two key alcoholysis processes [34].

4.7 Nutritional Properties

Oils and fats have been a significant part of the human diet for a long time. They act as an energy source and provide several essential compounds which are not synthesized by humans themselves like fatty acids. They carry several fat-soluble vitamins along with certain phytosterols. Fats and oil add texture and flavour to food and enhance its palatability [35].

4.7.1 Suggested Consumption Limits for Fats and Oils

Oils and fats have different fields inside our body ending up as membrane phospholipids, adipose tissue or as energy sources. Most of the essential fatty acids are ingested through diet. The intake of total fats produces approximately 15 to 40% energy globally. This high consumption of fats has led to increased incidences of chronic heart disease. Polik escents or to some extent palmitoleic acid being monounsaturated and in SSIS configuration are considered to be safe for consumption. DHA plays a major role in brain development of features; thus its intake should be increased in pregnant and lactating females [36].

4.7.2 Role of Fats in Health and Disease

Daily use of fats in diet should be well regulated with respect to quality and quantity in order to lead a healthy and long life. In today's lifestyle it is compulsory to choose the fats with minimal harmful compounds. DVD's disorders associated with increased consumption of facts are as follows:

4.7.2.1 Coronary Heart Disease

Heart disease includes atherosclerosis and thrombosis which leads to the major number of deaths in developed countries. In atherosclerosis there is narrowing of blood vessels due to deposition of lipids on the inner walls resulting in increased blood pressure leading to cardiac arrest. The enhanced LDL cholesterol and reduced HDL concentration is the major cause of coronary heart disease (CHD) followed by low plasma antioxidants and high plasma fibrinogen. The risk of CHD can be reduced by lowering fat content in diet which will further reduce the rate of mortality [37].

4.7.2.2 High Blood Pressure

The fact that non-vegetarian diets are more prone to higher blood pressure problems as compared to vegetarian diets is supported by the nature of dietary facts. Plant-based diets contain more monounsaturated and polyunsaturated fats and therefore reduce the risk of developing hypertension. Meat and meat products contain more saturated fatty acids, total fat and cholesterol, leading to high blood pressure [38].

4.7.2.3 Cancer

Studies have shown correlation between high-fat diets and rate of development of cancer globally. Breast cancer and colon cancer are more prone to be associated with high calorie consumption and obesity. Defaults which are helpful in cancer prevention like fruits, vegetables, fiber and complex carbohydrates are present in very low concentration in high-fat diets [39].

4.7.2.4 Inflammation

Inflammation is a sequence of events generated by our immune system in response to any foreign particle entering our body. During inflammation the immune cells secrete chemicals cytokines, interference and antibodies to generate swelling, redness, heat and pain. Omega 3derived fatty acids help in preventing inflammatory disorders by either directly inhibiting antigen antibody reaction or indirectly by modifying inflammatory gene expression [40].

4.8 Applications

4.8.1 Butter

Butter is an emulsion of fat and water which is produced mostly from cow milk. Milk is first centrifuged to convert it into cream which is kneaded and churned in order to convert it into butter [41]. It has major application in baking and frying where it provides the amazing flavour to the finished product. There are several disadvantages associated with butter like high cost, unhealthy due to high fat content, low spreading capacity after refrigeration storage, presence of trans fat, cholesterol and saturated acids. These disadvantages can be overcome by cow's diet modification, blending unsaturated vegetable oils or by fractionating anhydrous milk. Blended butters are produced globally due to its better spreading ability [43].

4.8.2 Margarine

Margarine is produced by processing various oils and fats including sunflower, canola, rapeseed, coconut, soybean, etc. The major processing steps involve fractionation, hydrogenation, blending and esterification. Different additives like salt, surface-active agents, protein, flavours, preservatives and vitamins are added to enhance the nutritional qualities and shelf life of margarine. The ratio of solid and liquid fat in margarine varies with temperature and affects the physical properties of the finished product [44].

4.8.3 Baking Fats and Shortenings

The baking industry is another major domain which utilizes the fats for the production of bread, biscuits, pie crusts, cookies, puff pastry, etc. Based upon their plasticity and melting behaviour, fats are selected for various products. To attain these desired properties various fats are blended to provide the final product with oxidative stability and nutrition. Fats used in the bakery industry to make various doughs are combinations of solid and liquid fats called plastic fats. At room temperature they are solid in nature but once the pressure is applied, they melt and turn liquid [42].

4.8.4 Frying Oil and Fats

Oils and fats are used in deep and shallow frying or food or food ingredients. This fried food is in too much demand, and pre-fried frozen and fried food products are growing extensively. While frying, the temperature of oil plays a major role in defining the texture and flavor of both finished food products. Heat and mass transfer operations take place during frying as the vapor from the inside of food travels outside the oil and oil tries to penetrate inside the food product. Low-temperature frying will lead to more oil penetration; high-temperature frying will cause undesirable modifications in oils. The ideal temperature for frying should be between 165 - 185 degrees Celsius, producing a crisp outer layer and soft inner region [45].

4.8.5 Ice Cream

Fat present in ice creams stabilize the aerated foam and contribute to creaminess and taste. It also enhances melting resistance and ratio of solid to liquid at different temperatures to provide better taste. Major fat used in the ice cream industry is milkfat but along with it a minimal amount of vegetable fat like palm, coconut, sunflower, etc., is also added [46].

Through the partially coalesced continuous three-dimensional network of homogenised globules that aids in the stabilisation of the air phase because fat globules surround air bubbles. Additionally, higher levels of fat aggregation are also correlated with improved melting resistance [47]. Ice cream's multipurpose ingredient, fat, enhances flavour, colour, texture, and mouthfeel. Due to the fact that fat is the primary carrier of flavour for many compounds and that low flavour intensity may not be able to be overcome by the addition of flavouring alone, a reduction in the fat content of the ice cream can cause a loss of texture and sensory properties [48, 49].

The "dryness" of ice cream, form retention, melting in the mouth, and texture creaminess—a highly desirable quality that is given by milk fat—are all explained by the presence of fat [50–52]. The body of the ice cream is determined by the milk fat since it interacts with other ingredients to create texture, mouthfeel, creaminess, and a lubricating effect. In order to produce goods that satisfy the demands of health-conscious consumers, some ice cream producers have recently reduced the amount of fat and replaced those fat solids with carbohydrates or proteins [53, 54]. The fat percentage in ice creams is typically between 10 and 16% [55].

In order for a fat replacer to transmit the required flavour profile and match the texture, mouthfeel, and functioning of the original fat, it is crucial to understand how to produce similar structures of fats [56, 57].

To minimise the amount of fat or to experiment with other sources, milk fat and vegetable fats can be substituted. In order to reduce the absence of fat in the structure, low-fat or fat-free formulations are typically tested with the addition of body agents. The availability of new sources, cost-saving measures, and the development of new goods have all led to the testing of fat substitutes. It has been investigated whether green coconut pulp could be used to make chocolate ice cream.

The outcomes of the physicochemical analyses demonstrated that coconut pulp has the potential to froth and emulsify, allowing it to be utilised to make ice cream even at low pH levels. By legislation, the frozen food item that doesn't contain milk solids is referred to as sorbet. However, using coconut pulp as a base, it was possible to create a product that had many of the same qualities as actual ice cream. Fresh pulp is ideal for making "ice cream" because it has minimal flavour and smell, making it a good ingredient to create goods with various flavours. The coconut pulp's high concentration of medium-chain fatty acids, particularly lauric acid, which are thought to be healthy [58], is another advantage.

To compensate for the absence or substitution of fat, product formulation must be improved in order to produce lower-fat ice cream [59], as modifications to a food system can cause an imbalance in the flavour profile through a variety of processes [60]. In order to address the expectations of consumers who are health-conscious, numerous researchers have examined tactics that have a strong emphasis on generating very pleasant low-fat ice cream [61]. The consumer market, which is increasingly looking for healthier options with more added value, would benefit greatly from the creation of products with good acceptance and perceived advancements in this field.

4.8.6 Chocolate and Confectionery Fat

Chocolate is dark, milk, and white; each has a unique amount of solid cocoa, milk fat, and cocoa butter. White chocolate, milk chocolate, and dark chocolate have fat contents of 28, 30, and 30.9%, respectively [62].

The quality of chocolate is most affected by its fat content among its various constituents. The rheological characteristics of fluid chocolate, release from the mould, snap, gloss, bloom prevention, melting properties, and flavour release are all impacted by the fat phase. The two primary types of fat utilised in chocolate production to give these qualities are milk fat and cocoa butter [63]. The primary element responsible for the rheological behaviour of chocolates is cocoa butter.

The primary acids in cocoa butter are palmitic acid (C16), stearic acid (C18:0), oleic acid (C18:1), and linoleic acid (C18:2), whereas lauric acid (C12) and myristic acid (C18:2) are present in relatively small amounts (C14). The composition of fatty acids may vary depending on where the product is produced, and cocoa butter made in Brazil contains more linoleic acid (3.5–3.6%) and less stearic acid (33.3–33.8%) than butter made in Ghana, Indonesia, Ecuador, or Malaysia. Approximately 34.5-36.5% of oleic acid and 25.1-27.9% of palmitic acid are also present in Brazilian cocoa butter [64]. Triglycerides make up 98% of milk fat, which also contains other trace amounts of lipids.

Due to the distinct stable crystal structures of milk fat and cocoa butter, the two fats cannot entirely mix when they are in a solid condition. In mixes of cocoa butter and milk fat, it tends to slow the pace of cocoa butter crystallisation and alters the temperatures at which the various crystals form. The physical and functional characteristics of chocolate, such as its melting point, tempering capacity, and hardness, are altered when milk fat content is increased. It takes around 50% more milk fat than cocoa butter to modify the crystal forms of the fat [65].

The cost of cocoa is influenced by many issues such as changing climatic conditions, presence of pests and rodents, and political issues in the major producing countries, making cocoa butter a costly raw ingredient for the chocolate business [66]. The difficulty of reducing fat in chocolate is to discover technological solutions to create a system with enhanced solid volume fraction and acceptable flow characteristics [67]. This is in addition to the present trend in the food industry to give consumers low-fat goods [68].

The triglyceride content of some vegetable fats is comparable to that of cocoa butter, hence cocoa butter equivalents can be mixed with chocolate [69]. The usage of substitutes depends on how closely their physical, chemical, and functional characteristics match those of cocoa butter; they don't create a eutetic combination and lower the melting point of the product [70].

In order to manipulate these kinds of items, new ingredients and knowledge are required [71]. However, lowering the fat level raises the viscosity of the molten chocolate, decreasing its eating quality and complicating processing. It also causes the chocolate to remain hard and difficult to swallow.

4.8.7 Cakes

From a technological as well as a sensory perspective, fat plays a crucial part in cakes and baked goods in general. According to [72] many bakery items need to have a significant amount of fat. Shortenings serve a variety of purposes in bread products, including lubrication, air entrapment, texture, softness, structure integrity, mouthfeel, and increased shelf life, according to [73].

Many systems, such as cake batters, incorporate air bubbles through the action of fat in whipping cream method, whereby the bubbles are trapped in the continuous phase of the emulsion at room temperature rather than continuing to exist in the aqueous phase [74]. The air bubbles are moved from the fatty phase to the aqueous phase structure as the batter is heated during cooking, creating a voluminous and foamed structure as a result. When heated, baking powder emits carbon dioxide. The mixture is under intense pressure from the gas that is created, increasing the volume of the finished product [75].

4.8.8 Bread

Wheat-derived lipids, fat (shortenings), and surfactants are crucial to the production of bread because they provide advantages at various stages of the baking process and during storage. According to the amount of wheat flour in the formulations, typically 2 to 5% fat is added. Generally speaking, the balance of fat and application depends on the ratio of the fat's solid phase to liquid phase at a particular temperature, with the volume of the bread having an impact on this ratio [76].

When manufacturing bread, fats are used to improve gas retention in the dough, increasing its volume and softness. They also lubricate, aerate, and aid in heat transfer, giving the dough a desirable texture. The quantity to be added will vary based on the type of flour (whole meal need higher levels than white). Until fermentation is complete (45 °C), a fat fraction should stay firm in the dough. Breads are made with a variety of shortenings, including butter, solid and liquid margarine, oils, and industrial fats. A plastic shortening's oil or oil fraction helps it become softer and gives it a moist mouthfeel.

The air bubbles used in the mixing process are a result of the solid fraction, which also adds to the structure of the dough and finished product. In the context of gas cells, fat crystals act as a cell wall or membrane for these cells during baking operations, causing them to enlarge. To lower the amount of solid fat in oils, certain emulsifiers can be added. In addition to boosting the dough rise, oven spring, and final loaf volume, the lubricating action permits more expansion during fermentation and bread baking, softening the finished texture and extending or maintaining the shelf life [77].

Wheat flour contains lipids that are derived from organelles, membranes, and spherosomes. They can be separated into starch lipids, free lipids, and lipids linked to non-starch components based on the lipid solubility. While raw starch does not go through gelatinization, starch lipids are tightly bonded in the starch granule and have no impact on how the dough is processed. Because they create a lipid layer at the gas/liquid interface that promotes higher gas retention of the dough, lipids or polar lipids have an impact on the production of breads by influencing the stability of gas bubbles in the dough. The bakery shortening needs to have a minimum of 20% solid fat content (SFC) at 25 °C and a minimum of 5% SFC at 40 °C for proper baking and the formation of the mass structure [78].

4.8.9 Biscuits

Biscuits are the major bakery products consumed globally and are produced in thousands of different flavours, shapes, and texture. The main ingredients for biscuits are Fleur sugar, fat/oils, milk, eggs and water but fats and sugars play a major role. Fats provide tenderness, texture, lubrication, extended shelf life, structural integrity and flavour to the final product. During the manufacturing of cookies, the fat should be either solid or semisolid at room temperature in order to provide the desired texture. This is achieved by adding more saturated fatty acids so that the shortenings will have more plastic properties. The solid to liquid fat index should be maintained adequately as high solid fat index will not be able to hold the required aeration due to less oil volume; on the other hand, low solid fat index shortenings will not be able to hold air till the mixing process is complete. The crystalline structure of fat is another feature. For the best creaming to occur, the fat must be in the crystal form.

A low-calorie product can be obtained by reducing the fat added to biscuits. Numerous writers have investigated the impact of various fat replacers partially replacing fat on the quality of biscuits (Bertolin). The functional and sensory qualities of fat can be mimicked by components of protein or carbohydrate origin in some diets [79].

References

1. Srilakshmi B (s2003) *Food Science*. New Age International Publishers. Hyderabad, India.
2. Breivik, H., ed. (2007) *Long-Chain Omega-3 Specialty Oils*, The Oily Press, Bridgwater, UK.

3. Christie, W.W. (1998) Gas chromatography-mass spectrometry methods for structural analysis of lipids. *Lipids*, 33, 343–353.
4. Christie, W.W. (2003) *Lipid Analysis*, The Oily Press, Bridgwater, UK.
5. Chumpitaz, L.D.A., Coutinho, L.F., and Meirelles, A.J.A. (1999) Surface tension of fatty acids and triglycerides. *J. Am. Oil Chem. Soc.*, 76, 379–382.
6. Coupland, J.N. and McClements, D.J. (1997) Physical properties of liquid edible oils. *J. Am. Oil Chem. Soc.*,74, 1559–1564.
7. Cuppett, S.L. (2003) Cholesterol oxides: sources and health implications, *Lipid Technology Newsletter*, 9, 9–13.
8. Diehl, B.W.K. (2002) 31P-NMR in phospholipid analysis. *Lipid Technol.*, 14, 62–65.
9. Dobson, G., ed. (2001) Spectroscopy and spectrometry of lipids. *Eur. J. Lipid Sci. Technol.*, 103, 815–840; 2002, 104, 36–68. *European Pharmacopoeia* 5th edition, 2004, Strasbourg, France.
10. Firestone, D., ed. (1999) *Physical and Chemical Characteristics of Oils, Fats, and Waxes*, AOCS Press, Champaign, USA.
11. Tripathy, D. B., Yadav, A., & Mishra, A. (2017). Applications of petrochemicals: A mini review. *Recent Adv. Petrochem.* Sci, 2, 67-69.
12. Tripathy, D. B., & Mishra, A. (2016). Renewable plant-based raw materials for industry. *Sustainable Inorganic Chemistry*, 353.
13. Gunstone, F.D. (2004) *The Chemistry of Oils and Fats*, Blackwell Publishing, Oxford, UK.
14. Hamilton, R.J. and Cast, J., eds. (1999) Spectral Properties of Lipids, Sheffield Academic Press, Sheffield, UK.
15. Hamm, W. and Hamilton, R.J., eds. (2000) *Edible Oil Processing*, Sheffield Academic Press, Sheffield, UK.
16. Keys, A. Andersen, J.T., and Grande, F. (1957) Prediction of serum–cholesterol responses of man to changes in the fats in the diet. *Lancet*, 2, 959–996.
17. Knothe, G. (2002) Structure indices in fatty acids. How relevant is the iodine value? *J. Am. Oil Chem. Soc.*, 79, 847–854.
18. Larsson, K.L., Quinn, P., Sato, K., and Tiberg, F. (2006) *Lipids: Structure, Physical Properties and Functionality*, The Oily Press, Bridgwater, UK.
19. Lawler, P.J. and Dimick, P.S. (2002) Crystallisation and polymorphism in fats, in *Food Lipids: Chemistry, Nutrition, and Biotechnology*, 2nd ed. (C.C. Akoh and D.B. Min, eds.), Dekker, New York, pp. 275–300.
20. McLean, B. and Drake, P. (2002) Review of Methods for the Determination of Fat and Oil in Foodstuffs, Review No. 37, Campden and Chorleywood Food Research Association Group, Chipping Campden, UK.
21. Mela, D.J. (2005) *Food, Diet, and Obesity*, Woodhead Publishing Limited, Cambridge, UK.
22. Rajah, K.K., ed. (2002) Fats in Food Technology, Sheffield Academic Press, Sheffield, UK.
23. Rossell, B., ed. (1999) *LFRA Oils and Fats Handbook Series*, Volume 1, *Vegetable Oils and Fats*, Leatherhead Food RA, Leatherhead, UK.

24. Sato, K. (2001) Crystallisation behaviour of fats and lipids – A review. *Chem. Eng. Science*, 56, 2255–2265.
25. Shahidi, F. (2005) *Bailey's Industrial Oil and Fat Products*, 6th ed., 6 volumes, John Wiley and Sons, Hoboken, USA.
26. Timms, R.E. (1978) Heats of fusion of glycerides. *Chem. Phys. Lipids*, 21, 113–129.
27. Timms, R.E. (1985) Physical properties of oils and mixtures of oils. *J. Am. Oil Chem. Soc.*, 62, 241–248.
28. Warner, K. (2007) Increasing gamma- and delta-tocopherols in oils improves oxidative stability. *Lipid Technol.*, 19, 229–231. Whitehurst, R.J. (2004) *Emulsifiers in Food Technology*, Blackwell Publishing, Oxford, UK.
29. Larsson *et al.*, *J Abnorm Psychol.* 2006 May;115(2):221-30. doi: 10.1037/0021-843X.115.2.221.
30. Folch *et al.* 1957. *J Biol Chem.* 226(1):497–509.
31. Potter NN, Hotchkiss JH (2012) *Food Science.* Fifth Ed. Springer Science & Business Media, Berlin, Germany.
32. Johnson S, Saikia N (2009) Fatty acids profile of edible oils and fats in India. Centre for Science and Environment. New Delhi, India.
33. Aloyor EO, Ozigagu CE, Oboh OI, Aluyor P (2009) Chromatographic analysis of vegetable oils: A review. *Scientific Research and Essay* 4: 191–197.
34. Mishra S, Manchanla SC (2012) Cooking oils for heart health. *Journal of Preventive Cardiology* 1: 123–131.
35. Gunstone FD (2002) *Vegetable oils in food technology: Composition, properties and uses.* Blackwell Publishing, New Jersey, USA.
36. Lawson HW (2013) *Food oils and fats: Technology, utilization and nutrition.* Springer Science & Business Media, Berlin, Germany.
37. Bonow, R.O. and Eckel, R.E. (2003) Diet, obesity, and cardiovascular risk, *New England Journal of Medicine*, 348, 2057–2058.
38. O'Brien RD (2008) *Fats and Oils: Formulating and processing for applications*, Third Ed. CRC Press, Florida, USA.
39. Ihekoronye AI, Ngoddy PO (1985) *Integrated food science and technology for the tropics.* MacMillan Education Ltd, London. UK.
40. Samarth NB, Mahanwar PA (2015) Modified Vegetable oil-based additives as a future polymeric material-A Review. *Open Journal of Organic Polymer Material* 5: 1–22.
41. Strayer D (2006) Food Fats and Oil. (9th Ed.). *Institute of Shortening and Edible Oils*, Washington, D.C., USA.
42. Achaya, K.T. (1997) Ghee, vanaspati and special fats in India.
43. Abhichandani, H., Bector, B.S. & Sharma, S.C. (1995).
44. S. F. Riepma, 1970. *The story of margarine*, Public Affairs Press, Washington, D.C.
45. E. Choe and D.B. Min, Chemistry of Deep-Fat Frying Oils, *Journal of Food Science*, 01 May 2007. https://doi.org/10.1111/j.1750-3841.2007.00352.x.

46. Achaya, K.T. (1997). Ghee, vanaspati and special fats in India. In *Lipid Technologies and Applications*, eds. F. D. Gunstone and F. B. Padley. Marcel Dekker Inc., New York, pp. 369390.
47. Afoakwa, E. O., Paterson, A., & Fowler, M. (2007). Factors influencing rheological and textural qualities in chocolate: a review. *Trends in Food Science & Technology*, 18, 290298. http://dx.doi.org/10.1016/j. tifs.2007.02.002
48. Beckett, S. T. (2009). *Industrial chocolate manufacture and use* (4th ed.). United Kingdom: Blackwell Publishing.
49. Jahurul, M. H. A., Zaidul, I. S. M., Norulaini, N. A. N., Sahena, F., Jinap, S., Azmir, J., Sharif, K. M., & Mohd Omar, A. K. (2013). Cocoa butter fats and possibilities of substitution in food products concerning cocoa varieties, alternative sources, extraction methods, composition, and characteristics. *Journal of Food Engineering*, 117, 467–476. http://dx.doi.org/10.1016/j. jfoodeng.2012.09.024
50. Haylock, S. J., & Dodds, T. M. (2009). Ingredients from milk. In S. T. Beckett, *Industrial chocolate manufacture and use* (4th ed.). Reino Unido: Blackwell Publishing.
51. Medeiros, M. L., & Lannes, S. C. S. (2009). Chemical evaluation of cocoa substitutes and study of formulated chocolate drink powder. *Ciência e Tecnologia de Alimentos*, 29(2), 247–253. http://dx.doi. org/10.1590/ S0101-20612009000200002
52. Do, T. A. L., Mitchell, J. R., Wolf, B., & Vieira, J. (2010). Use of ethylcellulose polymers as stabilizer in fat-based food suspensions examined on the example of model reduced-fat chocolate. *Reactive & Functional Polymers*, 70(10), 856–862. http://dx.doi.org/10.1016/j. reactfunctpolym.2010.07.012 Do, T. A. L., Mitchell, J. R., Wolf, B., & Vieira, J. (2010). Use of ethylcellulose polymers as stabilizer in fat-based food suspensions examined on the example of model reduced-fat chocolate. *Reactive & Functional Polymers*, 70(10), 856–862. http://dx.doi.org/10.1016/j. reactfunctpolym.2010.07.012
53. Suzuki, R. M., Montanher, P. F., Visentainer, J. V., & Souza, N. E. (2011). Proximate composition and quantification of fatty acids in five major Brazilian chocolate brands. *Ciência e Tecnologia de Alimentos*, 31(2), 541–546. http://dx.doi.org/10.1590/S0101-20612011000200040
54. Van Puyvelde, P., & Moldenaers, P. (2003). Biopolymer systems for low-fat foods. In B. M. McKenna, *Texture in food: semi-solid foods* (Vol. 1). Cambridge: Woodhead Publishing.
55. Sowmya, M., Jeyarani, T., Jyostna, R., & Indrani, D. (2009). Effect of replacement of fat with sesame oil and additives on rheological, microstructural, quality characteristics and fatty acid profile of cakes. *Food Hydrocolloids*, 23(7), 1827–1836. http://dx.doi. org/10.1016/j.foodhyd.2009.02.008
56. Zhou, J.; Faubion, J. M.; Walker, C.E. (2011). Evaluation of different types of fats for use in high-ratio layer cakes. *Food Science and Technology*, 44, 1802–1808.

57. Jacob, J., & Leelavathi, K. (2007). Effect of fat-type on cookie dough and cookie quality. *Journal of Food Engineering*, 79, 299–305. http:// dx.doi. org/10.1016/j.jfoodeng.2006.01.058
58. Goldstein, A., & Seetharaman, K. (2011). Effect of a novel monoglyceride stabilized oil in water emulsion shortening on cookie properties. *Food Research International*, 44, 1476–1481. http://dx.doi. org/10.1016/j. foodres.2011.03.029
59. Wilderjans, E., Luyts, A., Brijs, K., & Delcour, J. A. (2013). Ingredient functionality in batter type cake making. *Trends in Food Science & Technology*, 30, 6–15. http://dx.doi.org/10.1016/j.tifs.2013.01.001
60. Indrani, D., & Rao, G. V. (2008). Functions of ingredients in the baking of sweet goods. In S. G. Sumnu & S. Sahin, *Food engineering aspects of baking sweet goods*. New York: CRC Press.
61. Cauvain, S. P. (2003b). Bread making: an overview. In *Bread making improving quality*. Stanley: CRC Press. http://dx.doi. org/10.1201/9780203495001
62. Pareyt, B., Talhaoui, F., Kerckhofs, G., Brijs, K., Goesaert, H., Wevers, M., & Delcour, J. A. (2009). The role of sugar and fat in sugar-snap cookies: structural and textural properties. *Journal of Food Engineering*, 90(3), 400-408. http://dx.doi.org/10.1016/j. jfoodeng.2008.07.010
63. Manzocco, L., Calligaris, S., Pieve, S., Marzona, S., & Nicoli, M. C. (2012). Effect of monoglyceride-oil-water gels on white bread properties. *Food Research International*, 49(2), 778–782. http:// dx.doi.org/10.1016/j. foodres.2012.09.011
64. Smith, P. R., & Johansson, J. (2004). Influences of the proportion of solid fat in a shortening on loaf volume and staling of bread. *Journal of Food Processing and Preservation*, 28, 359–367. http://dx.doi. org/10.1111/j.1745-4549.2004.23079.x
65. Shahidi, F. (2005). *Bailey's industrial oil and fat products* (6th ed.). New Jersey: John Wiley & Sons. Retrieved from http://www.mrw. interscience. wiley.com/biofp
66. Kaur, A., Jassal, V., Thind, S. S., & Aggarwal, P. (2012). Rice bran oil an alternative bakery shortening. *Journal Food Science and Technology*, 49(1), 110–114. PMid:23572833 PMCid:PMC3550886. http://dx.doi. org/10.1007/ s13197-011-0259-6
67. Goesaert, H., Brijs, K., Veraverbeke, W. S., Courtin, C. M., Gebruers, K., & Delcour, J. A. (2005). Wheat flour constituents: how they impact Bread quality, and how to impact their functionality. *Trends in Food Science & Technology*, 16, 12–30. http://dx.doi.org/10.1016/j. tifs.2004.02.011
68. Latip, R. A., Lee, Y. Y., Tang, T. K., Phuah, E. T., Tan, C. P., & Lai, O. M. (2013). Physicochemical properties and crystallisation behaviour of bakery shortening produced from stearin fraction of palm-based diacyglycerol blended with various vegetable oils. *Food Chemistry*, 141, 3938–3946. PMid:23993569. http://dx.doi.org/10.1016/j. foodchem.2013.05.114

69. Bertolin, T. E., Centenaro, A., Giacomelli, B., Reinehr, C., & Gutkoski, L. C. (2013). Elaboration of biscuits with oatmeal and palm fat with added L-leucine and calcium for sarcopenia. *Food Science Technology*, 33(2), 345–354.
70. Zoulias, E. I., Oreopoulou, V., & Tzia, C. (2002). Textural properties of low-fat cookies containing carbohydrate- or protein-based fat replacers. *Journal of Food Engineering*, 55, 337–342. http://dx.doi. org/10.1016/S0260-8774(02)00111-5
71. Granger, C., Leger, A., Barey, P., Langendorff, V., & Cansell, M. (2005). Influence of the formulation on the structural networks in ice cream. *International Dairy Journal*, 5, 255–262. http://dx.doi.org/10.1016/j.idairyj.2004.07.009
72. Su, F., & Lannes, S. C. S. (2012). Rheological evaluation of the structure of ice cream mixes varying fat base. *Applied Rheology*, 22(6).
73. Ohmes, R. L., Marshall, R. T., & Heymann, H. (1998). Sensory and physical properties of ice creams containing milk fat or fat replacers. *Journal of Dairy Science*, 81(5), 1222–1228. http://dx.doi. org/10.3168/jds.S0022-0302(98)75682-6
74. Goff, H. D. (2008). 65 Years of ice cream Science. *International Dairy Journal*, 18, 754–758. http://dx.doi.org/10.1016/j.idairyj.2008.03.006
75. Lannes, S. C. S., & Ignácio, R. M. (2013). Structuring fat foods. In I. Muzzalupo (Org.), *Food industry* (Vol. 1). Rijeka: Intech Open Science.
76. Adapa, S., Dingeldein, H., & Schmidt, K. A. (2000). Rheological properties of ice cream mixes and frozen ice creams containing fat and fat replacers. *Journal of Dairy Science*, 83(10), 2224–2229. http:// dx.doi.org/10.3168/jds.S0022-0302(00)75106-X
77. Prindiville, E. A., Marshall, R. T., & Heymann, H. (2000). Effect of milk fat, cocoa butter, and whey protein fat replacers on the sensory properties of low-fat and nonfat chocolate ice cream. *Journal of Dairy Science*, 83, 2216–2223. http://dx.doi.org/10.3168/jds.S0022- 0302(00)75105-8
78. Santana, I. A., Ribeiro, E. P., & Iguti, A. M. (2011). Evaluation of green coconut (Cocos nucifera L.) pulp for use as milk, fat and emulsifier replacer in ice cream. *Procedia Food Science*, 1, 1447–1453. http:// dx.doi.org/10.1016/j.profoo.2011.09.214
79. Karaca, O. B., Guven, M., Yasar, K., Kaya, S., & Kahyaoglu, T. (2009). The functional, rheological and sensoru characteristics of ice creams with various fat replacers. *International Journal of Dairy Technology*, 62, 93–99. http:// dx.doi.org/10.1111/j.1471-0307.2008.00456.x

5

Oils and Fats as an Environmentally Benign Raw Material for Surfactants and Laundry Detergents

Subhalaxmi Pradhan[1], Chandu S. Madankar[2*] and Paridhi[1]

[1]*Department of Basic Sciences, School of Basic and Applied Sciences, Galgotias University, Greater Noida, U.P. India*
[2]*Department of Oils, Oleochemicals and Surfactants Technology, Institute of Chemical Technology, Mumbai, India*

Abstract
Surfactants/detergents have grown in popularity as a result of their widespread use in the home, industries, enhanced oil recovery, cosmetics, medicines, and food applications. Surfactants have low irritancy to skin and eyes, impart softness, good solubilization, and antimicrobial properties. Surfactants show basic surface activities like foaming, detergency, solubilization, emulsification, and wetting. Besides these properties, different modified raw materials based on fats and oils increase the demand and use of surfactants. The composition and formulation of the final product are based on the properties and function as well as the raw material used in the formulations. There are many raw materials based on the fats/oils that can be categorized based on their properties, such as surfactants, emulsifiers, wetting agents, softening agents, etc. Surfactants nowadays are present in many forms such as solid powder, tablets, pods, laundry balls, etc. A lot of research is going on towards the synthesis of advanced surfactants based on oils and fats like Gemini surfactants, biosurfactants, vegetable oils ethoxylates, and alkyl polyglucosides, polymeric surfactants (Polyquaternium, solidly based surfactants). This chapter describes the history of detergents/surfactants, raw materials based on fats and oils, synthesis methods, potential applications, formulations, environmental effects, and market analysis of detergents/surfactants.

**Corresponding author*: chandumadankar@gmail.com

Divya Bajpai Tripathy, Anjali Gupta, Pooja Agarwal, Anuradha Mishra, and Arvind Kumar Jain (eds.) Oils and Fats as Raw Materials for Industry, (117–144) © 2024 Scrivener Publishing LLC

***Keywords*:** Vegetable oil, fats, formulations, emulsifier, surface-active properties, biosurfactants, detergents

5.1 Introduction

Laundry detergents and the products based on them contain the surfactants mostly and other ingredients that will increase the cleaning and other properties. Laundry detergents are of many types, but two types are widely used, powdered granules and liquid detergents [1–3]. Liquid detergents have gained widespread attention from consumers, and the increasing demand for low foamer laundry detergents.

Detergents of any form can be formulated by addition of two or more surfactants; this is because of increasing synergistic of the complex molecules that will impart different effects and properties such as wetting, cleaning, foaming, etc. The complex bonding of soil with the fabric is removed by these detergents and the term used to distinguished percentage removal of soil is called "detergency". Detergency can be defined as solid unwanted substances in contact with liquid [4].

For the formulation of detergents, surfactants and the builders are the main key ingredients that play a significant role in the cleaning activity. They reduce the water hardness by the binding with the positively charged metal ions and all the ions form a complex with the binder during the cleaning process in the water which precipitates out the ions and increases the cleaning and the detergency [5, 6]. With the builders and the surfactants other ingredients are also used to boost the effects of this component, like whitening agents, stabilizers, bleaching agents, antibacterial agents, enzymes, fragrances, and color. Water is the major ingredient in the liquid detergent formulation [4]. The present chapter will deal with production of surfactants and laundry detergents from oils and fats, their application, market value and environmental sustainability for future generations.

5.2 History of Laundry Detergent

A Sumerian tablet from the 22nd century BC gives soap formulations composed of cassia oil, water, and alkali; however, it was unfamiliar during its discovery. Germany created the first synthetic detergent in 1916 as a result of the shortage of fats during World War I, even though the process for making industrial soap, which involves heating fats and oils with alkali, had largely not changed up until that time. Soaps have an advantage over surfactants in terms of redeposition of dirt and soil as well as retaining whiteness because they operate as self-water softeners when there is enough soap to form suds. The use of anti-redeposition chemicals was necessary due to the advent of multi-ingredient laundry detergents based on synthetic surfactants [7–9]. Even if chemical surfactants help to reduce dirt redeposition to some extent, the effect isn't powerful enough to eliminate the need for a whiteness retention aid. In the last 60 years since the first laundry detergent was released, it was found that the molecular structure can influence by chemical technology. It was recognized that decreasing surface tension was required for better cleaning, which was accomplished using chemical surfactants [10–12].

Prior to World War I, laundry soaps were mostly made of Na and K neutralised fatty acid soaps. During World War II, synthetic detergents were developed in Germany to replace the rare animal fats used in soap manufacture [13]. These chemical detergents are branched-chain alkyl benzene sulfonates and short-chain alkyl naphthalene sulfonates. They might soften water by removing hard minerals, similar to soap.

Long alkyl aryl sulfonates were employed as a detergent for the first time in 1930 and again in 1945. They were the surfactant that was most frequently employed in synthetic laundry detergents. The first produced synthetic detergent with sodium diphosphate as a builder was made and released in the United States [14].

Due to the inability of straight-chain detergents to perform in hard water, phosphates were mixed to make water soft. But further, phosphate was also replaced by Na_2CO_3 and EDTA as water softeners because phosphate was outstanding fertilizer for algae in aquatic systems; due to its activity algae blooms ensued, depleting the oxygen in the water, and thus it harms the aquatic animals [15–17].

Since the mid-1960s, complex phosphates have caused environmental problems due to their role in the eutrophication of streams. Since then, the detergent business has focused its efforts on developing an affordable replacement. The alkyl aryl group, which was used to make the surfactants present in the earliest synthetic detergents, was produced by combining benzene with propylene tetramer. These stiff alkyl benzene sulphonates, which are highly branched and non-biodegradable, were discovered in nature. They accumulated in the environment to the point where streams and sewage treatment facilities started to froth. Hard alkyl benzene sulphonates, which were previously used in the detergent industry in the USA, were gradually phased out in 1965 [18–20]. The discovery of some ion exchange materials and zeolites as detergent builders in the mid-1970s, led to a gradual move away from phosphate technology.

When synthetic powder detergents were converted to compacts, anionic linear alkyl aryl sulfonates (LASs), long-chain fatty alcohol sulphates (ASs), long-chain fatty acids, and nonionic alkyl ethoxylates were the most often utilised surfactants. The major U.S. manufacturers benefited from the recent change in production methods and formulas brought on by the changeover to compact discs.

This has been achieved without sacrificing value for customers or effectiveness. The switchover of our detergent manufacturing towards a phosphate-free formula is almost finished. About 32% of European detergent powder contains zeolite and Canada seems to be roughly halfway there; nevertheless, Latin America and several Pacific area geographies remain phosphate-based.

On the other hand, the transition to a phosphate-free construction approach increased the overall complexity of detergent powders. Furthermore, binding ions of hardness, and phosphate shows several important activities in the removal and cleansing of dirt. These activities are peptization and breakdown of soil, suspension and dispersion of soil. Hence phosphate was withdrawn from the detergent formula and the formulator needs to find out another alternative compound to replace the phosphates. Nowadays, to overcome complications like hardness management, alternate ingredients were added. Zeolites/layer silicates were used for managing the hardness of phosphate-based detergent. Then there is the rise of another complication, soil suspension, and to reduce this, poly-carboxylate polymers were used for maintaining the pH and peptization of soil citric acid [21–23].

Natural alcohol sulphates were first introduced in approximately 1932. Soap and other natural alcohol sulphates are made from natural oils. The first step includes converting natural oil to fatty acid ester, which are then reduced to alcohol with the help of a copper chromite catalyst [Scheme 5.1]. Conversion to natural alcohols in the current era entails converting fat to methyl esters; the esters are therefore shrunk to alcohol and methanol, which could be reused [24–27].

Dreft was the first laundry product to employ a natural alcohol sulfate in 1933, followed by Tide in 1947. Natural alcohol sulfates are still widely utilized as a high-volume surfactant in many laundry solutions today [24]. Other mainframe surfactants were quickly produced after the introduction of natural alcohol sulfates. Alkylbenzene sulfate (ABS) was the next significant low-cost surfactant technique, which was first used in 1947 [25]. Tetrapropylene, a highly branched compound, is one of the oligomers of propylene used for production of ABS [26]. Tetrapropylene is used frequently as petroleum industry raw materials have various applications

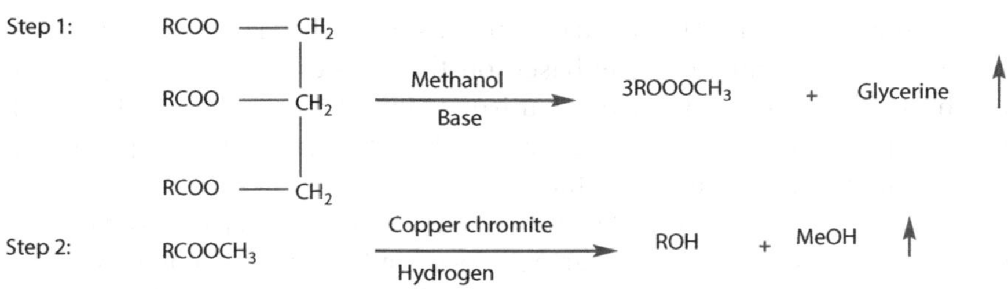

Scheme 5.1 Method of formation of natural alcohol from fats and oil.

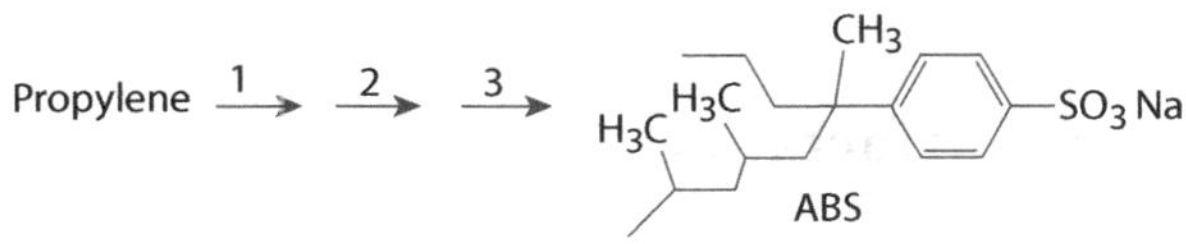

Scheme 5.2 Synthesis of ABS 1) Oligomerization of tetrapropylene, 2) Alkylation of Benzene using aluminium chloride, 3) Sulfonation using sulfuric acid.

and act as an octane booster in gasoline. The method of production of ABS surfactant is given in Scheme 5.2. First Tetrapropylene is produced by oligomerization of propylene followed by alkylation in presence of a Lewis acid catalyst and benzene to make the branched alkyl, such as aluminum chloride benzene. Then ABS surfactant is made by sulfating the alkyl benzene [27 A & 27 B].

After the discovery of natural alcohol sulfate in 1932 it was introduced into the detergent.

These alcohol sulphates still have been highly used in several laundry solutions today as a large surfactant [28–30].

5.3 Raw Materials in Laundry and Detergents

A surface-active agent is a distinct, long-chain molecule that contains both water-loving and water-hating components. The water-loving and water-hating components of a surfactant can be modified to affect a surfactant's wetting and emulsifying properties, foaming ability, foaming control ability, and dispersive ability [19, 20, 22–24].

The tendency of surfactants to adsorb at surfaces/interfaces in an orientated form is one of their distinguishing characteristics. The molecular behavior of surfactant molecules at the interface and surfaces estimates whether the surface or interfaces will become more hydrophilic or hydrophobic as a result of surfactant adsorption [31]. These qualities reveal the surfactant molecule based on the type of head charge present on the surfactants and their mechanism of interaction at the surface and the interface; these are the properties that help to distinguish properties of surface-active agents i.e., surfactants [32].

The cleaning temperature, fabric type, desired foam level, type of builder, state of product and manufacturing process all affect the type of surfactant used in a laundry product. Additionally, surfactants must adhere to stricter environmental compatibility criteria. Surfactants must also meet more demanding environmental compatibility standards [30].

5.4 Types of Surfactants

The surface-active compounds have various properties, and based on this surface-active property they play a key role in laundry formulations. They show properties such as:

a) They increase the wetting ability of water
b) They can emulsify and solubilize soils that are present in the wash solutions
c) They are a good cleanser and remove soil concerning washing action.

The surfactant is mainly classified based on the charge present in their head groups, such as anionic (-), Cationic (+), Amphoteric or non-ionic (neutral charge), etc. [32–34].

5.4.1 Anionic Surfactants

Anionic surfactants have hydrophobic fatty chain and polar groups, which prevents them from ionising in water. The solubility is achieved by the formation of hydrogen bonds between water molecules and some hydrophilic functions, such as the ether function of polyoxyethylene (hydration phenomenon), which are present in the hydrophilic component but are not ionizable for oxygen, nitrogen, or sulphur atoms [35]. Although sugar esters and alkanolamides are also present, polyoxyethylene (POE) or polypropylene (POP) derivatives make up the majority of the items in this category.

Anionic surfactants are one of the most important surfactants used in laundry and detergents because of their low cost of production and manufacturing; also, anionic surfactants are very effective and easily degradable, having a chain length of C-12 to C-18 atoms. The head shows containing the negative charge and shows the hydrophilic nature, e.g., Sulphates, phosphates and carboxylates, etc. [36, 37].

The anionic surfactants head group carries a negatively charged ion and during the washing, mechanism water contains calcium and magnesium and other metal ions which increases the hardness of water as we know negatively charged ion present on the surfactant reacts with the positively charged and precipitation of soil and clay formed which shows the cleaning action by anionic surfactants.

Some common types of anionic surfactants are used for the formulation of laundry and detergents.

5.4.1.1 *Alkylbenzene Sulfonates (ABS)*

The most common surfactants are alkyl benzene sulfonates (ABS). They may be branched (for example, tetrapropylene benzene sulfonate), or linear (for example, linear alkylbenzene sulfonates, or LAS). Because of their slow rate of microbial breakdown, the former are only employed in a few nations (biodegradability). This is because of the structural complexity of the molecule, the double bond present at the terminal position or inner side of the carbon chain, and isomers with the phenyl group present in the position of 1, 2, and 3 might be formed. LAS is the most sensitive anionic surfactant due to the existence of Ca^{2+} and Mg^{2+} ions along with other metal ions in water [35]. The number of builders will depend on the calcium precipitation in the water, lowering detergency. LAS detergency, on the other hand, increases with a concentration in the absence of precipitation. LAS is the most important surfactant for detergent powders in terms of volume. Sulfonates of paraffin or alkane (SAS: secondary alkane sulfonates) could be a source of anionic surfactants but these compounds are highly biodegradable in an aerobic environment. Due to their relatively expensive cost, these products have yet to be employed in detergents [34, 35].

5.4.1.2 *Paraffin- or Alkane Sulfonates*

Sulfates of primary alcohol (PAS) are anionic surfactants which are also from renewable materials and biodegradable in nature. Sulfonates of olefins (AOS) are made up of alkene sulfonates and hydroxy alkane sulfonates, represented in Figure 5.1 [36–38].

Unlike conventional surfactants, which have the most surface activity when C12 alkyl chains are utilized, olefin sulfonates have the highest

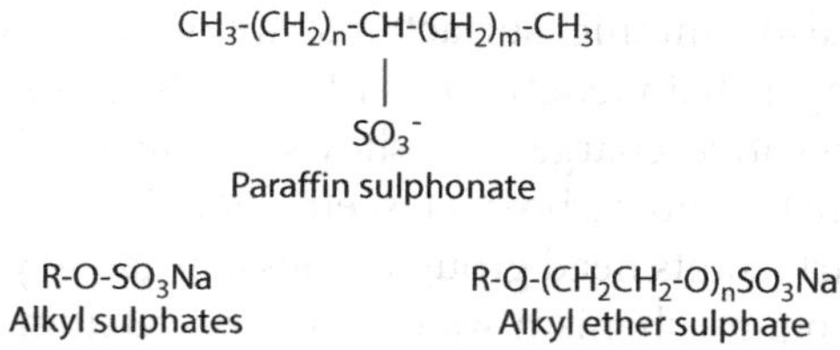

Figure 5.1 Structure of paraffin sulfonates.

activity by increasing the carbon chain to C-14 and C-16. Alkylbenzene sulfonates and fatty alcohol sulfates are more susceptible to water hardness than anionic surfactants. AOS is not utilized in washing goods in the United States because it can irritate the skin when combined with chlorine bleach, which is widely used for whitening [39]. AOS, on the other hand, has been utilized in laundry detergents and personal care products for many years in Asia, particularly in Japan and Korea, having excellent washing and effervescing abilities [39, 40]. This kind of surfactant can be found in shampoos, liquid detergents, and alkyl ether sulphates (fatty alcohol ether sulphates: FES). Alkyl ether sulfates vary from alkyl sulfates due to the presence of ether glycol units between the C- chain and the SO_4^{2-} group. The presence of ether glycol units between the carbon chain and the sulphate group distinguishes ether sulphates from alkyl sulphates. There are two methods for creating ether sulfates. The first is fatty alcohol infused with ethylene oxide molecules, and the second is sulfation of ethoxylated fatty alcohols with an air/SO_3 mixture (as for alkyl sulfonates), followed by alkaline neutralization with NaOH, NH_3, and amines. The features of ethoxy sulfates are determined by the number of ethylene glycol units in the initial ethoxy fatty alcohols. The most commonly used ether sulfates are Lauryl ether sulfates with n = 2 or 3, which have the best foam and detergency properties. HBO S 3390 - 2 (Hoechst) or Fenipon AC are the brands of these isethionates of acyl. They have excellent skin-friendly properties as well as aqueous solution stability. Since these are less soluble in water at room temperature, they are employed in creams and thickened liquids such as shampoos, foam baths and toilet bars. Because of its skin-friendliness and emollient characteristics, sodium acyl isethionate can also be employed in infant care accessories and face wash compositions, and these surfactants were studied thoroughly in the 1960s [41]. Tallow was the primary raw material at the time, hence expenses were considerable. Prices are getting more competitive now that palm oil is widely available. C16-C18 paraffin sulfonates give the finest results [42]. They are found in both powder and liquid detergents and are mostly popular in Southeast Asian countries.

5.4.2 Cationic Surfactants

Cationic surfactants are the surfactants synthesized by the "quaternization reaction". They react to the positively charged nitrogen and form quaternary salts. The word Cationic indicates that surfactants contain a positive charge on the head group. They show antibacterial properties due to the positive charge head group which can destroy microorganisms, as microbes have negatively charged cell particles due to carboxylate and peptidoglycan present on the cell wall. The cationic surfactants react with negatively charged particles on the microbes and increase the hydrophobicity resulting in the breaking of the cytoplasmic layer as result, the surface charge kills the microbes [43, 44]. Mainly cationic surfactants are used in fabric softeners due to their positive charge to reduce static effects present on the hair. C-8 to C-18 chain length is found in the alkyl chain of cationic surfactants. Cationic surfactants include alkyl trimethyl ammonium chlorides, such as dodecyl trimethyl ammonium chloride (DDTMAC, $C_{12}H_{25}(CH_3)_3NCl$) and others. Another popular cationic surfactant family is DADMAC, which comprises two long-chain alkyl groups of dialkyl dimethyl ammonium chloride [41].

5.4.2.1 Quaternary Ammonium Compounds

Regarding quaternary compounds, the word quaternary tells us that there are four chemical structures, which have a central positive charge surrounded by the same or different alkyl groups. Quaternary ammonium compounds are the cationic surfactants that contribute and use high in the class of cationic surfactants in the laundry and detergents fabric softeners formulations [45, 46].

Distearyl dimethyl ammonium chloride (DSDMAC), a quaternary ammonium molecule with two long alkyl substituents made from hydrogenated tallow, is the most well-known cationic surfactant. The vast majority of its long-chain alkyl substituents are the homologs of C16 and C18. [45]. There are very prominent properties of quaternary compounds. These properties mainly depend on the structure of quaternary compounds such as number of C-atoms, degree of saturation, and the number of oxygen atoms in the structure. The presence of the hydroxyl group in the structure will affect the solubility of the substance; hydroxy group increases the solubility and pourability in the formulation [47].

5.4.2.2 *Imidazolinium Compounds*

Imidazolinium compounds have a cationic head group and are comparable to quaternary compounds. The first carbon atom is a fatty aminoethyl chain, the second carbon atom is a long alkyl substituent, and a third carbon atom is a methyl group. Imidazolinium compounds, unlike quaternary ammonium compounds, have only one anion: methyl sulfate. This category's most widely utilized member is the tallow-alkyl derivative [48, 49]. It comes as a liquid and can be simply transformed into fabric softeners in the cold. The hydrogenated tallow chain imidazolinium derivative requires a more involved formulation to generate a finished item; it softens more readily than the non-hydrogenated equivalent. The importance of the oleyl derivative has recently attracted interest [50].

5.4.3 Non-Ionic Surfactants

Non-ionic surfactants are structurally composed of cationic and anionic components, and as a result, the head group of these substances is neutrally charged. Due to the presence of charge on non-ionic surfactants, it is not ionized in water. They cannot reduce water hardness but show excellent results in the oil and soil dirt removal by the solubilization and emulsification nature of the compound. Non-ionic surfactants are used in liquid detergents and powders. The most common class of non-ionic surfactants are ethoxylates, which have various derivatives such as alcohol ethoxylates, fatty acid ethoxylates, alkyl ethoxylates, alkylphenol ethoxylates, monoalkanolamide ethoxylates, ethylene oxide, and propylene oxide copolymers (Figure 5.2). Some polyoxyethylene esters mainly contain polyethylene glycols (PEGs) [46].

The most commonly and widely used non-ionic surfactants in the formulation of laundry and detergents are esters and alcohol ethoxylates. This type of surfactant is effective for reduction in hardness.

$$R\text{-}O(CH_2\text{-}CH_2\text{-}O)_n\text{-}H$$

Fatty alcohol ethoxylated

$$H\text{-}(O\text{-}CH_2CH_2)_m\text{-}O\text{-}(CH_2\text{-}CH\text{-}O)_n\text{-}(O\text{-}CH_2\text{-}CH_2)_m\text{-}H$$

Ethylene and proxylene oxids copolymer

Figure 5.2 Structures of non-ionic surfactants.

5.4.4 Amphoteric Surfactants

Zwitterionic surfactants are similar to amphoteric surfactants. Zwitterionic surfactants have a single molecule that is connected to both cationic and anionic centres (Figure 5.3). Zwitterionic surfactants are less frequent than anionic, cationic, or nonionic surfactants [32–34]. Due to their gentle nature, these surfactants are ideal for use in personal care and home cleaning products. They work well with all other surfactant classes and remain soluble and helpful even in the presence of high concentrations of electrolytes, acids, and alkalis. Two charged groups with different charges could be present in these surfactants. The positive charge is almost invariably ammonium, even if the negative charge could be anything (carboxylate, sulphate, or sulfonate) [40–44].

5.4.5 Chelating Agents

Removal of soil is a very complex process; it's not as simple as adding surfactant into the water and removing soil present on the fabric. During a cleaning cycle, the most important compound is water hardness. Water contains many ions such as calcium, magnesium, etc. These are the metal ions present in the water. These ions have a positive charge and the detergents have negative charge surfactant molecules. When the molecules, i.e., the negatively charged particle from the detergents and the positive charge ions present in the water interact with each other form a complex. The ions present in the water start consuming an excess of surfactants, as chelating agents bind with the ions, and precipitation is formed by forming a complex structure. As a result, positively charged ions get transferred their charge and converted into negatively charged particles. Phosphates, ethylenediaminetetraacetic acid (EDTA), sodium citrate, and zeolite compounds are some of the most popular chelating agents used in industrial cleaning products [40].

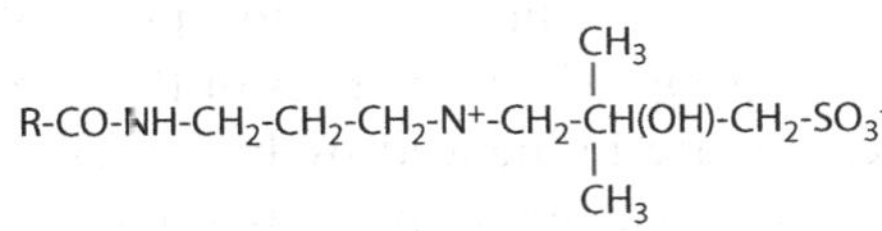

Figure 5.3 Structures of Zwitterionic surfactants.

5.4.6 Builders Detergents

Builders are the blends of chelating agents and the surfactants, as chelating agents are the most important to binding metal ions present in the water hardness and change the charge present on the metal ions. For the precipitation and to avoid excess consumption of surfactants, chelating agents are used. But using chelating agents is cost-consuming as an alternative. The formulators use builders to overcome the surfactants and chelating agents' problems [20–22]. Builders upgrade the cleaning efficacy of the surfactants. Because builders have different properties such as emulsifying, softening, and chelating properties, builders will soften the water by changing the charge present on the metal ions in the water. There are two ways, as follows.

A) Sequestration, builders can hold metal ion present in the water
B) Precipitation is the process that forms complex and removes metal ions from solution.

Builders can help to increase alkalinity so that the dirt and the oil stain can easily be removed while the cleaning action. There are some common builders we use in the laundry and detergents formulation: (i) phosphate derivatives, (ii) sodium carbonate, and (iii) sodium silicate. The phosphate-derived builders are sodium triphosphate (STPP), which is commonly used in detergents where the solution or water contains heavy metal ions [38–40]. Sodium carbonate is a builder used in the detergents' formulation which softens the water through the process of precipitation; when used in high amounts, sodium silicate acts as a builder in some detergents. It reduces rusting and lends crispness to detergent granules when used in lower quantities.

5.5 Synthesis Methods

Fatty acids having more number of C-atoms and their derivatives are amphiphilic substances with a lipophilic non-polar alkyl chain and a hydrophilic polar head group. The properties of the compound and, consequently, its application are influenced by how these two diametrically opposed sets of traits are balanced in the compound. Generally speaking, it is simpler to change the carboxyl/ester function than the alkyl chain. The choice of the starting material, which comes in three various forms depending on chain length: C12 (lauric oils), C16/18 (palm, tallow, and so on), and C22 (palm, tallow, and so on), is typically connected to changes in the alkyl chain (erucic oils). Ricinoleic acid is an example of an unsaturated hydroxy fatty acid present in castor oil. The carboxyl and ester functionalities play an important role in lipid chemistry [43–46]. Production of these functionalities can be carried out on a milligramme scale for analytical applications, such as producing methyl esters for gas chromatography or producing oxazolines and picoline esters for mass spectrometry. A specific triacylglycerol can be synthesised in-lab on a gramme scale, or acids, methyl esters, and other esters like olestra and monoacylglycerols can be produced commercially on a tonne scale. Methyl esters are increasingly employed as a source of biodiesel, biolubricant and biosurfactants. The different procedures for producing distinct triacylglycerol derivatives include hydrolysis, alcoholysis (especially methanolysis and glycerolysis), acidolysis, esterification, and interesterification (Figure 5.4). These triacylglycerol derivatives can be employed as biosurfactants, biolubricants, and biofuels in a variety of industries. These reactions might have as their starting point natural mixed triacylglycerols or acids or esters.

Nature contains both free molecules and, more typically, esters of long-chain alcohols (RCH_2OH) with structures that resemble well-known acids. Alcohol acetates make up a large portion of insect pheromones, while wax esters, which are present in jojoba and other vegetable and animal waxes, are made up of long-chain alcohols and long-chain acids. Acids, methyl esters, or triacylglycerols are used in the commercial manufacture of long-chain alcohols, which are essential oleochemicals. Depending on the catalyst employed, double bonds may be reduced or left unaltered throughout this catalytic process [9–12, 34–37]. Vegetable oils are frequently subjected to these processes, and the results are mixtures with varied chain lengths. C8 to C14 alcohols are produced by tallow, lard, palm oil, or palm stearin, whereas C16 and C18 compounds are created by erucic acid-rich oils. The mixed products can be fractionally distilled to produce different alcohols. By oligomerizing ethene, the petrochemical sector also creates dodecanol and other related alcohols. Production of long-chain alcohol has increased to about 1.5 million tonnes annually, with fat making up at least two-thirds of that amount. About 75% of these fatty alcohols are alcohol sulphates, alcohol ethoxylates, and alcohol ethoxylate sulphates. Although most of these processes begin with commodity oils and fats, the glycerol esters are rarely employed directly since the precious glycerol would be lost, among other things. Hydrogenolysis more frequently results in damage to acids, methyl esters, or wax esters made *in situ* from acids and alcohols. Esters,

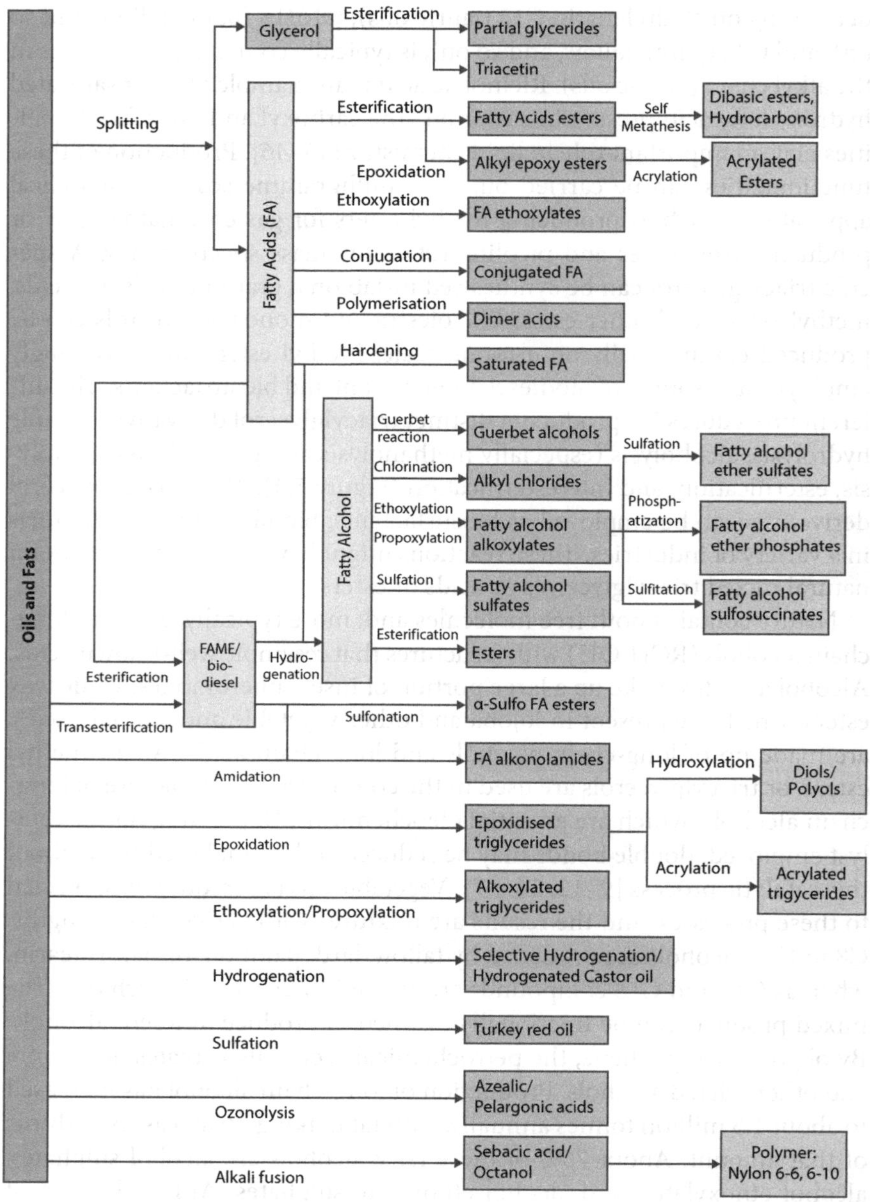

Figure 5.4 Synthesis of different functionalities of oils and fats.

aldehydes, alkanes, ethers, and acetals are all likely by-products of hydrogenolysis in addition to alcohols [49].

5.5.1 Synthesis of Methyl Ester Sulfonate Surfactant from Jatropha Oil

Esterification, transesterification, and sulfonation were three sequential processes used to produce the anionic surfactant methyl ester sulfonate (MES) from Jatropha oil. The free fatty acid content of the oil is first reduced to a minimal level utilizing photocatalytic esterification before continuing with alkali-catalyzed transesterification. If this step is neglected, soap will be made when oil is saponified in the presence of an alkaline catalyst, which will inhibit the production of alkyl esters from oil. Slamtet *et al.* (2018) [50] employed TiO_2 P_{25} catalyst to photocatalytically esterify Jatropha oil with a high free fatty acid content (5.4%) in a stirred box-size photoreactor with five 12V UV-A lamps at 352 nm. Alkali-catalyzed transesterification was carried out using oil and methanol in a 1:4 molar ratio with a 1% NaOH catalyst at 60 °C for one hour after the aforementioned procedure, which reduced the oil's free fatty acid concentration. The methyl ester product was then centrifuged separated from the reaction, rinsed with water to remove any remaining alcohol and catalyst, and dried at 110 °C to remove any moisture. After that the next sulphonation step was carried out taking methyl ester and sodium bisulphite in a Erlenmeyer two-necked flask fitted with reflux condenser at 100 °C for 4.5 hrs with stirring at 600 rpm. After completion of the reaction the product methyl ester sulfonate (MES) was purified by centrifugation at 4000 rpm and washed with 40% methanol at 50 °C for 1hr to remove excess alcohol. The product MES showed the lowest surface tension value of 33.12 Dyne/cm.

5.5.2 Synthesis of Non-Ionic Surfactants from Jatropha Oil

Non-ionic surfactants are synthesized from fatty acids of Jatropha oil. Fatty acids obtained after hydrolysis of Jatropha oil was refluxed with polyethylene glycol and further reaction with p-aminobenzoic acid (PABA) produced amino acid based non-ionic surfactant.

A combination of extracted fatty acids is typically esterified by recirculating polyethylene glycol in the solvent xylene at a temperature of 140 °C. The reaction is carried out until water is completely eliminated from the reaction mixture. About 27.4g of p-aminobenzoic acid are combined with the reactant and refluxed at a temperature of about 140 °C in order to completely eliminate the H_2O created during the reaction. When 0.2 mol of water is removed from the reaction, it is stopped, and several schiff bases are created by reacting the created diesters with the carbonyl groups in the ethanol (vanillin and salicylaldehyde) [51]. The reaction process is shown in Figure 5.5.

A surfactant's performance is often judged based on a number of factors, including emulsion stability, critical micelle concentration (CMC), and surface tension. Two qualities can be acquired by varying the surfactant content. By extrapolating the patterns in the two distinctive regions, CMC is determined. PEG species with short chains have low CMC, whereas PEG with longer chains has higher CMC. The CMC value hardly changes when the vanillin moiety is substituted for the o-hydroxy benzaldehyde moiety [51]. Surfactant molecules in an aqueous medium exhibit repellent and attractive forces that counterbalance one another. While drawing partially charged non-ionic chains and water, the polar phase repels hydrophobic chains. Surface tension at the CMC is used to calculate the surface activity of the surfactant molecules at the interface. Surfactants based on salicylaldehyde displayed a lower CMC than those based on vanillin [51].

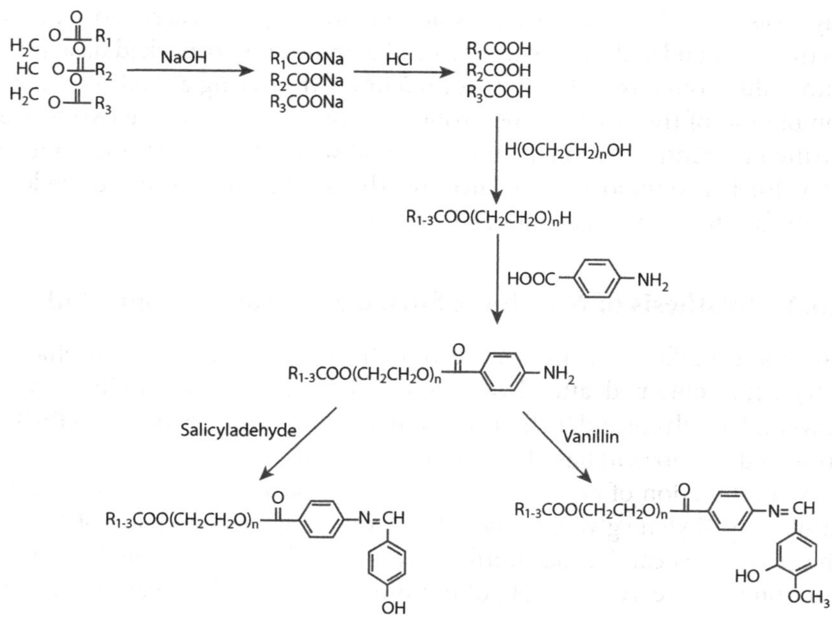

Figure 5.5 Synthesis of amino acid-based surfactants from Jatropha oil.

5.5.3 Synthesis of Anionic Surfactants from Castor, Sal and Thumba Oil

Mono and diethanolamide sulphates of Castor, Sal and Thumba were prepared by some researchers that can be used as anionic surfactants [52]. Thumba oil contains around 81% of unsaturated fatty acid, castor contains 90% of unsaturated hydroxy fatty acid-ricinoleic acid and sal oil contains 56.5% of saturated fatty acid [52]. The extracted oils of the above seeds were reacted with monoethanolamide and diethanolamide to produce respective ethanolamides of oil. Then sulfonation of ethanamides was done by reaction with chlorosulfonic acid. About 0.04 mol of oil was taken in a 250 ml two-necked RB fitted with a guard tube constructed of calcium chloride and a dropping funnel. The temperature of the medium was maintained between 110 and 115 °C. The mixture was then combined with 1% sodium methoxide w.r.t oil and stirred magnetically. Then 0.14 mol of monoethanolamine was placed in a dropping funnel and added dropwise to the reaction mixture within 30 minutes. The reaction mixture was stirred at 110-115 °C for 4 hrs. The progress of reaction was examined by Thin layer Chromatography in 5% chloroform-methanol solution as mobile phase. Then the cooled reaction mixture was dissolved in 150 ml of $CHCl_3$, and subsequently washed many times with a 15% aqueous NaCl solution till complete removal of CH_3ONa and monoethanolamine. Then it was concentrated using a rotary evaporator under reduced pressure and dried over anhydrous sodium sulphate. Purification of monoethanolamide from any traces of impurities, such as amino esters and free fatty acids was done by silica gel (60–120 mesh) column chromatography in 2% chloroform- methanol solution as eluent. After purification the yield of synthesised monoethanolamide obtained is around 92%. The above-mentioned procedure can be followed for synthesis of diethanolamide of different oils but diethanolamine was taken instead of monoethanolamine. The yield of Thumba oil diethanolamide obtained by the above method is around 89%. Characterisation of synthesized diethanolamides is generally done by IR, ^{1}H-NMR, ^{13}C-NMR and ESI–MS.

Sulphonation of Mono- and Diethanolamides of above oils were carried out by reacting with Chlorosulfonic acid. The scheme of above reactions is given in Figure 5.6.

About 0.109 mol of synthesized monoethanolamide of oil is taken and 0.12 mol of Chlorosulfonic acid added to it slowly dropwise in 20min. Then the mixture dissolved in 70 ml of chloroform at 5 °C by stirring using a magnetic stirrer. The reaction was continued for 3hrs at 5 °C and after that diluted using n-butanol and neutralisation of the mixture was done

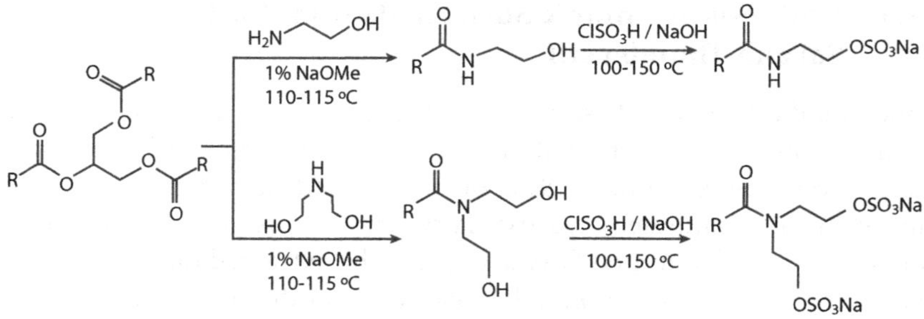

Figure 5.6 Synthesis of sulphated ethanolamide based surfactants from vegetable oils, where R=Fatty acid of vegetable oils.

using 18N NaOH solution. Butanol-water mixture is evaporated using rotavapour and drying of residue is done under reduced pressure at 80 °C. Then the unreacted mono- and diethanolamides of oil is removed by treating the product with diethyl ether. Drying of final sulfated sodium salt of the product is done by vacuum desiccator. The yield of sulfated mono- and diethanolamides of oil varies in the range of 85-90%.

The sulfated monoethanolamide of thumba and castor oil demonstrated superior CMC (critical micelle concentration) values, outstanding emulsifying performance, and calcium tolerance characteristics when compared to diethanolamides and SDS. These sulfated compounds can be used in a variety of detergent applications. However, due to their greater Ca tolerance, which is a vital detergency property, castor diethanolamides can also be used in detergent formulations.

5.5.4 Synthesis of Sugar-Based Surfactants from Vegetable Oils

Fatty acids isolated from non-edible oils react with sugar in the presence of biocatalyst lipase to manufacture sugar-based surfactants (Figure 5.7). The characteristics of the substrates affect the catalytic activity of lipase. Some lipases only catalyse branching and short-chain fatty acids, others are capable of catalysing medium and long-chain fatty acids as well [53]. When making fatty acid esters from sugar, organic solvents are crucial. They are also used as adjuvants in solid phase enzyme catalysed reactions [54–56]. Finding a solvent, however, that will solubilize both sugar and fatty acid and is also compatible with enzymes is the difficulty. Both fat and carbohydrates can be solubilized using organic solvents like pyridine and dimethyl formamide; however. lipase activity is limited when utilising these solvents [57]. Organic solvents are poisonous and non-volatile, so they are not used

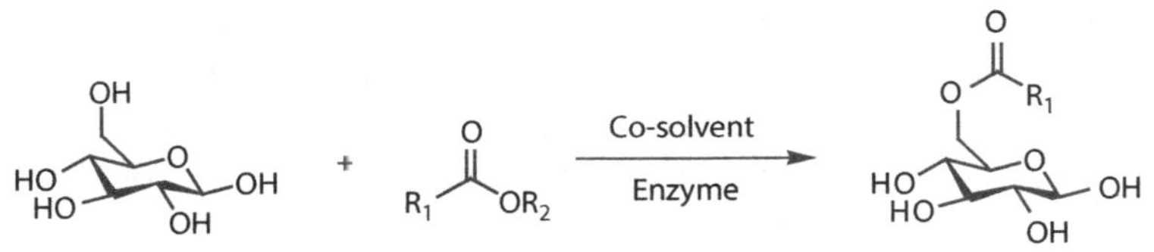

Figure 5.7 Synthesis of sugar-based surfactants from vegetable oils.

in the context of food applications. Ionic liquids (ILs), which improve the solubility of sugar, the reactivity of enzymes, and regioselectivity, will offer an alternative green reaction medium to conventional organic solvents for the manufacture of SFAEs to tackle the challenge [58].

5.6 Market Analysis

The global market for laundry detergent is anticipated to increase from \$61,711.0 million in 2020 to \$98,139.7 million in 2030, growing at a rate of 4.8% between 2021 and 2030 (Figure 5.8). In laundry detergent powder, popular constituents include bleach, surfactants, enzymes, dirt-preventing agents, optical brighteners, foam regulators, fragrances, colours, and corrosion inhibitors [12].

Factors affecting the growth of the laundry detergent market includ population growth, increasing urbanisation, novel product debuts, an

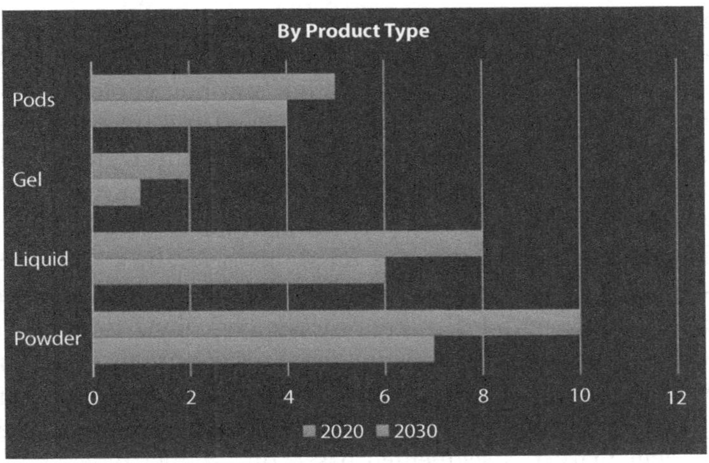

Figure 5.8 Trend of product type of laundry detergents.

increase in the adoption of automated washing machines, and rising consumer awareness of cleanliness. Boosted marketing efforts by top market players have increased consumer knowledge of gels, pods, and liquids. Due to customer demand for easily washable apparel, the market for liquid and pod laundry detergents has expanded. This is because liquid laundry detergents dissolve fast in cold water and offers an efficient clean while using less energy when used in washing machines. As a result, the exponential rise of the liquid laundry detergent industry was aided by the rising prevalence of automated washing machines in developed nations [59].

Laundry detergent market growth is likely to be boosted by the acceptance and rising popularity of natural or organic laundry detergent. Organic detergents are free of chemicals compared to conventional detergents because they are made from natural compounds and dissolve fast in water, so organic detergents are perfect for cleaning clothes. Because they are soft on fabrics, they last longer and require less water. Because it provides efficient cleaning and promotes sustainability, organic laundry detergent is anticipated to support the expansion of the laundry detergent market over the anticipated period [59, 60].

In addition to their ability to remove stains, laundry detergents are increasingly made to condition and give textiles a unique scent. In addition, industry participants are spending more on marketing and advertising in an effort to raise brand awareness among new target consumers, mainly for middle-class families. Due to the consequent rise in demand, manufacturers and retailers of laundry detergent have started to change their marketing strategies [46–49]. In the meantime, major companies are working heavily on their worldwide, digital, and social presence, as well as their on-shelf presence.

Asia-Pacific is the largest laundry detergent market in terms of value and volume due to its massive population and household size. Because individuals prefer to wash their clothing by hand and are not aware of the advantages of liquid and tablet laundry detergents, this region also has the largest market for powder laundry detergent. Growing urbanisation, rising disposable income, and rising washing machine usage seem to offer profitable expansion opportunities for industries. The COVID-19 pandemic's appearance sparked a statewide lockdown, which halted laundry detergent production and disrupted the supply chain. The rapid spread of the COVID-19 virus made people more conscientious of hygiene and sanitation; therefore the pandemic had little effect on the economics of laundry detergent industry. According to the market analysis, the laundry detergent market is segmented based on product type, application, distribution strategy, and geographical region. Depending on the type of product, the market is split into powder, liquid, gel, and pods/tablets. Applications may be employed both domestically and commercially. Based on the distribution channel, it is divided into supermarkets/hypermarkets, convenience stores, independent merchants, and others. North America is made up of the United States, Canada, and Mexico; Europe is made up of the UK, Germany, France, Italy, Spain, and the rest of Europe; Asia-Pacific is made up of China, Japan, India, South Korea, Australia, and the rest of Asia-Pacific, and LAMEA is made up of Brazil, Argentina, South Africa, Saudi Arabia, and the rest of Latin America, Middle East, and Africa [61]. In terms of product type, according to the laundry detergent market prediction, the powder segment contributed the most to the market with $27,039.4 million in 2020 (43.8%), and it is anticipated that this contribution will persist during the forecast period. Because of the convenience, good cleaning, and easy storage that gels provide, it is anticipated that the gel segment will grow quickly over the course of the forecast period. Due to its design and low likelihood of waste caused by the gel form's ability to avoid overdosing, it is also cost-effective.

The powder segment held the major share of 43.8% in 2020.

The household sector accounted for around 71.3% of the laundry detergent market share in 2020, based on industry statistics, the increased perception of automated washing machine, and expanding consumer spending on domestic items. According to the product's market trends, buyers may opt to buy detergents offline depending on the distribution channel. Examples of offline channels include supermarkets and hypermarkets, grocery stores, convenience stores, retail stores, and department stores. The convenience of finding favoured brands and laundry detergent types is what is driving the offline market's growth. People are increasingly buying laundry detergent online due to the globalisation of digital platforms. The supermarket/hypermarket segment dominated the market in 2020 with 35.3% of the total. Asia-Pacific dominated the global market for laundry detergent in 2020, and it is expected that this dominance will last the entire forecast period. The market growth for laundry detergent is being driven by the widespread usage of powder detergent, rising disposable income, and raised consumer awareness of hygiene and cleanliness.

5.7 Environmental Safety

There are many laundry detergents used for cleaning purposes and draining with water; the formulator has to consider and formulate the product according to a safety point of view. The used washed water should be treated by the municipal corporation to avoid environmental hazards. The amounts and effects of the substances used in laundry and detergents are taken into account during environmental assessment. The safety precautions cover two points. First is to decide the minimum limits of each ingredient added to the formulation of laundry and detergents and the second is to set a maximum threshold limit based on animal, water, human, and plant safety. These are the critical limits of each ingredient added into the formulation. These are the two higher limits and the lower ones that will help the formulator to develop a discharged system when the detergent shows cleaning actions. A prohibition on the use of phosphate detergents and the development of nonphosphate product substitutes has been enacted in response to public concerns, academic research, and governmental legislation. According to the (US) Soap and Detergent Association, manufacturers of cleaning products have taken the lead in reducing packaging waste and promoting proper waste disposal methods. Technological advancements have resulted in more concentrated products, products that combine two tasks in one, products with refill packages, and packages that utilize recycled materials. Products that are concentrated require less energy to make, transport, and package. With versatile products, separate packaging is no longer required. With refill packages, consumers can reuse primary packages more than once, using up less packaging overall and producing less waste. Recycling turns waste materials like plastic and paperboard into usable resources. Finally, detergents' environmental quality is being improved through life cycle assessment (LCA). LCA provides a whole assessment from manufacturing process to end user application, assessment of a product's environmental implications, usually from raw material acquisition to manufacturing and distribution, as well as customer use and disposal. LCA can be used to determine whether eliminating an environmental impact in one area, such as manufacturing, is worth it [62].

5.8 Future Trends

Consumers in the future want surfactants or laundry detergents with added values. There are lots of sustainable solutions incorporated into the formulation, and the goal of the chemist should be to synthesize the green surfactants which are much more effective in the future. Utilization of the waste and making the greener surfactants and final product would be more eco-friendly. The consumer needs a product that can utilize less water and energy concerning effective cost.

Customers continue to seek regionally and even locally specific goods; for instance, a product created to satisfy the needs of an older Western European population may not appeal to the youth of many new nations. Consumer fabrics are continuously changing, and not necessarily for the better when it comes to cleaning. For instance, despite a predilection for natural fibres, such as in Germany, synthetic microfibers are becoming more popular in industrialised nations, and the ratio of synthetic materials to cotton in garments produced in Asia is rising. These trends imply that we will need to clean fabrics more efficiently using less in the form of materials, energy, and water, as well as through more appropriate methods [63].

Not unexpectedly, there is ongoing interest in producing superior catalytic chemistry, which includes biomimetic systems as well as biocatalysts like enzymes. The rationale for formulation design is being strengthened by the study of soft condensed matter and mesoscale phenomena like self-assembly and interactions between surfactant and polymer, while formula development is being accelerated, especially for large and complex biomolecules, by tools like high throughput screening. But as can frequently be shown by comparing patent disclosures to textbook disclosures, most colloid research in the past lagged behind the formulator rather than outstripped them, sometimes by wide margins. Numerous theories, like the DLVO theory, continue to be unpredictable and correct, and there is a lack of experimental reference data for real systems, even simple ones with changeable counter-ions and pH. The effects of Hofmeister are still unknown. The importance of dispersion forces is usually overlooked because too many formulators only consider electrostatics. In the actual world, surfactants are frequently complex mixtures, yet a lot of academic study focuses on model molecules. Consumer habits and preferences are changing at a sometimes astonishing rate, with developing markets being able to outpace industrialised markets in terms of expectations and demands. All of this suggests that manufacturers of consumer washing products have a promising future.

5.9 Conclusion

Laundry detergents and surfactants have an important class of compounds and their unique trends toward sustainability gained widespread interest from consumers. A range of vegetable oil-based surfactants can be made by reacting fatty acids or the related methyl esters with various alcohols or amines. Surfactants have vast applications in personal care products, laundry detergents, and industrial cleansers, which include anionic, cationic, non-ionic, and amphoteric types. Currently researchers are more focused on synthesizing surfactants using renewable raw materials. Similar to this, relevant industries also evaluate the financial benefits of switching from petrochemicals to natural raw materials. The novel raw materials used in synthesizing surfactants and making a final product should be safer for consumers and the environment. The raw material used for the formulation of laundry surfactants and detergents has been modified and will help to optimize formulation concerning cost and transportation. As we see the trends of using detergents increasing day by day, vegetable oils will be an effective raw material for the formulation of laundry detergents with the optimized cost of production.

References

1. Schwartz, A.M and Perry J.W., *Surface Active Agents*, vol. 1. New York: NY Inter Science Publishers, Inc., 1949.
2. Linfield, W.M. (Ed.), *Anionic Surfactants. Surfactant Science*, vol. 7, part. 1. New York, M. Dekker, Inc., 1976.
3. Schink, M. (Ed.), *Nonionic Surfactants. Surfactant Science*, vol. 1. New York, M. Dekker, Inc., 1967.
4. Friedli, F., *Detergency of specialty surfactants*, CRC Press. 302 pp., Taylor & Francis, 2001.
5. S. N Augustin, Aqueous detergent-inorganic builder concentrates. U.S. Patent No. 3,350,319. 31. 1967.
6. O Brien, P Thomas, L Benjamin., Liquid detergent composition. U.S. Patent No. 4,268,406. 1981.
7. Wertz, H.M., Jean-Luc, Goffinet P.C.E., Near-neutral pH detergents containing anionic surfactant, cosurfactant and fatty acid. U.S. Patent No. 4,561,998. 31 Dec. 1985. Hinton, Gerald Thomas. "Detergent." U.S. Patent No. 6,465,413. 15. 2002.
8. Macıas-Zamora, J., Ramırez-Alvarez, N., Tracing sewage pollution using linear alkylbenzenes (LABs) in surface sediments at the south end of the Southern California Bight. *Environ. Pollut.*, 130, 229–238, 2004.
9. Sandhu, M.Y., Saleh, F.S., Afridi, S., The process for making low-density LAS surfactant detergent agglomerates using microwave heating. *Powder Technol.*, 326, 32–36, 2018.
10. Kocal, J.A., Vora, B.V., Imai, T., Production of linear alkylbenzenes. *Appl. Catal A.*, 221(1-2), 295–301, 2001.
11. Bordoloi, A., Devassy, B.M., Niphadkar, P., Shape selective synthesis of long-chain linear alkylbenzene (LAB) with AlMCM-41/Beta zeolite composite catalyst. *J Mol Catal A Chem.*, 253(1-2), 239–244, 2006.

12. Tolls, J., Kloepper-Sams, P., Sijm, D.T., Surfactant bioconcentration-a critical review. *Chemosphere*, 29(4), 693–717, 1994.
13. Johansson, I and Somasundaran, P. *Handbook for Cleaning/decontamination of Surfaces*, 1st ed. Elsevier, p. 992, 2001.
14. Clerck, K.D., Demeyere, H., Labeque, R., Laundry Cleaning of Textiles, *Handbook for Cleaning/decontamination of Surfaces*. 2007. p. 57.
15. Lomax. E.G., *Amphoteric surfactants*. CRC Press. 1996.
16. Schick. M.J., *Nonionic surfactants: physical chemistry*. CRC Press. 1987.
17. Sjöblom. K. J., Surfactants used in the food industry: a review. *J Dispers Sci Technol.*, 30(9), 1363–1383, 2009.
18. Kissa, E. Kinetics and mechanism of soiling detergency. *Detergency: Theory and Technology*. (Cutler, W. G.; Kissa, E. ed.) Marcel Dekker. 1, 331, 1987.
19. Spitz, L., Soap and Technology for the 1990s, American Oil Chemist's Society, Champaign, Illinois. 48–93, 1990.
20. Cutler, W.G., (Ed.), *Detergency: Part 1, Surfactant Science Series*, 5, 9–15, 1972.
21. Jakobi G., Lohr, A., *Detergents and Textile Washing: Principles and Practice*, VCH Velagsgesellschaft mbH, 1987.
22. Smulders, E., Rähse, W., Rybinski, W.V., Steber, J., Sung, E., Wiebel, F., Laundry *Detergents*, Published Online. 15 May (2003).
23. Kissa, E., Detergency: Theory and technology, in: *Surfactant Science Series*. W.G. Cutler (Ed.), Marcel Dekker. 20, 2, 1984.
24. Michael, S.S., Powdered detergents, in: *Surfactant Science Series*. Marcel Dekker. 71, 1–4, 1969.
25. Linfield, W.M, Cationic surfactants, in: *Surfactant Science Series*. E. Jungermann., (Ed.), Marcel Dekker. 4, 9, 1970.
26. Rubingh, N., Holland, P.M., Cationic Surfactants - Physical Chemistry, *Surfactants Science Series*. Marcel Dekker. 37, 41, 1991.
27. (A)Tripathy, D. B., & Mishra, A. (2016). Renewable plant-based raw materials for industry. *Sustainable Inorganic Chemistry*, 353.
27. (B) Tripathy, D. B., Yadav, A., & Mishra, A. (2017). Applications of petrochemicals: A mini review. *Recent Adv. Petrochem. Sci*, 2, 67–69.
28. Nace, V.M., Nonionic Surfactants, in: *Surfactant Science Series*. M.J. Schick (Ed.) 60, 32, 1981.
29. Schick, M.J., Nonionic surfactants, *Surfactant Science Series*. Marcel Dekker. 23, 22–25, 1987.
30. Schonfeldt, N. *Surface-active ethylene oxide adducts*, Pergamon Press, Oxford, 1970.
31. Leidreiter, H.I., Gruning, B., Kaseborn, D., Amphoteric surfactants: Processing, product composition, and properties. *Int. J. Cosmetic Sci.*, 19 (5), 239–253, 1997.
32. Tsujii, K., *Surface activity*, Academic Press, San Diego, 1998.
33. Board, N., *The Complete Technology Book on Detergents*. ISBN: 81-86623-78-7, Code: NI92, Publisher: NIIR.

34. M. Aoyagi, K. Takanashi, M. Yamamura, M. Murata, H. Yamada, H. Araki, T. Fukumoto, Quaternary ammonium salts of dicyano substituted teriary alkylene diamines as bleach activators, U.S. Pat. 4978770, 1990.
35. V. Au, G. Gurudev, B. Harirchian, M. Massaro, L. Khan, N. Abid, Compositions comprising nonionic glycolipid surfactants, U.S. Pat. 5389279, 1991.
36. Bronze, *Handbook of Detergents Part A: Properties, Surfactant Science Series*, Marcel Dekker. 82, 1999.
37. Miller, C.A., Raney, K.H., Solubilization-emulsification mechanisms of detergency. *Colloids Surf. A.*, 74, 169, 1993.
38. Sabatini, D., Scamehorn, J., Childs, J., Surfactant enhanced treatment of oil-contaminated soils and oil-based drill cuttings, online project report, University of Oklahoma (May 16, 2000, to May 15, 2001).
39. Scott, B.A., Mechanism of fatty soil removal. *J. Appl. Chem.*, 13, 133, 1963.
40. Huang, X., Yang, J., Zhang, W., Zhang Z., An, Z., Determination of the critical micelle concentration of cationic surfactants: An undergraduate experiment. *J. of Chem. Edu.*, 76, 93, 1999.
41. Shahidzadeh, N., Bonn, D., Meunier, J., A new mechanism of spontaneous emulsification: Relation to surfactant properties. *Europhys. Lett.*, 40 (4), 459–464, 1997.
42. Lange, H. General remarks on soiling and cleaning. Detergency, in: *Theory and Test Methods*. W. G. Cutler, R. C. Devis (Eds.) Marcel Dekker. 1, 66, 1972.
43. Bacon, O.C., Smith, J.E., Detergent action - mechanical work as a measure of the efficiency of surface-active agents in removing soil. *Ind. Eng. Chem.*, 40, 2361, 1948.
44. Schott, H., A kinetic study of fabric detergency. *J. Am. Oil Chem. Soc.*, 52, 225, 1975.
45. Kissa, E., Kinetics and mechanism of detergency. Part I: Liquid hydrophobic (oily) soils. *Textile Res. J.*, 45, 736, 1975.
46. Kissa, E., Kinetics and mechanism of detergency, Part II: Particulate soil. *Textile Res. J.*, 48, 395, 1978.
47. Kissa, E., Kinetics and mechanism of detergency. Part III: Effect of soiling conditions on particulate soil detergency. *Textile Res. J.*, 49, 384, 1979.
48. Vaughn, T.H., Vittine, A., Bacon, L.R., Detergent action of the system modified soda-soap water. *Ind. Eng. Chem.*, 33, 1011, 1941.
49. Yusran, S.S., Reza, F., Permadani, R.L., Utilization of Jatropha Oil as Feedstock for Synthesis of Methyl Ester Sulfonate Surfactant. *Proceedings of the 4th International Symposium on Applied Chemistry 2018*, AIP Conf. Proc. 2024, 020033-1–020033-8.
50. Tripathy, D. B., & Mishra, A., Waste Cooking Oil-Based Novel Gemini Imidazolinium Surfactants With Carbonate Linkage: Green Synthesis, Characterization and Properties Evaluation. *Journal of Surfactants and Detergents*, *20*(3), 553-564, 2017.

51. Kamalakar, K., Satyavani, T., Mohini, Y., Prasad, R. B. N., Karuna. M.S.L., Synthesis of Thumba, Castor and Sal Fatty Ethanolamide-Based Anionic Surfactants, *J Surfact Deterg.*, 2013 DOI 10.1007/s11743-013-1500-2.
52. Adachi S., Kobayashi T., Synthesis of esters by immobilized-lipasecatalyzed condensation reaction of sugars and fatty acids in water-miscible organic solvent. *J. Biosci. Bioeng.*, 99, 87–94, 2005.
53. Kobayashi, T., Lipase-catalyzed syntheses of sugar esters in non-aqueous media. *Biotechnol. Lett.*, 33, 1911–1919, 2011.
54. Cao, L., Bornscheuer, U.T., Schimid, R.D., Lipase-catalyzed solid phase synthesis of sugar esters. *Fett/Lipid*, 98, 332–335, 1996.
55. Otto, R.T., Bornscheuer, U.T., Syldatk, C., Schmid, R.D., Lipase-catalyzed synthesis of arylaliphatic esters of _-D(+)-glucose, n-alkyl- and arylglucosides and characterization of their surfactant properties. *J. Biotechnol.*, 64, 231–237, 1998.
56. Laane, C., Boeren, S., Vos, K., Veeger, C., Rules for optimization of biocatalysis in organic solvents. *Biotechnol. Bioeng.*, 30, 81–87, 1987.
57. Yang, Z., Huang, Z.L., Enzymatic synthesis of sugar fatty acid esters in ionic liquids. Catal. Sci. Technol., 2, 1767–1775, 2012.
58. Cutler, W.G., Kissa, E., *Detergency: Theory and Technology*, Surfactant Science Series. Marcel Dekker. 20, 193-201, 1987.
59. Cutler, W.G., Davis, R.C., *Detergents: Theory and Test Methods*, Surfactant Science Series. Marcel Dekker. 5, 1972.
60. Janicke, W., Hilge, G., Biodegradability of anionic/cationic surfactants under aerobic and anaerobic conditions of wastewater and sludge treatment. *Tenside Surf. Det.*, 16, 472–482, 1979.
61. Steber, J., Wie vollständig sind Tenside abbaubar? *Textilveredelung*. 26, 348–354, 1991.
62. Tripathy, D. B., Mishra, A., Gupta, A., & Yadav, A. (2017). Biodegradability of laundry detergent surfactants. *Int J*, 5(1), 130.
63. Larson, R.J., Federle, T.W., Shimp, R.J., Ventullo, R.M. Behavior of LAS in soil infiltration and groundwater. *Tenside Surf. Det.* 26, 116, 1989.

6

Oils and Fats as Raw Materials for Cosmetics

Shilpi Bhatnagar* **and Shilpi Khurana**

Department of Chemistry, Deshbandhu College, University of Delhi, Kalkaji, New Delhi, India

Abstract
The cosmetic industry has been growing rapidly in the past few years. Cosmetics play an important role in our day-to-day life for cleaning our body, enhancing personal hygiene, preventing body odour, protecting our body from harsh environmental conditions and making the skin look more beautiful and attractive. Oils and fats are natural products widely distributed in nature in both plant and animal kingdom and some are of mineral origin. They have been used in cosmetic formulations and personal care products as solubilizers, emollients to moisturize, nourish and lubricate the skin; super-fatting agents, and dispersing agents. Essential oils extracted from different parts of plants form the basis of perfumes, fragrances and flavours in the cosmetic industry. The quality of cosmetics is improving day by day using new technologies and new ingredients. To reduce the impact on the environment due to hazardous synthetic chemicals added in cosmetics there is an increasing awareness and demand to use safe, sustainable raw materials of natural origin.

***Keywords*:** Emollients, hydrate, moisturize, formulation, flavour, fragrance

6.1 Introduction

The surface of the skin is exposed to various pollutants, sun rays, UV rays, humidity, hot and cold air so a regular skin care routine should be followed to keep the skin radiant and healthy. Depending on the type of skin

**Corresponding author*: sbhatnagar@db.du.ac.in

Divya Bajpai Tripathy, Anjali Gupta, Pooja Agarwal, Anuradha Mishra, and Arvind Kumar Jain (eds.) Oils and Fats as Raw Materials for Industry, (145–168) © 2024 Scrivener Publishing LLC

a variety of creams are available in the market like cold creams, vanishing creams, day cream, night cream, cleansing cream, foundation, anti-aging creams, suntan lotion etc. Oils and fats are essential components in skin care and cosmetic formulations [1, 2]. Fats and oils applied to the skin in the form of creams and lotions penetrate the epidermis to some extent via the pilosebaceous unit. Naturally occurring fats and vegetable oils were recognized as earliest emollients and their action on the skin is called emollience. Emollients are materials used for prevention of dryness, protection of skin, improving the skin barrier function and to enhance the appearance of skin by smoothening flaky skin cells [3]. The primary reason of dry dull skin is the decrease in water content and excessive loss of water [4].

6.2 Theoretical Aspects of Emollients

Emollients work by occluding the surface and forming an oily film, thus conserving the skin water content and allowing rehydration for a longer time [5]. Due to their occlusivity and insensitivity of bacterial contamination they are being successfully used as key components in various cosmetic products. The chemical structure, molecular weight, properties such as spread ability, viscosity, volatility, stability, solubility, penetrating, hydrating, and lubricity contribute to the use of emollients in cosmetic formulations.

The action of emollient is different from lubrication of skin arising from oiliness or greasiness [6]. A good emollient should contain some solids in addition to pure liquid components to increase the melting point of the film laid down on the skin to a point little greater than the body temperature. Creams are semi-solid emulsions of either oil in water or water in oil. Oil in water creams (O/W) are more preferred; oil emulsion (W/O) are more greasy and not easily washable.

The action of emollients can be achieved in two ways:

- ☐ Forms an occlusive barrier to retain the water content inside the skin.
- ☐ Trapping water from the atmosphere by use of a humectant material to enhance the moisture holding capacity. Some of the different types of materials that can be used as emollients are as follows:
  1. Hydrocarbon oils & waxes (mineral oil, petrolatum, paraffin, ceresin, ozokerite, polyethylene, squalene)

2. Silicone oils (Dimethyl polysiloxanes, methyl phenyl polysiloxanes)
3. Triglyceride esters: vegetable and animals fats and oils
4. Alkyl esters & alkenyl esters
5. Fatty acids and alcohols
6. Lanolin and its derivatives
7. Waxes

Out of the various types, hydrocarbons and silicones are completely non-polar imparting characteristic roles in emollient action. Hydrocarbons used widely as emollients are Petrolatum, Paraffins and mineral oils. Mineral oils are isolated from a mineral source particularly a distillate of petroleum. They are a complex combination of highly refined saturated branched chains of different molecular weights and naphthenic hydrocarbons. Mineral oils are chemically stable, inexpensive and act as an efficient moisturizer imparting emollience and occlusivity [7, 8]. Highly refined and purified mineral oils are used in lotions, creams and ointments, lip care products and other skin care products as they are free from rancidity and keep the skin hydrated [9]. Mineral oils remain on the skin for a longer time so its poor washability and excessive occlusive nature sometimes may be desirable and sometimes undesirable as it may clog the skin pores, increase the risk of blackheads and acne and suffocate the skin. Mineral oils may remove appreciable quantities of natural oils from the surface of the skin so spermaceti, cetyl alcohol, cocoa butter, vegetable oils and lanolin may be added. The hydrocarbons such as petrolatum, paraffin wax, ozokerite have been employed in hand creams and lotions. They act as emollients by forming a hydrophobic occlusive film thereby helping hydration of the corneum by allowing water diffusing from the underlying tissues to accumulate in the corneum. Their sticky and waxy effect has restricted their use to barrier hand creams where a temporary, functional protective film is desired.

Silicon chemistry is very vast due to the variety of structures possible. Silicone oils are purely synthetic and are formed by a complicated process terminating with the hydrolysis of a mixture of different chlorinated compounds of Silicon. They contain a long hydrophobic non-polar chain owing to its water repellency. They are chemically inert and are added to barrier creams with a vanishing cream basis to leave a water repellant film on the skin. The lower polymers have been used as antifoaming agents, e.g., in perfumes solubilized with surface active agents. The higher polymers have been added to shampoos as hair conditioners. Silicones are less occlusive than mineral oils and are considered excellent emollients by lubricating and protecting the skin [10].

Squalene is a natural oil that was first isolated from shark liver oil [11, 12]. It is present in adult pure human sebum (about 10%) in epidermal lipids (3%) and in skin surface lipids (from 1.3 to 17.3%). It consists of six nonconjugated double bonds which permit oxidation. The complete hydrogenation of squalene produces a stable, colorless and odorless hydrocarbon oil called perhydro squalene and squalane [13]. Squalane works as an excellent emollient for hydrating and improving the skin texture and acts as a skin lubricant. Since squalane is a saturated hydrocarbon, it does not have the rancidity potential of squalene. It is readily emulsified and is compatible with most of the components of cosmetic products. It is a natural antioxidant and protects the skin from carcinogens. It helps to brighten the skin giving it a youthful appearance.

Structure of Squalene

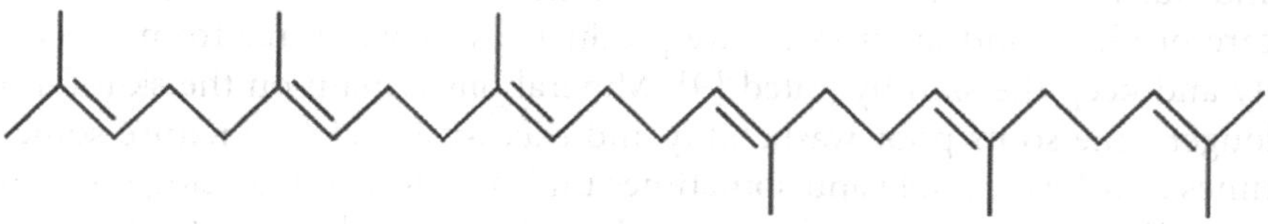

6.3 Commonly Used Vegetable/Plant Derived Oils

Vegetable oils play an important role in cosmetics as they are inexpensive, easily available and safe to use. Due to their various properties, they have been used to hydrate the skin and are present in high concentration in the seeds and fruits of plants. They are majorly composed of triglycerides. They have been derived from nuts (almonds, pistachios, walnuts), plants (primrose, perilla), fruit seeds (cherry, berries), and vegetables (carrot) etc. The presence of saturated or unsaturated fatty acids is responsible for the texture of the oil, its resistance to oxidation and its skin-healing properties [14, 15]. They are used in the treatment of many dermatological disorders due to their broad-spectrum antimicrobial activity. Vegetable oils that are lightweight, fast absorbing and do not leave a sticky effect on the surface are called "dry oils" [16]. They are rich in antioxidants and are used in skin care and hair products. Dry oils mainly consist of omega-3 and omega-6 fatty acids that enter into the skin with more ease. Sunflower and safflower oils are examples of dry oil that contain linoleic acid that repairs the skin and keeps it moisturized.

Vegetable oils are used in ayurvedic massage known as Abhyanga for treatment of various disorders, for relieving stress, for improving blood circulation and for maintaining a healthy lifestyle [17]. The selection of oil depends on the "prakrithi", "vikrithi" of the patient and on the "doshas" in a person related to Pitta, Vata, Kapha, Agni and Ama. Coconut oil, sesame oil, and mustard oil are the most used oils for Abhyanga [18].

- Coconut oil: The raw material for this oil is copra or coconut kernel. The coconuts as they come from a tree contain 30-40% oil, while dried copra contains 65-70% oil. It mainly consists of Vitamin E so it is used to moisturize and smoothen the dry skin and helps in wound healing. It contains antimicrobial, antioxidant and anti-inflammatory properties.

  It forms a protective layer on the top of the skin to shield the skin from environmental toxins and pollutants. It has a calming effect and can soothe irritated skin [19]. Coconut oil being low molecular weight, straight linear chain made up of saturated fats penetrates into hair strands and is absorbed quickly, preventing dandruff, hair loss, and split ends, leaving the hair smooth, long and shiny. It protects the hair scalp and helps in hair conditioning [20].
- Castor oil: It is obtained from the seeds or beans of castor plants found in most tropical regions. The beans contain 30-35% of oil and may be expressed or solvent extracted to recover oil [21]. It is in great demand and is produced globally nowadays. A refined grade castor oil is odorless, stable, richer in Vitamin E and contains above 90% ricinoleic acid. It is a good plasticizing agent and an excellent pigment dispersing agent due to its unique property of high viscosity (due to the presence of a hydroxyl group in the acid part) and is an important ingredient of modern lipsticks. It moisturizes the skin and prevents wrinkles due to its antioxidant property.
- Argan oil: Cosmetic grade argan oil is prepared from unroasted argan kernels and contains over 80% unsaturated fatty acids and bioactive molecules [22]. It has powerful antioxidant properties and has an effective role in skin and hair cosmetics. It has skin-healing properties and antibacterial, antifungal properties due to which it prevents pigmentation and is useful for treating acne [23]. It is an important

ingredient of anti-aging creams. It has good astringent properties, making its use as toners and in makeup removal. Due to its moisturizing effect, it is used in lotions, soaps and for hair conditioners, shampoos and oil. It penetrates into the hair follicle, keeping the hair moisturized and hydrophobic.

- Olive oil: It is prepared from the fruit of an olive tree. The ripe fruit contains 40-60% oil. Olive oil mainly consists of glycerides of oleic acid (70-80%) and unsaturated acids. It has no trans fatty acids and contains natural antioxidants that contribute to a long-lasting effect for a healthy skin. It is safe to use and can be applied on the skin directly as a moisturizer. It is also an important component of face wash, body wash, soaps and lotions. It is non-toxic, rich in Vitamin E, Vitamin K and helps in maintaining skin elasticity. It can be used for make-up removal, thus acting as a cleanser. It is used for controlling hair loss and dandruff. Extra virgin olive oil is also a good choice for massage and nourishment of the delicate skin of newborn babies [19].
- Cottonseed oil: It is derived from the seeds of cotton plants. It has excellent properties for hair and skin care products due to the high content of Vitamin E and Linoleic acid, which is one of the fatty acids present in cottonseed oil. It helps in moisturizing the hair, preventing protein loss and protecting hair from pollutants. It is also a common ingredient for anti-dandruff shampoos.
- Jojoba oil: Jojoba oil is extracted by cold pressing from jojoba desert shrub. It is a stable, long-chain linear liquid wax ester that is rich in vitamins and antioxidants. It is an effective emollient for skin and retains moisture on skin thus replenishing the lost skin. It acts as a brilliant humectant that forms a protective layer on the skin. It prevents premature aging and improves skin tone. Jojoba oil is rich in palmitic acid and stearic acid that helps in hair conditioning, prevents dandruff in hair, thus promoting hair thickness.
- Almond oil: It is extracted from kernels of almond and contains almost 62% of oleic acid, 24% of linoleic acid and 6% of palmitic acid. It is rich in Vitamins A, E and D and can also be used in baby skin care products. It helps to clean the pores of the skin, thus giving glow and brightness to the skin. It prevents skin damage from harmful pollutants and UV rays

and helps to rejuvenate the skin. It helps in reducing the stretch marks thus softening the skin.

- Cocoa butter: It is extracted from cocoa beans. It has a melting point lower than body temperature so it easily glides on the skin, making it free from blemishes, and helps to heal skin irritation. It helps to reduce stretch marks and is a wonderful moisturizer for body and lips. It is rich in oleic, palmitic, stearic, linoleic acids and other bioactive components so it is used as an additive in many cosmetic products like hair shampoo, lipsticks, lip balm, massage oils, lotions, creams, etc. Due to its antioxidant and photoprotective properties it has the potential to reduce skin diseases and prevents premature aging [24].
- Shea butter: It is extracted from nuts of the Shea tree, mainly found in West Africa. It has been used as a star ingredient in beauty products for many years. It is used as a refatting agent and helps to minimize dryness. It is most valued as a skin emollient. It contains different kinds of fatty acids and Vitamins A, E, and F which aid in maintaining oil balance of the skin and help in regenerating healthy cells. It is used in hair conditioning of dry hair and reduces breakage and dandruff. It has a high concentration of unsaponifiable lipid content, making it the most appreciated natural choice for the cosmetic industry.

6.4 Lanolin and Its Derivatives

Lanolin is a natural wax and is mainly composed of esters formed by a combination of different alcohols and higher fatty acids [25]. It is also called wool wax. The hydrophobic and adhesive character of lanolin makes it an excellent emollient, emulsifier, moisturizer, and stabilizer. It is widely used in cosmetics and personal care formulations and is rising in demand across the globe. It is used in hand treatment products and lip care products to moisturize lips. It is used in foundation creams and oil-based lotions as a skin hydrator and to soften rough skin. Lanolin is used in hair sprays as a plasticizer, in shampoos as a conditioner and in hairstyling products.

Dewaxed lanolin or liquid lanolin is more effective in hand cosmetics over lanolin due to its less stickiness and greater solubility in hydrocarbons at low temperatures also.

Lanolin waxes in which the lower melting esters are removed by solvent extraction can be used in small amounts as an emollient in hand preparations and to increase the body of creams and lotions.

Lanolin alcohols mainly consisting of sterols, triterpene alcohols and aliphatic alcohols have been applied advantageously in the formulation of hand creams and lotions.

Modified lanolin products such as acetylated lanolins, acetylated lanolin alcohols, polyoxyalkylene lanolins, and alcohol lanolin esters are used in hand treatment products. They are exceptionally non-tacky and extremely hydrophobic and have the property of depositing on the skin as emollient in hand creams and lotions.

6.5 Lecithin

Phospholipids are complex fat-soluble substances that contain, in addition to fatty acids and glycerol, a nitrogenous base (such as choline or ethanolamine) and phosphoric acid. It is found in egg yolk, marine sources, soyabeans, rapeseed, cottonseed and sunflower oil. It is a phospholipid and is known as phosphatidylcholine. It is used as an emollient in hand creams and lotions. It is used as an excellent emulsifier that stabilizes and gives consistency to cosmetic products. In addition, it has found application as a dispersing agent and surfactant. The concentration of lecithin generally employed has not exceeded 5% and is usually found in 1-2%. By adding hydrogen to lecithin hydrogenated lecithin is obtained which works well for the skin to hydrate it and form a protective barrier, thus improving the effectiveness of cosmetic products.

Structure of Lecithin

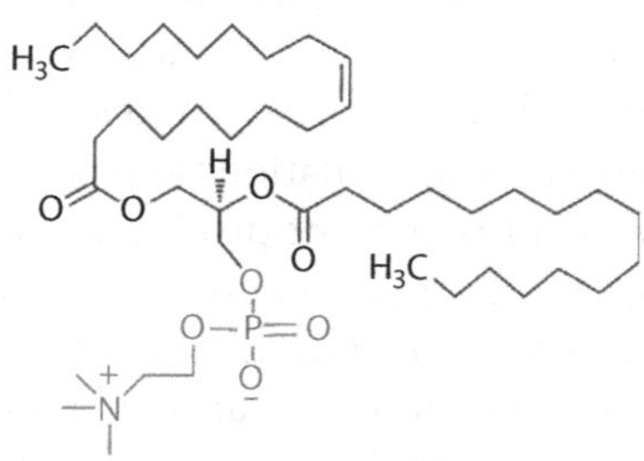

6.6 Essential Oils

The demand for essential oils is growing rapidly as they play an important role in the food and beverage market, for fragrances and aromatherapy and in the cosmetic and pharma industry. Fragrances play an important role in making cosmetic products attractive. Essential oils are of great interest nowadays in several cosmetic products, perfumes due to their unique pleasant odor in addition to their therapeutic properties [26, 27]. Essential oils are volatile liquids soluble in organic solvents but insoluble in water. They show diverse biological properties such as antioxidant, antiseptic, antimicrobial, anti-inflammatory, and emollient and are popular ingredients nowadays to keep the skin fresh and youthful [28, 29]. Essential oils are the key ingredients in many cosmetic preparations of creams, gels, perfumes, ointments, massage oils, deodorants and many other beauty products [30–32]. Essential oils are environmentally friendly so there has been increased interest in the use of essential oils by beauty and wellness companies. It has been reported that there are 300 essential oils commercially available from about 3,000 plant species [33] and they are produced globally for the flavour and fragrance market. These oils are extracted from buds, flowers, petals, seed, leaves, fruits, bark roots, etc., of plants. These parts of plants are processed to yield the essential oils devoid of cellulose, glycerides, starch, sugars, tannins, salts minerals present in plants. The chemical composition and quality of essential oils vary widely depending on the location of crop, part of plant and the extraction method used [34, 35]. Due to the complexity of active agents, they act as preservatives and are vital assets for incorporating into various skin care products. Their chemical nature also varies due to different functional groups present in them, e.g., Linalool (sandal oil, lavender oil) contains alcoholic group; citral (lemongrass oil) contains aldehyde group; eucalyptus oil of wintergreen contains ester group; cymene contains hydrocarbon and Eugenol (Bay oil) contains phenolic group.

- ☐ Eugenol, also called as clove oil, is commonly obtained from cloves, cinnamon, nutmeg, basil and other plants. It is used in perfumes and in dentistry due to its antiseptic and anesthetic properties [36]. It is used in many cosmetic formulations of aftershave lotions, bath products, fragrances, moisturizers, shampoos and skin care products.
- ☐ Geraniol is derived from plants such as geranium and lemongrass. It occurs naturally in rose oil, citronella oil and

other essential oils. Due to its rose-like pleasant aroma, it is used in perfumes, lotions, creams and other personal care products. It is widely used to induce the flavor of fruits. Due to its antiseptic, antibacterial and anti-inflammatory properties it is efficiently used to treat various skin problems and infections.

- ☐ Sandalwood essential oil has a classic scent and is used in perfumes, talcs and air fresheners. It has excellent anti-inflammatory, astringent and antidepressant properties [37]. It is used for cooling, relaxing, improving the skin tone and hydration of dry skin making the skin supple, refreshed and bright. It also reduces the blemishes and skin rashes. It is also used in hair products to maintain soft texture and lustre of hair. It has an amazing influence on the senses and is very popular for its use in aromatherapy. It enhances mental and physical health and well-being.
- ☐ Rose oil is extracted from the most popular flower, Rose flower, also called the king flower. Due to its diverse use in the cosmetic industry, it is also called "Liquid Gold" [38]. It has a pleasant rose scent and is used as a fragrance in different kinds of body lotions, creams, rose water and other skin care and hair products. It contains Vitamins A, C, D and E to heal the texture of skin, reducing wrinkles and dark spots and stimulating blood circulation, giving a soothing and hydrated skin [39].
- ☐ Tea Tree Essential oil has anti-inflammatory and antimicrobial properties and is widely used to treat acne and promote wound healing. It is also used for treating fungal skin conditions. Due to its antibacterial properties, it is used as a natural alternative to deodorants as it reduces underarm odour. It is very powerful to treat acne and blemish-prone skin. It is also used to fight bad breath and can be used as an effective mouthwash to fight germs and plaque causing bacteria [40].
- ☐ Lemon oil is essential oil obtained from citrus fruits that are in high demand and are produced on a large scale globally. Lemon essential oil is an ideal ingredient for various skin problems, skin elasticity, acne and stretch marks [41]. It has high content of vitamin C and acts as an antioxidant, helps to slow down the aging process and makes the

skin look brighter. It also possesses antibacterial and anti-inflammatory properties [42]. Lemon oil has good antiseptic properties and is effective for controlling dandruff and preventing infections on the scalp, keeping the hair healthy and nourished.

- Lavender essential oil has an amazing relaxing and soothing effect and has notable benefits on the skin. It helps to kill acne-inducing bacteria, heals the skin tissues and dull-looking skin. It has antioxidative effects [43] and dramatically reduces wrinkles and fine lines on skin, thus slowing the aging process. It can be diluted or mixed with coconut oil and aloe vera for enhancing its effect on skin and to treat inflammation on a burn or wound. It stimulates hair growth and prevents scalp infections.
- Neem essential oil has regenerative properties and promotes hair follicle growth. Neem oil suppresses inflammation and has antifungal properties [44] for treating irritation on scalp and resulting in healthier hair. Neem oil has high content of essential fatty acids and is used to treat skin diseases like eczema and acne and is thus a popular ingredient in skin care products. It is deeply absorbed in the skin and is used for anti-aging treatment.

6.7 Use of Waxes in Cosmetics

Waxes have been used extensively in cosmetics for make-up products, creams, lipsticks, mascaras, etc., due to their versatile properties. They occur naturally or are synthetic compounds made from high fatty acid esters (typically C36-C50) or from polymers [45, 46]. They are harder and less greasy than oils and fats. They are used as thickeners, as emollients and provide stability to cosmetics [47, 48]. They are resistant to oxidation and microbes. They have excellent molding properties and give gloss in cosmetic products. They also act as emulsifiers and are used to create a blend of water and oil which are normally immiscible. Natural or synthetic emulsifiers like cetearyl olivate (from olive fruit), PEG, PPG are chosen for use in cosmetics. Waxes are water insoluble and form water repellant films because of the hydrophobic carbon chains in their structure [49].

Classification of commonly used waxes:

Natural	Semi-synthetic	Synthetic
Beeswax	Paraffin	PP (polypropylene)
Carnauba	Microcrystalline	PE (polyethylene)
Jojoba	Montan	PTFE (polytetrafluroethylene)
Candelilla		Polyamide
		Fischer-Tropsch

Some Constants of waxes:

Wax	Melting point (°C)	Acid value	Saponification value	Iodine value
Spermaceti	41-49	0.5-2.8	121-135	3-9
Bees Wax	61-63	16.8-20.6	90-102	8.8-11
Candelilla	65-70	16-19	46-65	14.37
Ceresin	54-77	0	<2	0
Carnauba	80-87	4-8	78-88	13.5
Ozocerite	58-100	0	0	7.8

Beeswax is the best-known natural wax obtained from honeycombs by solvent extraction, expression or boiling in water. It is a mixture of esters, cetyl and myricyl myristate and palmitate with some free acids, free alcohols and hydrocarbons [50]. There may be changes in its consistency and physical characteristics, i.e., amorphous or crystalline nature of beeswax due to their source. It is commonly used in most emollient creams to maintain its consistency. The major use of beeswax in cosmetics is to stiffen the preparation in which it is used without hardening it. Its flexibility and plasticity facilitate its application. It acts as a natural exfoliator and removes dead cells of the skin and helps in repairing skin damage. It is used for softening of hair and promotes hair growth.

Carnauba wax is the hardest natural wax, having a high melting point of 85°C. It is the most expensive commercial wax and is produced from the leaves of carnauba palm which grows in Brazil. It contains myricyl cerolate with free myricyl alcohol, free cerotic acid and hydrocarbon. It is too brittle but when combined with other waxes especially beeswax gives lustre and consistency to cosmetic products. It raises the melting point of gels, has emollient and hypoallergenic properties. It is a preferred additive to thicken lipstick, lip balm, foundations and other skin care preparations. It is capable of forming a water repellant film and it counteracts the solubilizing action of soaps in water, so it is preferred in mascara formulations.

Ceresin is a blend of hydrocarbons of complex composition. It has a relatively high melting point and is available in various grades with a range of melting points. It is used in various cosmetic preparations as it assists the formulator to stiffen the preparation without making it too brittle or too hard.

Candelilla wax is a vegetable wax obtained from the leaves of a Euphoria plant. It is harder than beeswax but not so hard or brittle as carnauba wax. It acts as an emollient, thickener to give firm texture in cosmetic products.

Paraffin Wax consists of 20-40 hydrocarbons in length and is derived from crude oil. It can be mixed with other types of waxes to form products for several applications. It has a melting point close to body temperature so it acts as a natural emollient and is used in skin moisturizing products removing the dead cells thus making the skin look soft and silky. Paraffin wax bath treatment is an effective remedy to improve blood flow and relax muscles. It acts as a form of heat therapy to minimize muscle spasms, inflammation and helps in relieving the pain of people with joint mobility issues [51].

6.8 Use of Oils, Fats and Waxes in Lipsticks and Eye Care Products

Lips do not contain any sweat glands but salivary glands are present on the inner surface that maintain moisture on the lips. In cold and dry weather conditions, the thin corneal layer of lips dries and produces cracked lips. Lipstick is a popular cosmetic product that is used by women to impart an attractive color, gloss and hydration to the lips. In recent years the choice of shades of color, texture, and lustre has become wider. Lipsticks change the appearance and size of the lips. The stick mainly contains a colored material dispersed and suspended in a suitable base prepared by mixing oils and waxes in required proportions to form the desired melting point and viscosity [52].

The main ingredients of raw materials for a lipstick are:

- ☐ Mixture of oils
- ☐ Mixture of wax
- ☐ Colors
- ☐ Preservatives
- ☐ Fragrances
- ☐ Surfactants and other additives

Various oils and fats are used in lipsticks. Earlier, olive oil and sesame oil were used but their propensity for turning rancid and poor solvent power for staining the dye has restricted their use. Mineral oils are resistant to rancidity but are poor solvents for color and they make the lipstick run off readily. They are used in small proportions to enhance gloss. Paraffin oil is used not more than 5% to give a glossy appearance after application. It acts as a lubricant to help remove sticks from the molds after pouring. Castor oil (40-50%), oleyl alcohol and liquid paraffin (5%) are used to give glossy appearance to lips. Lanolin and its derivatives are added as plasticizers. They assist in dispersion of colors. Isopropyl myristate, isopropyl palmitate and butyl stearate helps in smooth application. They have an excellent wetting effect on insoluble dyes and pigments due to their low surface tension and act as a solvent for oils and waxes. Lecithin is used in small proportions to improve smoothness, emollience and ease of application. A refined-grade castor oil is odorless, tasteless and a good plasticizing agent. It is one of the most useful oils and an important ingredient of modern lipsticks. The high viscosity of castor oil is of advantage in delaying the settling of pigments from the molten lipstick mass and in lessening the tendency of the applied lipstick to smear and run off. Sometimes fats are added in the lipsticks for softening of lips and for dispersion of insoluble pigments. Cocoa butter and hydrogenated vegetable fats have been used.

Waxes are added in lip care products as they act as an emulsifier for ingredients. They give shape and structure to the lipstick and keep it solid even in warm climates. The ideal wax blend will maintain the stick to at least 50°C and will hold the oil phase so that it does not separate from the product and keep other substances together. The gloss and hardness of lipstick are largely dependent on the quality and type of wax used [53].

Beeswax is the traditional stiffening agent for lipsticks and is extensively used. Its property of shrinking on cooling helps in synthesis of molded products. If used as the only wax the lipstick will lack gloss and will tend to drag on application. It usually binds oils and high melting point waxes.

It is usually used in low concentration of 3-10% as at higher concentration it results in a dull waxy appearance and the stick crumbles during use.

Candelilla wax is used in 5-10% concentration. A mixture of candelilla wax and beeswax is a good combination for making lipstick.

Carnauba Wax increases the melting point of the base and hardens the lipstick; it is used in small quantities. A small percentage of this wax raises the softening point and the strength of the lipstick. It brings attractive lustre to the lipstick.

Ozokerites or amorphous hydrocarbon waxes are available in a wide range of melting points and textures. The harder ones confer strength and thermal stability. They are mainly used to increase the melting point of the base.

Paraffins are too weak and brittle to be of much value in lipsticks. If used in small proportions they may improve the gloss.

Eye make-up is one of the oldest and most extensively used cosmetics for enhancing the eyes. It makes the eyes look larger and attractive. For the formulation and manufacture of eye cosmetics the purity, safeness, and toxicity of raw materials must be tested because of the sensitivity of the area. Mascara is used to darken and thicken the eyelashes. It can be made as a cream or liquid. Waxes such as beeswax and carnauba wax are used as thickeners and provide firmness to mascara. A cosmetic-grade anhydrous lanolin melting point 38-40°C is used in eye shadows and mascara for its lubricating and adhering properties. Eye pencils are wax-based containing lubricants and emollients that are soft to use and ease the application.

6.8.1 Formulation of Lipsticks Using Wax

Emulsifiers, emollients, binders and colorants contribute to the properties of a fine lipstick. Waxes enable the mixture to be formed into easily recognized shapes of cosmetics. Fragrance and colour are added as antioxidants and preservatives to prevent lipstick from becoming rancid.

1. Carnauba wax 25 gram
2. White Petroleum 30 gram
3. Beeswax 40 gram
4. Liquid Paraffin 5 gram
5. Colour As per requirement
6. Perfume

Put the boiler on an oven with Carnauba wax, Beeswax, Liquid Paraffin and White Petroleum jelly. Stirred it well with a spoon to dissolve at 60°C.

Then added color and perfume and transferred the hot mixture into the dice and wait till it cools down to normal temperature [54].

Formulation of Eyebrow Pencil using wax:

1.	Beeswax	150grams
2.	Stearic acid	100 grams
3.	Triethanolamine	25 grams
4.	Burnt umber	25 grams
5.	Lanolin anhydrous	13 grams

Heated beeswax, stearic acid and lanolin in an oven at 80°C till all the ingredients are mixed properly. Now added burnt umber and triethanolamine with stirring. Allowed it to cool down and transferred the mixture into the pencil dice [54].

6.9 Cleansing Creams

Cleansing cream or lotion is required for removal of facial make-up, surface grime and oil from the face. A good cleansing cream should be conveniently used, spread with ease and should efficiently remove applied cosmetics without irritation. It should be effective in removal of make-up, surface griminess and oil from the skin especially the face and neck, thus keeping the skin soft after application.

Cleansing creams are of two types:

a) White, emulsified cold cream (beeswax borax type)
b) Translucent, liquefying, anhydrous type consisting of a mixture of hydrocarbon oils and waxes.

The beeswax-borax-type creams are usually white, have high lustre and are free from graininess. They have a firm consistency and are easy to spread. They contain a high percentage of mineral oil for cleansing efficacy. They are made as oil in water type emulsion and when applied on the skin much of the quantity of water evaporates, leaving the remaining oil which imparts the cleansing action on the skin. In this type of cream borax acts as an important ingredient. When borax is dissolved in water it produces boric acid and sodium hydroxide. Sodium hydroxide reacts with the free fatty acids in the beeswax, mainly cerotic acid producing soft soap that acts as the emulsifying agent and emulsifies the oil phase containing beeswax, mineral oil, paraffin, etc., in the aqueous phase.

6.9.1 Formulation of Beeswax-Borax Cold Cream Type

The raw materials with the least melting point are melted. Other oils and waxes are added in order of increasing melting points with continuous stirring. The aqueous phase ingredients are mixed separately and then heated to the same temperature as oil phase. Emulsifying agents are added to specific phases. The two phases are then mixed continuously to form a smooth mixture [55].

A. Mineral oil 28gms
   Isopropyl myristate 14gms
   Acetoglyceride 2.5 gms
   Petroleum jelly 7.5 gms
   Beeswax 15gms

B. Borax 1gm
   Water 32gms
   Preservative

C. Perfume
   Heated materials of A and B separately at about 75°C. Added the contents of B to the contents of A slowly with continuous stirring until a thick stable emulsion is formed. Add perfume when the temperature cools to 35°C. Stirred and stored it in a suitable container.

6.9.2 Liquefying Cleansing Cream

They are prepared by a combination of oils and waxes and have a consistency that they liquefy when rubbed on skin. Mainly, mineral oil, paraffin, petrolatum and waxes are used.

Mineral oil 80gms
Petroleum jelly 15gms
Ozokerite wax 5gms
Preservative
Perfume

Heated all the oils and waxes together at 65°C. Cooled with stirring. Added preservative and perfume after cooling at 40°C. Stored in a suitable container.

6.10 Oil Shampoo

Increased use of synthetic products leads to dry and frizzy hair. Frequent use of shampoos removes natural oil from the hair and reduces the lustre. Use of oil-based shampoos helps in hair conditioning and nourishment of the scalp [56, 57]. Oil shampoos basically consist of detergents made of sulphonated oils. They can remove oil and dirt from the hair without forming foam. Sulfonated vegetable oils are prepared by reacting vegetable oils with sulphuric acid or other sulfonating agents, and are good detergents. Oils such as coconut oil, almond oil, castor oil, argan oil, and sunflower oil are beneficial for moisturizing the hair and are used for making oil-based shampoos [58, 59].

6.11 Conclusion

The cosmetic industry is prospering globally. Due to enormous innovations in the cosmetic industry, there is a diverse range of cosmetics flooding the market. With the increase in demand, awareness among consumers and sophistication of technology, scientists are still searching for superior quality of oil and fats keeping in mind the safety of the environment and thus reducing the use of hazardous chemicals in cosmetic formulations. Researchers are looking for natural green active ingredients in cosmetics which are more effective and safer considering the sustainability principles.

References

1. Ahsan, H. (2019) The biomolecules of beauty, biochemical pharmacology and immunotoxicology of cosmeceuticals. *J. Immunoassay Immunochem.*, 40, 91-108.
2. Siti Zulaikha, R., Sharifah, N.S.I., Praveena, S.M. (2015) Hazardous ingredients in cosmetics and personal care products a health concern: A Review. *Public Health Research*, 5(1), 7-15.
3. Blank, I.H. (1957) Action of emollient creams and their additives, *J. Am. Med. Assoc.* 164, 412-415.

4. Blank, I.H. (1956) Mechanism of action of agents used for the relief of dry skin. *Am. Perf.*, 67, 35-39.
5. Flesch, P. (1963) Chemical basis of emollient function in horny layers, *Proc. Sci. Sec. TGA*, 40, 12-16.
6. Frazier, C.N. and Blank, I.H. (1954), A formulary for external therapy of the skin, Charles C Thomas, Springfield, III, 55-69.
7. Tripathy, D.B., Yadav, A., & Mishra, A. (2017). Applications of petrochemicals: A mini review. *Recent Adv. Petrochem. Sci*, *2*, 67-69.
8. Franks, A.J. (1964) White mineral oils and petrolatums in cosmetics, *SPC*, 37, 319-326.
9. Lachenmeier, D.W., Mildau, G., Rullman, A., Marx, G., Walch, S.G., Hartwig, A., Kuballa, T. (2017), Evaluation of mineral oil saturated hydrocarbons (MOSH) and mineral oil aromatic hydrocarbons (MOAH) in pure mineral hydrocarbon based cosmetics and cosmetic raw materials using 1HNMR spectroscopy, *F1000 Research*, 6, 682.
10. O'Lenick, A.J. & O'Lenick, K.A. (2007) Silicon Polymers in Skin Care, *MRS Bulletin, The Materials Science of Cosmetics*, 32(10), 801-806.
11. Gopakumar, K. (2012) Therapeutic Applications of Squalene - A Review, *Fishery Technology Journal*, 49(1), 1-9.
12. Gershbein, L.L., Singh, E.J. (1969) Hydrocarbons of dogfish and cod liver and herring oil. *J. Am Oil Chem.Soc.*, 46, 554-557.
13. Ciriminna, R., Pandarus V., Beland F. & Pagliaro M. (2014) Catalytic Hydrogenation of Squalene to Squalane, *Organic Process Research & Development,* 18(9), 1110-1115.
14. Ahmad, A. & Ahsan, H. (2020) Lipid based formulations in cosmeceuticals and biopharmaceuticals. *Biomedical Dermatology*, 4(12), 1-10.
15. Alvarez, AMR & Rodriquez, MLG. (2000) Lipids in pharmaceutical and cosmetic preparations, *Int. J. Fats Oils*, 51(1-2), 74-96.
16. Bonte, J.C.A.F. (2021) Vegetable fats in cosmetology. *Revista Boliviana De Quimica*, 38(2), 68-79.
17. Tripathy, D. B., & Mishra, A. (2016). Renewable plant-based raw materials for industry. *Sustainable Inorganic Chemistry*, 353.
18. Roshy, J.C., Anu, C., Joseph, C.T. (2012), Role of abhyanga (oil massage) to lead a healthy life. *Ayur Pharm Int. J. Ayur Alli Sci*, 1, 163-167.
19. Verallo-Rowell, V.M, Dillague, K.M., Syah-Tjundawan, B.S. (2008) Novel antibacterial and emollient effect of coconut & virgin olive oils in adult atopic dermatitis. *Dermatitis*, 19, 308-315.
20. Rele, A.S., Mohile, R.B. (2003) Effect of mineral oil, sunflower oil and coconut oil on prevention of hair damage. *J. Cosmet. Sci.*, 54, 175-192.
21. Damirchi, S.A., Achachlouei, B.F., Alirezalu, K. and Alirezalu, A. (2011) Physiological and Medicinal Properties of Castor Oil, *Recent Progress in Medicinal Plants*, pp. 323-337.
22. Charrouf, Z. Guillaume, D. (2008) Argan oil: occurrence, composition and impact on human health. *Eur J. Lipid Sci Technol*, 110, 632-636.

23. Monfalouti, H.E., Guillaume, D., Denhez, C. (2010) Therapeutic potential of argan oil: a review. *J. Pharm Pharmacol.* 62, 1669–1675.
24. Scapagnini, G., Davinelli, S., Renzo, L.D., Lorenzo, A.D., Olarte, H.H., Micali G., Cicero A.F. & Gonzalez, S. (2014) Cocoa Bioactive Compounds: Significance and Potential for the maintenance of Skin Health. *Nutrients*, 6(8), 3202-3213.
25. Sengupta, A. & Behhera, J. (2014) Comprehensive view on chemistry, manufacturing & applications of lanolin extracted from wool pretreatment. *American Journal of Engineering Research*, 7(3), 33-43.
26. Herman, A; Herman, A.P.; Domagalska, B.W.; Mlynarczyk, A. (2013) Essential oils and herbal extracts as antimicrobial agents in cosmetic emulsion. *Indian J. Microbiol*, 53, 232-237.
27. Manou, I.; Bouillard. L.; Devleeschouwer, M.; Barel, A. (1998) Evaluation of the preservative properties of Thymus vulgaris essential oil in topically applied formulations under a challenge test, *J. Appl. Microbiol.*, 84, 368-376.
28. Mishra, A; Mishra, A.; Chattopadhyay, P. (2012) Assessment of *in vitro* sun protection factor of Calendula officinalis L. (Asteraceae) essential oil formulation. *J. Young Pharm.*, 4, 17-21.
29. Winkelman, W.J. (2018) Aromatherapy, botanicals and essential oils in acne. *Clin. Dermatol.*, 36, 299-305.
30. Sharmeen, J.B., Mahomoodally F.M., Zengin, G. and Maggi, F. (2021) Essential oils as natural sources of fragrance compounds for cosmetics and cosmeceuticals. *Molecules*, 26(3), 666.
31. Sarkic, A. and Stappen, I. (2018), Essential oils and their single compounds in cosmetics - a critical review, 5, 11.
32. Happy, A. A., Jahan, F. Momen, M.A. (2021) Essential oils: Magical Ingredients for skin care; *Journal of Plant Sciences* 9(2), 54-64.
33. Hussain, H.; Al-Harrasi, A.; Green, I.R. Frankincense (Boswellia) Oils (2016) In Essential Oils in Food Preservation, Flavor and Safety; Elsevier, Amsterdam, The Netherlands, pp. 431-440.
34. Salgueiro, L., Martins, A., Correia, H. (2010) Raw Materials: The importance of quality and safety: a review, *Flavour Fragr.*, 25, 253-271.
35. Sanli, A., Karadogan, T. (2017) Geographical impact on essential oil composition of endemic Kundmannia anatolica Hub-Mor. (Apiaceae), *Afr. J. Tradit. Complementary Altern. Med.*, 14, 131-137.
36. Nejad, S.M., Ozgunes, H. & Basaran, N. (2017) Pharmacological and Toxicological Properties of Eugenol, *Turkish J. of Pharmaceutical Sciences*, 14(2), 201-206.
37. Sharma, M., Levenson, C. Browning J.C., Becker, E.M., Clements, I., Castella, P. and Cox, M.E. (2018), East Indian sandalwood oil is a phosphodiesterase inhibitor: a new therapeutic option in the treatment of inflammatory skin disease. *Frontiers in Pharmacology*, 9, p. 200.

38. Mohamadi, M.; Shamspur, T.; Mostafavi, A. (2013) Comparison of microwave assisted distillation and conventional hydrodistillation in the essential oil extraction of flowers Rosa damascene *Mill, J. Essent. Oil Res*, 25, 55-61.
39. Mohebitabar, S., Shirazi, M., Bioos, S., Rahimi, R., Malekshahi, F. and Nejatbakhsh, F. (2017), Therapeutic efficacy of rose oil: A comprehensive review of clinical evidence. *Avicenna Journal of Phytomedicine*, 7(3), 206.
40. Chandradas, D; Jayakumar, H.L.; Chandra, M.; Katodia, L.; Sreedevi, A; (2014). Evaluation of antimicrobial efficacy of garlic, tea tree oil, cetylpyridinium chloride, chlorhexidine and ultraviolet sanitizing device in the decontamination of toothbrush. *Indian J. Dent.*, 5, 183-189.
41. Maurya, A.K.; Mohanty, S., Pal, A., Chanotiya, C.S. and Bawankule, D.U. (2018), The essential oil from citrus limetta risso peels alleviates skin inflammation: *in vitro* and *in-vivo* study. *Journal of Ethnopharmacology*, 212, 86-94.
42. Hsouna, A.B., Halima, N.B., Smaoui, S. and Hamdi, N. (2017), Citrus lemon essential oil: Chemical composition, antioxidant and antimicrobial activities with its preservative effects against Listeria monocytogenes inoculated in minced beef meat. *Lipids in Health and Disease*, 16(1), 146.
43. Marin, I.; Sayas-Barbera, E.; Viuda-Martos, M.; Navarro, C.; and Sendra, E. (2016), Chemical composition, antioxidant and antimicrobial activity of essential oils from organic fennel, parsley and lavender from Spain. *Foods*, 5(1), 18.
44. Kumar, V.S. and Navaratnam, V. (2013). Neem (Azadirachta indica): Prehistory to contemporary medicinal uses to humankind. *Asian Pacific Journal of Tropical Biomedicine*, 3(7), 505-514.
45. Mark, B., Overberger M. (1988), *Encylopedia of Polymer Science and Technology*. John Wiley & Sons, p. 784.
46. Tinto, W.F., Elufioye, T.O., Roach, J. (2017), Pharmacognosy, Chapter 22: waxes, 443-455.
47. Chuberre, B., Araviskaia, E., Bieber, T., Barbaud, A. (2019) Mineral oil and waxes in cosmetics: an overview mainly based in the current European regulations and the safety profile of these compounds, *J. of the European Academy of Dermatology and Venereology*, 33(57), 5-14.
48. Pirow, R., Blume, A., Hellwig, N., Herzler, M., Huhse, B., Hutzler, C., Pfaff, K., Thierse, H.J., Tralau, T., Vieth, B., Luch, A. (2019), Mineral oil in food, cosmetic products, and in products regulated by other legislations, *Crit Rev Toxicol*, 49(9), 742-789.
49. Petry, T., Burry, D., Fautz, R., Hauser, M, Huber, B., Markowetz, A., Mishras, S., Rettinger, K. Schuh, W., Tiechert, T. (2017), Review of data on the dermal penetration of mineral oil and waxes used in cosmetic applications, *Toxicology Letters*, 280, 70-78.
50. Hossain M. E., Khan, M.I., Ketata, C., Islam, M.R. (2010) Comparative pathway analysis of paraffin wax and beeswax for industrial application, *Journal of Characterization and Development of Novel Materials*, 1(14), 1-13.

51. Sibtain, F., Khan, A., Rehman, S.S. (2013) Efficacy of Paraffin wax bath with and without joint mobilization techniques in rehabilitation of post traumatic stiff hand, *Pak J. Med Sci*, 29(2), 647-650.
52. Balsam, M.S., Sagarin, E. (2008) *Cosmetic Science and Technology*, Second Ed., Wiley Interscience Publication, NY, USA, 365 pp.
53. Kasparaviciene, G., Savickas, A., Kalveniene, Z., Valziene, S., Kubiliene, L.J. (2016), Evaluation of beeswax influence on physical properties of lipstick using instrumental and sensory methods, *Evidence Based Complementary and Alternative Medicine*, 1-8.
54. Gaur, G. & Sharma, Y.K. (2018) Preparation and Properties of Wax Emulsions, Wax Dispersion and their applications in coatings. *International J. of Engineering & Scientific Research*, 5(6), 118-124.
55. Deakers, T.W. (1936) Beeswax in creams, *DCI*, 39, 448.
56. Dias, M.F.R.G. (2015) Hair Cosmetics: An Overview, *International J. of Tricology*, 7(1), 2-15.
57. Alessandrini, A. & Piraccini, B.M. (2016) Essential of Hair Care Cosmetics, *Cosmetics*, 3(4), 34.
58. Arora, P., Nanda, A. & Karan, M. (2011). Shampoos based on synthetic ingredients vis-a vis shampoos based on herbal ingredients A review, *Int. J. Pharma Sci Rev Res*, 7, 41-46.
59. Rele, A.S. & Mohile, R. (2003). Effect of mineral oil, sunflower oil and coconut oil on prevention of hair damage. *J.Cosmetic.Sci.*, 54, 175-192.

7

Oil and Fats as Raw Materials for Coating Industries

Monika Kaurav[1]*, Kantrol Sahu[2], Ramakant Joshi[2], Wasim Akram[3], Pooja Mongia Raj[4], Rakesh Raj[5] and Sunita Minz[6]

[1]Department of Pharmaceutics, KIET School of Pharmacy, KIET Group of Institutions, Ghaziabad, UP, India
[2]Institute of Pharmaceutical Research, GLA University, Mathura, UP, India
[3]Department of Pharmaceutics, Shriram College of Pharmacy, National Express Way, A.B. Road Banmore, Morena (M.P), India
[4]Department of Pharmaceutics, Delhi Pharmaceutical Sciences and Research University, Govt. of NCT of Delhi, Pushp Vihar, Sector III, New Delhi, India
[5]DESU, Maharani Bagh Campus, Delhi Skill and Entrepreneurship University, New Delhi, India
[6]Department of Pharmacy, Indira Gandhi National Tribal University, Amarkantak, MP, India

Abstract
Recently, coating industries are focusing on more environmentally friendly coating materials with the help of various uses of oils and fats. Paint, varnish, lacquer, enamel, synthetic resins, oilcloth, linoleum, printing ink, insulation, and waterproofing are all products of the coating industry. Various vegetable oils including polymeric constituents such as polyols, alkyds, polyetheramides, polyetheramides, epoxies and polyurethanes are the ultimate materials as protective coatings along with them constitute the single, largest, easily available, low cost, non-toxic, non-depletable, biodegradable. The paint and varnish sector consumed 17% of all non-edible fat. Linseed oil is still used more than any other oil. Large-scale joint studies of raw materials and their applications have been conducted, and this has encouraged knowledge sharing. Linseed oil has now been blended with different oils, and synthetic resins and cellulose derivatives have been used to create

**Corresponding author*: monika11kaurav@gmail.com

Divya Bajpai Tripathy, Anjali Gupta, Pooja Agarwal, Anuradha Mishra, and Arvind Kumar Jain (eds.) Oils and Fats as Raw Materials for Industry, (169–194) © 2024 Scrivener Publishing LLC

a wide range of novel and useful coatings. This chapter provides an overview of recent advancements and future prospects in the use of oils and fats in coatings technology.

***Keywords*:** Fats, vegetable oils, coating industry, animal fats, coating applications

7.1 Introduction

Oils and fats come from a variety of different places, including plants and animals (such as tallow and lard), with soybean, palm, rapeseed, and sunflower oil seeming to be the most significant in terms of quantity. The vast majority of the nearly 10.1 crore heaps of oils and fats produced globally in 1998 found their way into human diets. There were 14 million tonnes available for oleochemistry. Since 2010, annual growth in output has averaged 3%. This tendency is expected to persist in the near future and develop further in the near and long term. The fatty acid spectrum of an oil is what ultimately defines how it will be used. Coconut oil and palm kernel oil require extra care since they contain a disproportionately large number of fatty acids compared to short-chain length (mostly 12 and 14 carbon atoms: C_{12}, C_{14}).

These, for instance, can be refined into effective surfactants for use in laundry detergents, dishwashing liquids, and even cosmetics. Raw ingredients for polymer applications and lubricant, long-chain fatty acids (e.g., C_{18}, saturated and unsaturated) are found in oils including rapeseed, sunflower, soybean, and palm oil, as well as animal fats like tallow [1].

Triglycerides like oils and fats are composed of glycerine along with saturated and unsaturated fatty acids. However, there are a few exceptions to this rule, including castor oil (12-hydroxyoleic acid glycerol triester) (ricinoleic acid). Triglycerides, from a chemical perspective, provide two reactive sites: presence of = bond and acid group in the unsaturated fatty acid chain [2–4]. More than 90% of derivatization processes in triglyceride-based product development occur at the carboxylic group, while less than 10% occur via various alkyl chain or double bond including oleochemical reactions via which fats and oils are converted into the other oleochemical by-products such as fatty acid methyl esters, fatty acids, glycerol, and their hydrogenation products [1].

Coating is the process of applying a thin, continuous layer of coating material to a dosage form or intermediary in the pharmaceutical industry.

Tablets, pellets, pills, beads, and spherules are coated for aesthetics, drug stabilization (from light, moisture, oxygen), to separate reactive components in formula, modified drug release, identification, gastric acid and enzyme protection, and increased mechanical strength. Organic solvents or water are used to coat films [5, 6].

Fats and vegetable oils (VO) have also been used for food and pharmaceutical coating from ancient times and again get attention in the modern pharmaceutical industry. Fats and VO offer many advantages such as hindering direct contact with surroundings and moisture absorption thus delaying browning of medicaments and foods, masking unpleasant taste and preserving flavours, inhibiting depletion of active ingredients, providing pressure and acid resistance, improving granulation and tableting capacity of drugs; coating itself controls dissolution of dosage forms and provides sustained release characteristics [7, 8].

7.2 Vegetable Origin Oils and Fats

Even in the days of cave paintings, VO and fats were utilised as the main component in paints and coatings. Today, polymer chemists and scientists have returned to the widespread use of materials derivative from VO in paints and coatings owing to a number of environmental and health hazards emerging from products derived from fossil fuels, as well as concern over the reduction of petroleum resources by the end of the twenty-first century [7]. Owing to lack of oil crops and the concomitantly low variety in fatty acid content, industry has been more interested in creating novel crops that are tailored for certain uses. These novel crops include those with greater concentrations of fatty acids that are desired as well as those with rare, uncommon fatty acids. Since existing sources only include around 10 distinct types of fatty acids, the oleochemical and fine chemicals sectors have indicated interest in novel fatty acids with unexpected characteristics and functions for non-food uses [9].

Since the time of the first cave paintings, VO have been utilised as binders or additions in paints and coatings. Vegetable oil serves mostly as a drying oil in coatings. Highly unsaturated oils called drying oils will oligomerize or polymerize when showing to oxygen in the air, often with the help of a catalyst. Because of the crosslinking, the molecular weight rises as a result. This resulted in the creation of coating formulations based on renewable materials that performed better [10].

7.2.1 Types of Vegetable Oils and Fats

Vegetable or plant oils are an example of a renewable resource that may be utilised as a trustworthy foundation for the creation of new items with a broad range of structural and functional variants. Plant oils are a desirable industrial raw material for the coating industry owing to their wide readiness and comparatively low cost [11]. In 1999/2000, the world produced 84.6 million tonnes (Mt) of the principal VO, which come as palm trees, soybeans, rapeseeds, cotton, sunflower, palm kernel, olives, and coconuts. By 2009/2010, that number had risen to 137.3 Mt. (an increase of 62%). The largest oleochemical process by volume, the synthesis of fatty acids accounts for around 52% of the oils and fats utilised in industry. Between 2001 and 2008, the global supply of fatty acids approximately doubled [12]. Here below in Figure 7.1 and Table 7.1 a few examples of VO and fats obtained from plants and vegetables are mentioned.

7.2.1.1 Castor Oil

Among all VO, castor oil is considered to be the best substitute because of its naturally occurring hydroxyl groups, low toxicity, biodegradability, affordability, benefits to industry, and ease of supply [13]. Castor oil, which is made from the nut of the castor nut *Ricinus communis*, has a very high concentration (90%) of ricinoleic acid, a fatty acid with a hydroxy group (12-hydroxy, 9c-octadecenoic acid). Castor oil is an intriguing oil for the coating business even though it doesn't have any drying capabilities on its own.

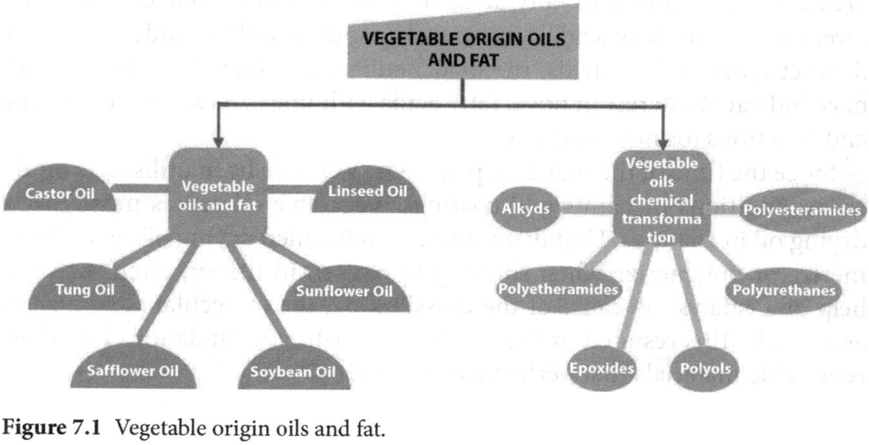

Figure 7.1 Vegetable origin oils and fat.

Table 7.1 Oils and fats obtained from plant source.

Plant oil	Plant sources	Major fatty acid	% of major fatty acid	Applications
Sunflower oil	Helianthus annuus L.	Linoleic/oleic	69% and 18.3%	Surface coating, thin film material
Linseed oil	Linum usitatissimum	Linolenic/ linoleic	52% and 17%	Film coating
Soybean oil	Glycine max	Linoleic/oleic	53% and 24%	Protective coating
Castor oil	Ricinus communis	Ricinoleic/ linoleic	89% and 4%	Antibacterial coating
Safflower oil	Carthamus tinctorius	Linoleic/oleic	78% and 13%	Paints, protective coats
Tung oil	Aleurites fordii	Elaeostearic/ oleic	80% and 8%	Paints, varnishes
Dehydrated castor oil	Ricinus communis	Linoleic/ conjugated oleic	65% and 22%	Gloss and Paints

This is partially because the oil's hydroxyl activity enables it to be employed in, for instance, polyurethane coatings [14].

Due to their diverse variety of characteristics, polyurethanes have several industrial uses. The development of materials made of both linear and cross-linked polyurethanes that provide the design flexibility needed for engineering applications is gaining more and more attention. Many of these polyurethanes have a wide range of chemical and mechanical qualities that make them helpful in biomedical applications. They are also biodegradable polymers. Castor oil in particular, which is both affordable and environmentally beneficial, has recently been related to the development of polyurethane elastomers with environmentally favourable characteristics. One of the main components of castor oil, a fatty acid that contains hydroxyls, may be utilised directly as a raw material for the production of polyurethanes. Additionally, castor oil and polyethylene glycol have shown high compatibility, which might result in better qualities when creating polyurethane elastomers [15].

Several research studies recommended substituting petro-derived goods such coatings, foams, sealants, adhesives, and encapsulating material with castor oil as a source of polyol [16]. As a result, castor oil attracted a lot of interest in the synthesis of polyurethanes; previously other VO were produced for the synthesis of polyol. Castor oil is a good choice for using a cross-linking monomer to create polyurethane materials since it has three hydroxyl groups and about 3.04 unsaturation/triglycerides [17].

7.2.1.2 Linseed Oil

In order to make linseed oil, flax seeds are used, often via mechanical expulsion of *Linum usitatissimum*. Many regions of the world, especially those with a temperate temperature, cultivate this oilseed crop. The oil contains a high a-linolenic acid concentration, which results in a high level of iodine 175-204, depending on, for example, cultivar and climatic circumstances throughout the growing period. Due to its high unsaturation level, linseed oil is extremely prone to autoxidation and polymerization, which leads to crosslinked and durable films when exposed to air. Linseed oil has been a main component of paints and varnishes for ages because of this [14].

The majority of linseed oil produced globally is used for food production, with the remaining 30% going toward technical uses. Linseed oil is a desirable substitute for petroleum-based polymers due to its characteristic fatty acid content, cheap cost, sustainability, availability, and simplicity of handling. As a result, it has previously been utilised as a raw material for the production of biopolymers as well as in paints and coatings. Additionally, it is beginning to be used in biomedicine, food packaging, and agriculture [18].

7.2.1.3 Tung Oil

Aleurites fordii or A. montana's nuts are used to extract tung oil, also known as (Chinese) wood oil (oil content approx. 18%). In the beginning planted in South-East Asia, these trees are now grown in Malawi, Argentina, and the United States. More than 70% of the unique fatty acid eleostearic acid is present in the oil (SC, 11 t, 13t-octadecatrienoic acid) [19].

Another drying oil with some similarities to linseed oil is tung oil. It is easily accessible as a primary product made from the Tung tree's seeds. For thousands of years, this oil has been used to make paints, waterproof coatings, cement, and caulk. Tung oil is particularly useful in paints, varnishes, linoleum, oilcloth, and printing inks because of its capability to dry fast and polymerize into a durable, glossy, impermeable covering. Pure tung oil is resistant to acids and water. It is flexible and unlikely to check, resists marring, and penetrates well. A conjugated triene called an elaeostearic acid glyceride makes up its main component. This oil's remarkable drying abilities and quick polymerization are primarily due to its highly unsaturated, conjugated structure. The molecules of tung oil bind together to form a compact complex structure when it dries and cures. The reason for tung oil's success as a finish is this procedure. The surface is resistant to numerous substances and waterproof thanks to the crosslinking of the oil's molecules. Additionally, the bonding makes the surface flexible, enabling it to tolerate wear and tear [20].

7.2.1.4 Sunflower Oil

The communal sunflower *Helianthus annus L.*, a member of the Asteraceae family, is cultivated for commercial purposes all over the world. Helianthus seed is naturally high in linoleic acid (55–70%) and weak in oleic acid (20–25%). It contains 35–42% oil by weight. A particularly good source of polyunsaturated fatty acids is sunflower seed (approximately 31.0%) [21]. Standard sunflower oil's thermal and oxidative stability is inadequate for biolubricants due to its high content of polyunsaturated fatty acids. Saturated triacylglycerols, on the other hand, have subpar low temperature characteristics. These factors contribute to the acceptable oxidative stability, good low temperature capabilities, and superior lubricating qualities of high-oleic sunflower oils. The essential thing to keep in mind is that basic oil works well as a lubricant under moderate circumstances (lower temperature and smaller load), but additives are required for larger loads. Sunflower oil has also been investigated for use in the manufacturing of paints, soaps, detergents, varnishes, agrichemicals, surfactants, adhesives, polymers, fabric softeners, and coatings [22].

7.2.1.5 Safflower Oil

Although it has a greater linoleic acid content and rather fewer saturated fatty acids than soybean oil, safflower oil, which is derived from the thistle *Curthamus tinctorius*, is quite comparable to soybean oil. As a result, it dries a little bit more quickly and has a somewhat greater iodine content than soybean oil is 140–150. It has good anti-yellowing qualities, much like soybean oil, and is utilised in the production of light-colored, anti-yellowing alkyd paints [14]. Safflower oil is well suited to the coatings industry due to its exceptional high linoleic acid concentration of 78%. Safflower oil's use in alkyd coating resins may have been initially documented by Thurmond,

Hempel, and Marling. Currently, the narrative is very different. Safflower oil has drawn additional customers as a result of its sustained success in surface coatings. Safflower oil-based protective coatings have great initial and through-dry performance, outstanding gloss and gloss retaining, outstanding original colour with minimal after-yellowing, low inclination to wrinkle, good exterior durability, and consistent polymers that contribute to superior film flexibility. Safflower oil can be employed where light-coloured finish products are essential due to its outstanding colour and heat-bleachability properties. Progresses through safflower oil consumption kept up with the periods when it comes to the usage of drying oils in coating formulations. This suggests that safflower oil is likewise utilised as a home paint vehicle in water-thinned systems [23].

7.2.1.6 Soybean Oil

The primary output of the crop soybean *Glycine max*, which is widely grown in places like the USA and Europe, is oil in addition to protein. The flaked beans, which have a dry weight oil content of 15–18%, are used to extract the oil using hexane. The crude oil needs to be refined before being put to use, especially to get rid of the significant number of gums (lecithin, or phospholipids). In the coatings business, soybean lecithin is used as flocculation-controlling and pigment-wetting agents due to its suitability as a surfactant [14]. However, because it is readily accessible on a lot of scale and has enough double bonds, soybean oil is frequently used to make polyols (4.6 per triglyceride). High-grade bio-based polyols were created by the ozonolysis/reduction of soybean oil and glycerin, which were combined to act as an esterification agent. These polyols were then utilised to create hard and flexible foams as well as high-performance coatings. Foams, elastomers, varnishes, and adhesives there have been created by responding to a polyol derived from soybean oil through various isocyanates [24].

Genetic engineering can also be utilised to generate soybean oil with high quantities of linolenic acid, a polyunsaturated fatty acid with limited oxidative stability, in contrast to the oxidative stability of high oleic acid oils. By boosting the expression of the FAD3 gene, which encodes the enzyme that transforms linoleic acid to linolenic acid, soybean seeds have been produced with linolenic acid content exceeding 50% of the total oil. Conventional soybean oil, in comparison, contains around 10% of the oil's total as linolenic acid. These genetically altered soybean varieties are similar to linseed oil in terms of their quantities of linolenic acid. For drying oils used in coating applications, such as paints, inks, and varnishes, high linolenic acid oil has a low oxidative stability, which is a desired quality [25].

7.2.2 Vegetable Oils Chemical Transformation

Triacylglycerols of fatty acids are what make up natural VO. They have functional groups that may undergo a variety of chemical reactions, including double bonds, epoxies, hydroxyls, esters, and additional functional groups. One such example is the natural process known as "drying oils," which is the usual occurrence of VO drying out. In their virgin forms, VO have a propensity to dry out or form films; this behaviour is directly tied to the fatty acid makeup of the oil. VO that are drying or semi-drying are typically employed in surface coatings. VO that are non-drying may also be used for this determination by adding appropriate substances such as hydroxyl group or modernizers (such as vinyls, acrylics, and acrylic co-polymers) to the VO' chemical structure, which transforms them into film formers (conflicting to drying or semi-drying VO, wherever drying is a normal occurrence.) [26]. Longer drying durations are needed for virgin VO. Virgin vegetable oil-based films do not perform as well in terms of physico-mechanical stability and corrosion resistance as desired. As a result, several chemical processes take place at the functional groups found in VO. Hydrogenation, tranesterification, auto-oxidation, epoxidation, hydroxylation, acrylation, isocyanation, and other chemical responses are often carried out in VO. Eventually, these processes produce monomers or polymers that are actively employed as coating materials' binders [7].

A polyol (often glycerol, trimethalolpropane, or pentaerythritol), a multifunctional acid (phthalic acid or trimellitic acid), and an unsaturated fatty acid produced through a polycondensation process make up alkyds, which are oil-modified polyesters. They play a significant role as binders in surface coatings [27]. Polyesteramides are amide modified alkyds made when a vegetable oil amide diol reacts with an acid or anhydride in an esterification process. They differ from alkyd in that they have repetition ester and amide components in their backbone and better qualities such hardness, ease of drying, resistance to water vapour, and resistance to chemicals, especially alkalis [28].

The amide and ether moieties in polyetheramides alternate. The ether and amide groups in polyetheramides' backbones are what give their coatings their strong adherence and chemical resistance. In the area of coatings, polyetheramides is still in its infancy and has a plethora of potential uses [7]. Toluene 2,4-diisocyanate [TDI], diphenylmethane diisocyanate, naphthylene 1,5-diisocyanate, hexamethylene diisocyanate, and isophorone diisocyanate are some examples of the (di- or poly-) isocyanates that are used as the raw ingredients for single also dual packet polyurethanes. This reaction results in the development of polyurethanes. Polyurethane coatings have the advantage over the materials indicated above in that they often exhibit curing or drying at room temperature. A coating material uses an isocyanate treatment to bring its curative temperature down from raised to ambient temperature if it exhibits high temperature curing or drying (>100 °C) [29].

Vernolic acid, a fatty acid with an obviously one of the naturally occurring oxirane rings elements of its triglyceride support, is present in large amounts in VO such as those produced by the vernonia, Euphorbia, and Cephalocroton species. Epoxidation, a chemical process that incorporates the oxirane ring at the unsaturation sites of VO, is another option. Epoxidation can be accomplished via a variety of techniques, including chemo-enzymatic epoxidation, hydrogen peroxide, the use of dioxirane, phase-transfer catalyst, and others [30]. Dietary fats epoxies frequently include hydroxyls, as shown by spectrum analyses. The usage of vegetable oil epoxies in coatings is anticipated to lessen our reliance on petro-based chemicals and lower the price of coatings and paints deprived of sacrificing their performance and service lifespan. Applications for paint and coatings have employed epoxidized fatty acids [31].

One of the significant derivatives of VO is polyol. These lengthy aliphatic chains, often referred to as oleochemical polyols, typically include hydroxyls, double bonds, active methylene groups, and occasionally -OCH3, -Cl, -Br, and oxirane rings, among other functional groups [32]. Castor oil, a natural polyol, has been employed as a raw ingredient in hyperbranched polyurethane coatings in order to develop advanced surface coating constituents.

7.3 Animal Origin Fats & Oils

Lipids are significantly vital. Without lipids, there seems to be no thought, no sensation, and no life. Lipids are crucial for the formation of vitamins and glycolipids, which are used as intracellular messengers, as well as cell membranes. Triglycerides, or fats derived from plant, marine, and animal sources, are the most prevalent types of lipids. In addition to serving as a significant source of energy, fats play an important role in the taste, flavour, and texture of foods we eat every day [33]. A sizable portion of the fats used in the world today, both for foods and for a variety of non-food applications, are produced by animals. Around 170 million tonnes of vegetable

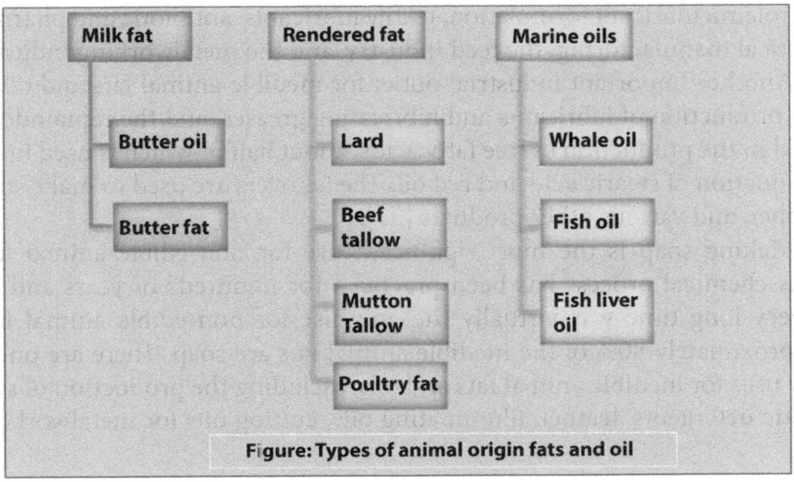

Figure 7.2 Types of animal origin fats and oils.

and animal oils and fats are produced annually worldwide, with nearly 25 million tonnes coming from animals (14%) [34]. The most popular animal fats are butter, lard, fish oil, and tallow [35] (Figure 7.2).

Fish oils, rendered fats, and milk fats are also examples of animal fats. One of the most common type of milk fat is cow's milk fat. In terms of fat content, all other milks are virtually insignificant. They are valuable both on their own and as a base for dairy items like cheese. Additionally, when consumed in large quantities, their fat is transformed into a part of the food. Butter production has conventionally been a subject of dairy technology. Rendered fats are by-products of meat production, even when produced in large quantities. There isn't any livestock raised for fat. Rendering fats are distinct from at least some oilseeds in this regard because they are grown primarily for the oil they produce.

Marine oils are typically by-products from fish that were captured to be refined as a source of protein. Additionally, some fish are primarily caught for the aim of producing oil. There are numerous species are also available as sources, ranging from small fish like sardines to large mammals like whales. Milk fats, rendered fats, and marine oils all have different fatty acid compositions. When compared to other fats, milk fats have a greater density of short-chain fatty acids, which range from C4 to C10. Saturated fatty acids are found in adipose tissue, whereas marine oils contain a lot of polyunsaturated fatty acids, which literally translate to mean more than three [36]. Animal fats and oils have numerous applications, which include petroleum lubricant formulation, textile lubricants, antibiotic and pharmaceutical manufacturing, the feed industry, and the metalworking industry.

Another important industrial outlet for inedible animal fats and oils is the production of lubricants and lubricating greases, and the remainder is used in the production of free fatty acids, about half of which is used in the production of stearic acid and red oil. The leftovers are used to make soap, rubber, and various other products [33].

Making soap is the most significant use for non-edible animal fats. This chemical process has been practiced for hundreds of years and for a very long time was virtually the only use for non-edible animal fats. Approximately 80% of the inedible animal fats are soap. There are only a few uses for inedible animal fats and oils, including the production of synthetic detergents, leather, illuminating oils, cutting oils for metalworking,

Table 7.2 Various type of animal-derived fats and oils.

S. no.	Fats & oil	Source	Production/ year	Application
01	**Milk fat** • Butter oil • Butter fat	Ruminants	6.0 Million metric ton	Cheese, creams, yoghurt and various dairy products, cosmetics
02	**Rendered fat** • Lard • Beef tallow • Mutton Tallow • Poultry fat	Pigs, Cattle Sheep's, lamb Hen, egg, turkey, goose	12.5 Million metric ton	Soap, lubricants, lubricating greases, rubber, printing inks, paints and varnishes, protective coatings, food, cosmetic & pharmaceutical industry
03	**Marine oil** • Whale oil • fish oil • Fish liver oil	Whale, fish, Fish liver	1.2 Million metric ton	Food, cosmetic & pharmaceutical industry
04	**Neatsfoot oil**	Feet & shinbones of cattle	3-4 million pounds	Lubricant, dressing for leather

printing inks, paints and varnishes, and about 5% of the total, or about 100 million pounds, of inedible animal fats and oils [37] (Table 7.2).

7.4 Various Applications of Oils and Fats in Coating Industry

Recently the application of renewable raw materials is in growing mode for the food and chemical industries. Use of oils and fats in the coating industry balances the economic and ecological problems worldwide. Oil and fats are one of the important raw materials in the coating industry. It provides elegant appearance and economical balance to the film. VO are more commonly used for coating with their slight substitution based on petrochemical raw materials.

Biohybrid coating material is in trend to produce enhanced quality solutions which are flexible, scratchproof, transparent and impact resistant. Coating quality of castor oil with isocyanate alkyl triethoxy silane can be improved by incorporating silica nanoparticles in it. This mixture is then hydrolyzed so that organic-inorganic hybrid coating is formed for extra improved quality [38]. Another enhanced coating material is castor oil mixed with DGEBA epoxy in the ratio of 30:70 and also with the addition of 10 wt. % of prehydrolyzed TEOS in it. This results in formation of better quality organic-inorganic hybrid coating material [39].

China holds the earliest record of having used natural oils as a waterproofing agent for caulking, boats and furniture. "Drying oil" is used to form the crosslinked or polymerized waterproof films which is a mixture of conjugated oils, like tung, linseed and various fish oils. Pigments are generally added to these materials for proficient use as paints and varnishes. Annual consumption rate of about 1.2 billion pounds are reported by this material until the 1950s. Synthetic materials such as epoxy and urethane resins are completely replaced by this natural products and today nearly 450 million lbs of fatty acids are being utilized by industries where esters and alkyds of synthetic compound leads.

Soybean oil and sunflower oil which have multiple non-conjugated double bonds are known as "semi-drying" oils. These oils are generally formed by partial oxidation to produce modified oils containing reactive epoxides that have the ability to quickly crosslink. Polymeric coating materials, foams and plastics are formed by oxygenation of triglycerides as well as highly unsaturated fatty acids. This leads to addition of hydroxyl groups in epoxides which is further treated with crosslinked agents or isocyanates. However, the rate of reaction can be increased by addition of metallic catalysts such as manganese and cobalt to the oil. Nowadays, drying oils are in use with the aid of tighter regulatory constraints on volatile organic solvents to be efficiently used as latex paints. Biodegradable plastics are made up of plant fatty acids with long chain glucans called Graft copolymers [40–42].

Among the wide range of potential application of the oil-derived polymer composites, coatings as well as adhesives have received much consideration. Seed oil of Ceylon ironwood (Mesua ferrea L.), have been prepared into a polyester resin by Konwar *et al.* They worked on improving the hardness and chemical resistance of coating material by reinforcing it with nanoclays. This provides high performance as well as improved glossiness by coating material [43].

Highly branched molecules of oil can be obtained by glycerolysis reaction followed by anhydride reaction and then, with a hydroxy-acid. This is then subjected for reaction with poly(amido-amine)-epoxy mixture in the presence of the nanoclay so that hydrogen bonding occurs between resin and nanoplatelets to make it more flexible. Nanoclays play an important role to increase the tensile strength of oil coating from 2.7 to 7.1 MPa only by addition of 2.5 wt. % in it. Elongation point at break is also increased by about 24% to 145% by nanoclay addition with hardness impact of 19%. Glossiness, thermal strength and chemical protection against acid and water is also improved by nanoclay concentration of 2.5 wt.%. All these features of nanoclay addition make it a wonderful ingredient for coating material to incorporate in it. However, broth culture technique containing the bacterial strains is used to evaluate the biodegradability or quality of films. Similar outcomes were reported by use of clay-nanocomposite obtained by reactive mixtures of epoxy-amine as well as bio-based polyurethane derived from Mesua ferrea L. seed oil which showed effective thermal stability, hardness, biodegradation and tensile properties [44].

Plant oils are also used as surfactants to improve the stability and spreading property of coating materials. Surfactant nature of oils and fats are also needed for its cleansing capabilities. Sulfonated oils and fatty acids are the good surfactants as it showed the higher cleansing capabilities. In the early 1900s, structural similarity between fatty acids and petrochemicals was recognised and applied for the new possibilities of usage such as coating ingredients. Generally, hydrogenation of fatty acids is needed for the sulphate salts for its improved stability, whereas, alkylated sulphates show excellent surfactant qualities with improved wetting properties which implements better coating material formulation.

Cationic surfactant with the addition of amines and ammonium salts in combination with aliphatic groups has tremendous advantages. It possesses superb stability properties along with high surface activity. Additionally, these materials also showed the antimicrobial properties where it is applied. Chemical modification in cationic surfactants is required to explore its usage. It is conjugated with the polyamines, imidazolines, and amino esters in combination with various alkyl and aromatic ligands. Specially in shampoo and cosmetics, triethanol amine derivatives are extensively used. It is also to be noted that use of plant oil–based coating material replaces the synthetic polymers in a partial or minimum way. So the goal of using plant oils and fats should be well filtered and future researchers will further design in this coating system [41–44].

7.5 Regulatory and Safety Issues of Vegetable Oil and Fats Coatings

Vegetable oil and fat coatings should fulfil all food ingredient requirements since they are edible [45]. Druchta and Johnston [46] released a technical assessment report on edible coating materials. For product quality and safety and nutritive value, all coating components and functional additives should be GRAS and utilised within FDA limits (FDA). If the coating material is not GRAS, but the maker can establish its safety, the manufacturer may file a GRAS Affirmation Petition or commercialise the material without FDA consent (self-determination). Vegetable oil/fat coatings may contain antioxidants, antimicrobials, colorings, etc. According to EU laws, additives must always be labelled with its name or E-number and functional category.

Regulatory statutes cover another crucial topic. Many vegetable oil and fat coatings contain allergens. Milk, soy, fish, peanuts, almonds, and wheat are common allergies. Milk (whey, casein), wheat (gluten), soy, and peanut proteins generate edible films and coatings. No matter how little the amount, if coating materials have a well-known allergy they must be properly mentioned for consumers' awareness [47].

To be commercially successful, all coating materials and chemicals used to boost food quality and shelf life must be accepted by customers. Further research should evaluate sensory qualities, including sensory analysis on actual products to establish if customers can identify coated products and preference tests to determine which products are favoured [48].

Higher doses of natural substances can cause oral toxicity. VO and fats dose must balance effectiveness and toxicity. Since vegetable oil and fat coatings are consumed, all ingredients must be listed on the label. Antimicrobials, antioxidants, essential oils, colour, and other additives are authorized before addition [49].

7.6 Patents of Oils and Fats Used for Industrial Coating

About 90% oil produced is of vegetable source, generated by the seed-processing industry, intended for human use. This sector has seen a rise in demand for waxes and oils from a variety of natural resources, particularly for use in food and drug applications. When VO are in their virgin forms, they have a propensity to dry out or form films; this behaviour is correlated with the makeup of their fatty acids. Drying is a built-in property of VO and is influenced by the unsaturated component, or the quantity of carbon-carbon double bonds. In surface coatings, VO that is drying or semi-drying is typically employed. By incorporating appropriate substances like vinyl, acrylics and hydroxyls into the VO structure and undergoing chemical procedures to change such oils into film formers. Non-drying VO can also be used for this purpose. The inherent fluidity and flexibility retention properties of VO and several of its derivatives are a significant advantage as well. As a result, VO can be utilized as diluents or, in certain circumstances, can convert into free-flowing materials for coating devoid of the need for a solvent. VO and its derivatives are used in coatings because of their distinctive structural characteristics.

By selecting the proper solid fat content profile that is not excessively brittle, hard, soft, or waxy, we can achieve requirements by coating fats on dosage forms. The fat used for coating products is malleable and does not break off while being consumed. The procedure of coating with fats doesn't involve any heating, water, or solvents. This feature enables the physical qualities of delicate functional ingredients that are challenging to prepare to be improved. Examples include useful substances with low melting points, such as alpha-lipoic acid, and advantageous microbes like Lactobacillus. Individual particles can be coated without causing any stickiness. The quantity of powdered fats used as coating agents can also be adjusted (coating rate). Such distinct properties of VO gain interest of most the researcher and patenting of oil/wax-based coating. Patent 5162057 was awarded to Yohko Akiyama *et al.* for coatings for reliable continuous release preparations in which the fatty acid esters of polyglycerols are used as coating materials for coating of solid dosage forms like tablets, granules, pills and fine granules. Examples include behenic acid hexa(tetra)glyceryl ester, stearic acid penta(tetra)glyceryl ester, oleic acid di(tri)glyceride, lauric acid mono(deca)glyceryl ester. It is possible to achieve reliable continuous release of the preparations' active ingredients even after a lengthy period of storage by coating with coating agents [50].

Chia oil containing an edible nanoemulsion was the subject of WO2018029626, an Argentine patent application. This patent illustrated nanoemulsion of chia oil made of water, at least one antioxidant, between 0.05 and 0.2%, 2 - 5% polysorbate, 0.5 - 5% at least one emulsifier other than the polysorbate, and 10 - 20% chia oil (Salvia hispanica L.). Beverages and desserts, like fruit juices and jellies, edible chia oil nanoemulsion formulations were revealed [51]. Nanoemulsion of cinnamon oil in order to prevent the growth and expansion of harmful food microbes was patented by the Republic of Korea, KR20160005182. The pharmaceutical and cosmetic industries could also benefit from this development, in addition to using it as food additives, food coating agents, preservatives, etc. [52]. According to Houbin *et al.* (2013), a mustard oil nanoemulsion application patent from China, CN103315956, was created for lessening odour of oil and preventing volatilization. This oil can also be utilized to increase microbial resistance in edible products and medications [53]. A palm oil nanoemulsion developed by Wang Weichun Feng Wei was described in a patent application from China with the number CN103750050. This nanoemulsion addressed issues with existing young animal feeds, such as their high production costs, long production times, high equipment costs, and low oil content. The process used to create the palm oil nanoemulsion included palm oil with an emulsifier, and breakdown of cells in mixture through ultrasonic wave. The method was straightforward; the whole method was simple to control. The time of production was low, the equipment investment and production costs were low, while created nanoemulsion has high oil content, and the livestock's ability to digest the product improved [54].

7.7 Recent Approaches for Coating

7.7.1 Vegetable Oil as Corrosion Inhibitors

Vegetable oil and its derivatives are used in coatings because of their distinctive structural characteristics. In addition to providing strength in paints and coatings as binders, oils from vegetables and plant extracts are also employed as organic material for inhibition of corrosion. Raja and

Sethuraman (2008) have created an outstanding evaluation in this regard. They also discovered that both Annona aquamosa and Pongamia glabra had demonstrated protection of steel from corrosion under acidic conditions [55]. Castor oil, commonly known as Ricinus communis, has been used by Srivastava and Sathiyanathan *et al.* to prevent steel from corroding in acidic environments [56, 57]. Recently, Dar published a summary reflecting the application of VO and other plant extracts to inhibit corrosion. The anticorrosion activity of numerous plant extracts and seed oil has also been studied in acidic and alkaline media as well as in chloride ion by researchers [58]. They report heterocyclic molecules like tannins, alkaloid, cellulose, flavonoids and some others are thought to be responsible for the anticorrosion activity.

7.7.1.1 Coatings Made of Polymeric Vegetable Oils

Triacylglycerols of fatty acids make up natural VO. They have functionalities in their backbone that are appropriate, including hydroxyls, double bonds, esters, epoxides and other functional groups which may participate in a variety of chemical processes. One such instance is the drying of VO, which generally happens as a natural process in VO and is referred to as "drying oils". When VO are in their virgin forms, they have a propensity to dry out or form films; this behaviour is correlated with the makeup of their fatty acids. Drying is a built-in property of VO and is influenced by the unsaturated component, or the presence of carbon-carbon double bonds, and bonds either conjugated or nonconjugated, and the position of the attachment of substituent around the carbon-carbon double bond, viz. cis or trans. In surface coatings, VO that is drying or semi-drying is typically employed. By incorporating appropriate substances like vinyls, acrylics and hydroxyl into the VO backbone and chemically modifying them into film-forming agents, non-drying VO can also be used for this purpose. Methylene group a- or b- attached to the ester link can be oxidised to create hydroperoxide in completely saturated fatty tristers. Even though it could potentially be a site of reaction for film breakdown or degradation, it is a slow and inconsequential film-forming reaction. When VO is dried, it can react with oxygen in the air to create polymeric polymers having cross-linked assemblies [59, 60]. Upon removal of hydrogen from a bisallylic group at double bonds in a polyunsaturated fatty acid chain, drying VO is oxidised by air. Epoxidation, peroxidation, crosslinking, hydroperoxidation and peroxidation reaction through radical recombination can be achieved. The inherent fluidity and flexibility retention properties of VO and several of its derivatives are a significant advantage as well. Resultant, VO could also be utilised as filler or, in certain circumstances, can convert into free-flowing coating powder without using a liquid. As a result, utilization of volatile organic compounds (VOC) is eliminated by making it easier to manufacture film-forming agents without solvent.

A few modified polymers originated using VO are

1. Alkyds
2. Polyesteramides [PEA]
3. Polyetheramides [PEtA]
4. Polyurethanes [PU]
5. Epoxies
6. Polyols

7.7.2 Non-Edible Vegetable Oil-Based Polyurethane Coatings

Polyurethane coatings prepared from non-edible vegetable oil
Jojoba oil, thevetia peruviana seed oil, tung oil, algae oil, linseed oil, neem seed oil, pongamia glabra seed oils, castor oil, cashew jatropha oil and nutshell oil are among the most widely utilised non-edible VO included [61–63].

7.7.2.1 Non-Edible Vegetable Oil-Based Anticorrosive Polyurethane Coatings

Polyurethane coatings that prevent corrosion: adhesion
For anticorrosive performances, it is essential to prevent the absorption of corrosive media. Due to the heterocyclic rings in its molecular structure, like urethane linkages containing nitrogen and oxygen, the polyurethane coating is uniform and well-adhered over the metal substrate, making it an outstanding anticorrosive coating material by nature.

Coordination bonds are created as a result of interactions between the numerous lone pairs of electron on heteroatoms and the metal's empty d-orbital [64]. Simply put, the well-adhered PU coating's barrier effect serves as the protection mechanism [65].

Polyurethane Coatings That Prevent Corrosion: Crosslinking
An examination of several PUs helped to clarify how the three-dimensional network formed by crosslinking affected the polyurethanes characteristics. The coating structure's vacant space, free volume and pores are reduced by the crosslinked structure, making it difficult for the corrosive

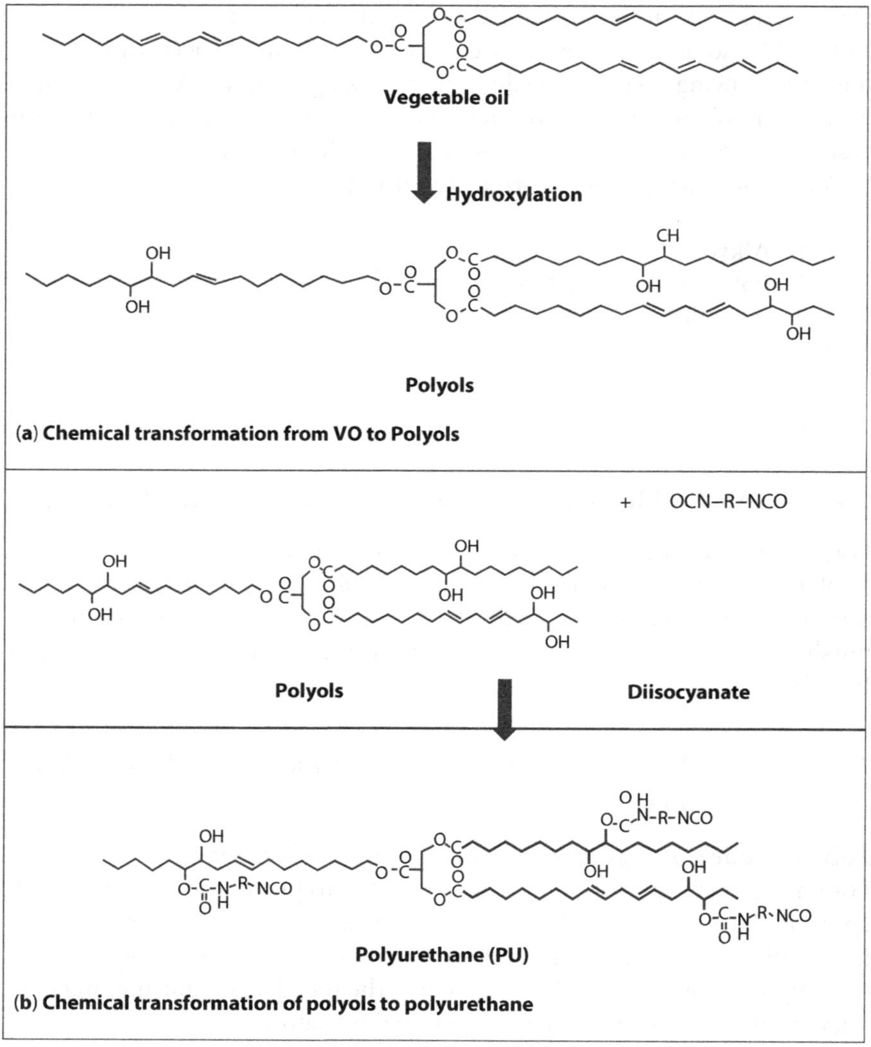

Figure 7.3 Chemical modification of vegetable oil as coating material [68].

electrolyte to penetrate [66, 67]. A groundbreaking study by Ahmad *et al.* on the polyurethane coatings made from linseed oil via hydroxylation and epoxidation was published in 2002 [68] (Figure 7.3). These coatings exhibit good anticorrosive capabilities because of their extensively crosslinked structure. Gite *et al.* used an immersion approach to assess the anticorrosive characteristics of neem seed oil-based alkyd modified polyurethane coatings; they found that the anticorrosive nature of these coatings was improved as a result of an enhanced crosslinking reaction [69]. In order to improve the cross-linking of polyurethane coating and reduce flow of electrolyte mass via the coating, Ranade and co-workers investigated integration of polyurethane [70].

The combination of castor oil: Gite *et al.* demonstrated that the combination of castor oil monoglycerides with isophorone diisocyanate trimer results in superior crosslinked polyurethanes which is highly resistant to alkali, acids and fluids [69]. Poly(epichlorohydrin-triol) crosslinker was added by Siyanbola *et al.* to castor oil-based polyurethane coatings. Increased cross-linking and better adhesion on mild steel resulted in greater chemical resistance [70]. Recent research by Benedetti *et al.* stated that tannin modified castor oil-based polyurethane coating, that also favours crosslinking owing to multitude of -OH groups in the structure increased corrosion protection of aluminium alloy AA2024 [71].

7.7.3 Organic-Inorganic Hybrid Anticorrosive Polyurethane Coatings

For polyurethane coatings to have certain features, such anticorrosive capabilities, admixing of organic and inorganic chemicals were investigated. A resistant barrier phase was formed at the edges between metal substrate and corrosive media, that prevents penetration of electrolyte ions via coatings, can be used to explain how inorganic filler prevents corrosion [72].

7.7.4 Antimicrobial Polyurethane Coating Based on Nonedible VO

Over the past few decades, the risk of bacterial infection has become a more widespread worry. To avoid this, antibacterial coating has provided a method to prevent spreading of microbial infection in many settings, including the production of healthcare product, medical devices, dental and office equipment, and drug as well as food products [73, 74].

7.8 Conclusions

VO and fat coatings provide many food sector benefits, but few pharmaceutical industrial applications till date have been developed. These materials are preferable over synthetic polymers. Several newer natural materials have been studied for VO and fat coatings. VO and fat coating materials supply is the most critical component in determining their qualities. Plant-based fats and oils are low-cost, readily available medicinal coating materials that supply nutrition. VO and fat-based coatings have many benefits. Reduced weight loss and longer shelf life reduce food wastes and manufacturing costs. Edible films and coatings need further research to develop breakthrough materials and procedures that benefit world economies and the environment.

References

1. Tripathy, D. B., & Mishra, A., Renewable plant-based raw materials for industry, *Sustainable Inorganic Chemistry*, 353, 2016.
2. H. Eierdanz, (Ed.). Perspektiven nachwachsender Rohstoffe in der Chemie, VCH, Weinheim, pp. 209-217, 1996.
3. G. Kreienfeld and G. Stoll. In *Alkyl Polyglycosides: Technology, Properties and Applications*, K. Hill, W. von Rybinski, G. Stoll (Eds.), p. 225, VCH, Weinheim, 1997.
4. K. Hill, Fats and Oils as Oleochemical Raw Materials, *Journal of Oleo Science*, Vol. 50, pp. 433-444, 2001.
5. Remington's. *The Science and Practice of Pharmacy*, Volume I. 21st ed. Indian Edition, Lippincot Williams and Wilkins, pp. 929-938, 2005.
6. L. Joseph L. Johnson Aqualon Company, *Coatings Technology Handbook*, Third Edition. In: *Pharmaceutical Tablet Coating*. pp. 112-1,112-3, 2005.
7. M. Alam, D. Akram, E. Sharmin, F. Zafar, S. Ahmad. Vegetable oil based eco-friendly coating materials: A review article, *Arabian Journal of Chemistry*, 7(4), pp. 469-479, 2014.
8. https://www.nof.co.jp/contents/english/business/food/coating/summary/index.html
9. P. Samyn P. Active Barrier Coating for Packaging Paper with Controlled Release of Sunflower Oils. *Molecules*. Vol. 26, p. 3561, 2021.
10. G. Lligadas, J.C. Ronda, M. Galia and V. Cadiz, V. Renewable polymeric materials from vegetable oils: a perspective. *Materials Today*, Vol.16, p. 337-343, 2013.
11. Y. Xia and R.C. Larock. Vegetable oil-based polymeric materials: synthesis, properties, and applications. *Green Chemistry*, Vol.12 (11), pp. 1893-1909, 2010.
12. U. Biermann, U. Bornscheuer, M.A. Meier, J.O. Metzger, and H.J. Schäfer. Oils and fats as renewable raw materials in chemistry. *Angewandte Chemie* Vol. 50, pp. 3854-3871, 2011.

13. P.M. Paraskar, M.S. Prabhudesai, V.M. Hatkar and R.D. Kulkarni. Vegetable oil based polyurethane coatings–A sustainable approach: A review. *Progress in Organic Coatings*, Vol. 156, p. 106267, 2021.
14. J.T. Derksen, F.P. Cuperus, and P. Kolster, P. Paints and coatings from renewable resources. *Industrial Crops and Products*, Vol. 3, pp. 225-236, 1995.
15. Oprea S. Effect of the long chain extender on the properties of linear and castor oil cross-linked PEG-based polyurethane elastomers. *Journal of Materials Science*. Vol. 46(7), pp. 2251-8, 2011.
16. Y. Hu, C. Liu, Q. Shang and Y. Zhou. Synthesis and characterization of novel renewable castor oil-based UV-curable polyfunctional polyurethane acrylate. *Journal of Coatings Technology and Research*, Vol. 15, pp. 77-85, 2018.
17. N.J. Vickers. Animal communication: when I'm calling you, will you answer too. *Current Biology*, Vol. 27, pp. R713-R715, 2017.
18. Díez-Pascual AM. Antimicrobial Coatings Based on Linseed Oil/TiO2 Nanocomposites. *Handbook of Antimicrobial Coatings*. 2018:411.
19. J.T. Derksen, F.P. Cuperus and P. Kolster, P. Renewable resources in coatings technology: a review. *Progress in Organic Coatings*, Vol. 27(1-4), pp. 45-53, 1996.
20. M. Samadzadeh, S.H. Boura, M. Peikari, A. Ashrafi, and M. Kasiriha. Tung oil: An autonomous repairing agent for self-healing epoxy coatings. *Progress in Organic Coatings*, Vol. 70(4), pp. 383-387, 2011.
21. S. Guo, Y. Ge and K. Na Jom. A review of phytochemistry, metabolite changes, and medicinal uses of the common sunflower seed and sprouts (Helianthus annuus L.). *Chemistry Central Journal*, Vol. 11(1), pp.1-10, 2017.
22. R. Garces, E. Martínez-Force, J.J. Salas and M. Venegas-Calerón. Current advances in sunflower oil and its applications. *Lipid Technology*, 21(4), pp. 79-82, 2009
23. A.E. Rheineck and L.O. Cummings. Safflower oil utilization in surface coatings. *Journal of the American Oil Chemists Society*, Vol. 43(6), pp. 409-412, 1996.
24. D.P. Pfister, Y. Xia and R.C. Larock. Recent advances in vegetable oil-based polyurethanes. *ChemSusChem*, Vol. 4(6), pp. 703-717, 2011.
25. Cahoon EB. Genetic enhancement of soybean oil for industrial uses: prospects and challenges. *AgBioForum*, 6 (1&2), pp.11-13, 2003.
26. Y. Lu and R.C. Larock. Novel polymeric materials from vegetable oils and vinyl monomers: preparation, properties, and applications. *ChemSusChem*, Vol. 2(2), pp. 136-147, 2009.
27. A. Hofland. Alkyd resins: From down and out to alive and kicking. *Progress in Organic Coatings*, Vol. 73(4), pp. 274-282, 2012.
28. S.S. Mahapatra and N. Karak. Synthesis and characterization of polyesteramide resins from Nahar seed oil for surface coating applications. *Progress in Organic Coatings*, Vol. 51(2), pp. 103-108, 2004.
29. S. Ahmad, S.M. Ashraf, E. Sharmin, M. Nazir and M. Alam, M. Studies on new polyetheramide-butylated melamine formaldehyde based anticorrosive

coatings from a sustainable resource. *Progress in Organic Coatings*, Vol. 52(2), pp. 85-91, 2005.
30. A. Köckritzand A. Martin. Oxidation of unsaturated fatty acid derivatives and vegetable oils. *European Journal of Lipid Science and Technology*, Vol. 110(9), pp. 812-824, 2008.
31. D.L. Trumbo, J.T. Otto. Epoxidized fatty acid-derived oxazoline in thermoset coatings. *J. Coat. Technol. Res.* Vol. 1–8, pp. 107-111, 2011.
32. A.S.A Hazmi, M.M. Aung, L.C. Abdullah, M.Z. Salleh, and M.H. Mahmood. Producing Jatropha oil-based polyol via epoxidation and ring opening. *Industrial Crops and Products*, 50, pp. 563-567, 2013.
33. D. Swern, W.C. Ault, A.J. Stirton. *Animal Fats and Oils in Industry*, pp. 538-543.
34. https://www.rea.co.uk/download/companies/reaholdingsplc/Presentations/Investor_presentation_July_2017.pdf
35. Woodgate SL, van der Veen JT. 21 Fats and Oils–Animal Based. Principles and Applications. 2014:481.
36. Michael Bockisch. *Fats and Oils Handbook*. AOCS Press, 1998, pp.121-173.
37. F.D. Gunstone, F.D. Production and Trade of Vegetable Oils. In *Vegetable Oils in Food Technology*, F.D. Gunstone (Ed.), 2011.
38. Y. Mülazim, E. Cakmakc, M.V. Kahraman. Preparation of photo curable highly hydrophobic coatings using a modified castor oil derivative as a sol-gel component. *Progress Org Coat*, Vol. 72, pp. 394-401, 2011.
39. E. Sharmin, D. Akram, A. Ghosal, O. Rahman, F. Zafar, S. Ahmad. Preparation and characterization of nanostructured biohybrid. *Progress Org Coat*, Vol. 72, pp. 469-72, 2011.
40. H.S. Kwatra, J.M. Caruthers, B.Y. Tao. Surface chemical modification of natural cellulose fibers. *Ind Eng Chem Res* Vol. 31, pp. 2647-2651, 1992.
41. P. Wang and B.Y. Tao. Synthesis and characterization of long-chain fatty acid cellulose ester (FACE). *J. Appl. Polym. Sci.* Vol. 52, p. 755-761, 1994.
42. B.Y. Tao, Chapter 24 - Industrial Applications for Plant Oils and Lipids, Editor(s): Shang-Tian Yang, *Bioprocessing for Value-Added Products from Renewable Resources*, Elsevier, 2007, pp. 611-627.
43. U. Konwar, N. Karak, M. Mandal, L. Mesua ferrea. Seed oil based highly thermostable and biodegradable polyester/clay nanocomposites. *Polym Degrad Stab* Vol. 94, pp. 2221-30, 2009.
44. S. Dutta, N. Karak, J.P. Saikia, B.K. Konwar. Biocompatible epoxy modified bio-based polyurethane nanocomposites: mechanical property, cytotoxicity and biodegradation. *Biores Technol.* Vol.100: pp. 6391-7, 2009.
45. S. Guilbert, N. Gontard,. Edible and biodegradable food packaging. In *Foods and Packaging Materials—Chemical Interactions*; Ackermann, P., Jägerstad, M., Ohlsson, T., (Eds.); Royal Society of Chemistry: England, UK, pp. 159-168, 1995.
46. J.M. Druchta, C.D. Jonhston. An update on edible films-scientific status summary. *Food Technol.* Vol. 51, pp. 60-62, 1997.

47. L.R. Franssen, J.M. Krochta. Edible coatings containing natural antimicrobials for processed foods. In *Natural Antimicrobials for Minimal Processing of Foods*; Roller, S., Ed.; CRC Cambridge: Boca Raton, FL, USA, pp. 250-262, 2003.
48. K. Kraśniewska, M. Gniewosz. Substances with antimicrobial activity in edible films—A review. *Pol. J. Food Nutr. Sci.* Vol. 62, pp. 199-206, 2012.
49. R. Ribeiro-Santos, M. Andrade, N. Rados de Melo, A. Sanches-Silva. Use of essential oils in active food packaging: Recent advances and future trends. *Trends Food Sci. Technol.* Vol. 61, pp. 132-140, 2017.
50. Y. Akiyama, N. Nagahara, M. Yoshioka Coatings for stable sustained release preparations US Patent 5158608. 1991.
51. M.J. Martínez, A.M.R. Pilosof, C. Arzeni, V.M. Pizones Ruiz-Henestrosav, F.A. Bellesi, M. Von Staszewski. Chia oil edible nanoemulsion. WO Patent201802962-6 A1. 2016.
52. M.S. Gi, C.M. Jung, C.J. Yeon, J.Y. Ji. Nano-imulsion comprisingcinnamon oil and method therefor. KR Patent 2016000518-2 A. 2016.
53. L. Houbin, Z. Huaming, L. Chengjin, L. Liang. Mustard oil nanoemulsion and preparation method thereof. CN Patent 10331595-6 A. 2013.
54. W. Weichun, F.Wei, Z Aixia. Preparation method of palm oil nanoemulsion. CN Patent 10375005-0 A. 2014.
55. P.B. Raja, M.G. Sethuraman. Natural products as corrosion inhibitor for metals in corrosive media--a review. *Mater. Lett.* Vol. 62, pp. 113-116, 2008.
56. K. Srivastava, P. Srivastava. Studies on plant materials as corrosion inhibitors. *Br. Corr. J.* Vol. 16, pp. 221-223,1981.
57. R.A.L. Sathiyanathan, S. Maruthamuthu, M. Selvanayagam, S. Mohanan, N. Palaniswamy. Corrosion inhibition of mild steel by ethanolic extracts of Ricinus communis leaves. *Ind. J. Chem. Technol.* Vol. 12, pp. 356-360, 2005.
58. M.A. Dar. A review: plant extracts and oils as corrosion inhibitors in aggressive media. *Ind. Lubr. Tribol.* Vol. 63, pp. 227-233, 2011.
59. N. Lahhit, A. Bouyanzer, J.M. Desjobert, B. Hammouti, R. Salghi, J. Costa, C.Jama, F. Bentiss, L. Majidi. Fennel (Foeniculum Vulgar) essential oil as green corrosion inhibitor of carbon steel in hydrochloric acid solution. Portugal. *Electrochim. Acta* vol. 29, pp. 127-138, 2011.
60. Y. Lu, R.C. Larock. Novel polymeric materials from vegetable oils and vinyl monomers: preparation, properties, and applications. *ChemSusChem* Vol. 2, pp. 136-147, 2009.
61. M.A.R. Meir, J.O. Metzger, U.S. Scubert. Plant oil renewable resources as green alternatives in polymer science. *Chem. Soc. Rev.* Vol. 36, pp. 1788-1802, 2007.
62. P.K. Sahoo, L.M. Das, M.K.G. Babu, S.N. Naik. Biodiesel development from high acid value polanga seed oil and performance evaluation in a CI engine. *Fuel* Vol. 86, pp.448-454, 2007.

63. A. Kumar, S. Sharma. Potential non-edible oil resources as biodiesel feedstock: An Indian perspective. *Renew. Sustain. Energy Rev.* Vol. 15, pp.1791-1800, 2011.
64. M.S. Gaikwad, V.V. Gite, P.P. Mahulikar, D.G. Hundiwale, O.S. Yemul. Eco-friendly polyurethane coatings from cottonseed and karanja oil. *Prog. Org. Coat.* Vol. 86, pp.164-172, 2015.
65. M. Alam, S. M. Ashraf, A.R. Ray, S. Ahmad. Development of anticorrosive poly(ether-urethane) amide coatings from linseed oil: A sustainable resource. *J. Polym. Environ.* Vol. 18, pp. 208-215, 2010.
66. A. Chaudhari, A. Kuwar, P. Mahulikar, D. Hundiwale, R. Kulkarni, V. Gite. Development of anticorrosive two pack polyurethane coatings based on modified fatty amide of Azadirachta indica Juss oil cured at room temperature—A sustainable resource. *RSC Adv.* 2014, 4, 17866-17872.
67. S. Ranade, L. Neelakantan. Exploring *in situ* integration of pongamia oil to improve barrier properties of polyurethane coatings. *J. Appl. Polym. Sci.* Vol. 137, p. 49553, 2020.
68. S. Ahmad, S.M. Ashraf, E. Sharmin, F. Zafar, A. Hasnat. Studies on ambient cured polyurethane modified epoxy coatings synthesized from a sustainable resource. *Prog. Cryst. Growth Charact. Mater.* Vol. 45, pp. 83-88, 2002.
69. V.V. Gite, R.D. Kulkarni, D.G. Hundiwale, U.R. Kapadi. Synthesis and characterisation of polyurethane coatings based on trimer of isophorone diisocyanate (IPDI) and monoglycerides of oils. *Surf. Coat. Int. B Coat. Trans.* Vol. 89, pp. 117-122, 2002.
70. T.O. Siyanbola, R. Enishetty, R. Kumar, O.O. James, G.I. Olasehinde, S. Kaki, R. Narayan, K.V.S.N. Raju. Specific crosslinking effects of poly(epichlorohydrin)-triol on urethane polymer matrix of castor seed oil-based coatings. *J. Coat. Technol. Res.* 2021, 18, 129-141.
71. J.V. Nardeli, C.S. Fugivara, M. Taryba, M.F. Montemor, S.J.L. Ribeiro, A.V. Benedetti. Novel healing coatings based on natural-derived polyurethane modified with tannins for corrosion protection of AA2024-T3. *Corros. Sci.* Vol. 162, p. 108213, 2020.
72. S. Ahmad, F. Zafar, E. Sharmin, N. Garg, M. Kashif. Synthesis and characterization of corrosion protective polyurethane fattyamide/silica hybrid coating material. *Prog. Org. Coat.* Vol.73, pp. 112-117, 2012.
73. J.C. Tiller. Antimicrobial Surfaces. In *Bioactive Surfaces. Advances in Polymer Science*; H.G. Börner, J.F. Lutz, Eds.; Springer: Berlin/Heidelberg, Germany, V. 240, 2010.
74. K. Vasilev, J. Cook, H.J. Griesser. Antibacterial surfaces for biomedical devices. *Expert Rev. Med. Devices* 6, pp. 553-567, 2009.

8

Oil and Fats as Raw Materials as Corrosion Inhibitors and Biolubricants

Anurag Bapat[1], Subhalaxmi Pradhan[2] and Chandu S. Madankar[1*]

[1]Department of Oils, Oleochemicals and Surfactants Technology, Institute of Chemical Technology, Mumbai, India
[2]Department of Basic Sciences, School of Basic and Applied Sciences, Galgotias University, U.P., India

Abstract

Petroleum crude oil is depleting very fast throughout the world. Researchers are focusing on alternative sources that can reduce our dependence on petroleum feedstock. Naturally occurring vegetable oils and natural fats have long been and continue to be the most alternative renewable feedstock for industrial applications such as foods, fuels, corrosion inhibitors and lubricants. Natural extracts of vegetable seeds are an incredibly important feedstock for the industry since they are renewable, sustainable, non-toxic, non-polluting, and cost effective in comparison to synthetic feedstock. Eco-friendly vegetable oils, particularly plant oils, have excellent anticorrosion, lubrication, and viscosity indices due to their inherent affinity for metal surfaces, making them suitable as corrosion inhibitors and lubricants. Corrosion has both economic and social consequences, affecting the safety and health of people who work in industries or live in communities. As an outcome, researchers are exploring and implementing naturally occurring and eco-friendly corrosion inhibitors. In view of the environmental impacts and rapidly depleting resources of mineral oil and synthetic oil lubricants, their replacement with vegetable oil lubricants (biolubricants) that are biodegradable and nontoxic is gaining pace. The current chapter will focus on production of fatty alkyl esters, epoxides of different vegetable oils and their potential use as biolubricant.

Keywords: Vegetable oils, fatty alkyl esters, transesterification, epoxidation, biolubricant

---

**Corresponding author*: chandumadankar@gmail.com

Divya Bajpai Tripathy, Anjali Gupta, Pooja Agarwal, Anuradha Mishra, and Arvind Kumar Jain (eds.) Oils and Fats as Raw Materials for Industry, (195–230) © 2024 Scrivener Publishing LLC

8.1 Introduction

Since man constructed machines, lubricants and lubrication have been built into moving parts. Water, animal origin fats, vegetable origin oils were utilized for lubricating in the early days of machines. Natural lubricants were supplanted by mineral oil and petroleum products as the petrochemical industry grew in the late 1800s. Due to worries about the petrochemical-derived lubricants' perishability, more damaging impacts, and environmental challenges, a search has developed for lubricants of biological origin [1]. The lubricant industry's environmental consciousness favors products that use naturally accessible eco-friendly lubricants rather than depleting finite sources of mineral hydrocarbons. Generally, three basic raw materials are used for the biolubricant production.

Vegetable origin oils, animal-derived fats and synthetic esters are all examples. Petrochemical-based lubricants have been around for a long time, and prior to that, lubricants were largely made up of vegetable oils. Throughout World War II, many synthetic esters were created and used as lubricants for many years [2].

Nowadays, about 80–85% of the raw materials used in synthetic lubricants come from various sources, like synthesised hydrocarbons (50–55%), organic esters (40–45%). Polyglycols (10%) and phosphate esters (5%) are two further forms of synthetic lubricants [90]. Synthetic esters provide a number of benefits as compared to mineral oils; they can be used with mineral oils to improve production. They are made up of basic and pure raw materials for getting the appropriate molecular architectures for excellent lubrication. Synthetic esters have good viscosity indices, are oxidatively and thermally stable, and are free of impurities found in petroleum-based lubricants.

They act as basic feedstocks for lubricant preparation for a long period of time, despite resistance to their usage due to their high cost. Synthetic esters are flexible, allowing them to replace biodegradable esters in 90% of all lubricants without affecting effectiveness. Some tribological characteristics, such as the thermal and oxidative tolerance, may be greatly enhanced when compared to products with mineral bases.

Internal combustion engines, automobile and industrial gearboxes, compressors, turbines, and hydraulic systems all demand the usage of specially formulated lubricants. To meet more than 90% of all lubrication uses, between 5,000 and 10,000 distinct lubricant formulas are required. On average, lubricating oils are made up of 93% basic oils and 7% chemical additives and other components, accounting for over 90% of lubricant usage.

Efforts are being made all around the world to precisely leverage the benefits of agriculture renewable resources. The European Union (EU) and Asia Pacific are working together to explore lubrication-related scientific, industrial, and commercial potential. Several corporations have become involved due to the ecologically benign characteristics of vegetable origin derived lubricants as well as the marketing advantages. Mobil Chemical, for example, was planning to establish a neat lubricant formulation line. Shell and British Petroleum teamed up with the French National Railways to synthesize biodegradable railway lubricants. A variety of greases, metalworking fluids, and hydraulic fluids based on vegetable origin oil were developed by the Lubrizol Corporation of the United States. The Indian Oil Corporation is oriented towards developing vegetable oil–based lubricants for agricultural and forest machines, as well as the transportation industry.

8.1.1 Consumption of Lubricants in India and Worldwide

The worldwide lubricant market has seen substantial demand and rising usage during the last decade. As indicated in Figure 8.1, India utilised roughly 5% of the total worldwide consumption of 36.3 millions tonnes of lubricants in 2010. Automotive lubricants accounted for 53% of all lubricants consumed globally, followed by industrial lubricants (32%), marine based oils (5%), and processed oils (10%). Hydraulic oils accounted for 37% of all industrial lubricants consumed, followed by industrial gear oils at 7%, metal working fluids at 16%, and greases at 9% [3]. In India, the automobile industry consumed 47% of all lubricants in 2010. Industry lubricants came in second with 17%, followed by marine and aviation with 15%, hydraulic oils with 11%, and processing oils with 10% (Figure 8.2).

Strong economic development will increase lubricant necessity by more than 2% yearly to 41 million tonnes in 2015 as countries recover from the effects of the global economic crisis that started in 2009. The region with the greatest growth will still be Asia, followed by Africa and the Middle East. Looking at the worldwide lubricant market, ongoing demand for lubricants reached 36.3 million tonnes in 2010, and is likely to continue to increase. In contrast to all other regions, Western Europe's lubricant consumption in 2010 had generated effective rates of growth of 1.7 to 2.5% over past years. The worldwide lubricant industry is anticipated to develop at a compound annual growth rate of 16.8% over the next ten years due to growing demand in emerging regions [4, 5].

Approximately 50-60% of lubricants utilized worldwide will end up in the environment and harm it. Therefore, spills and evaporation of

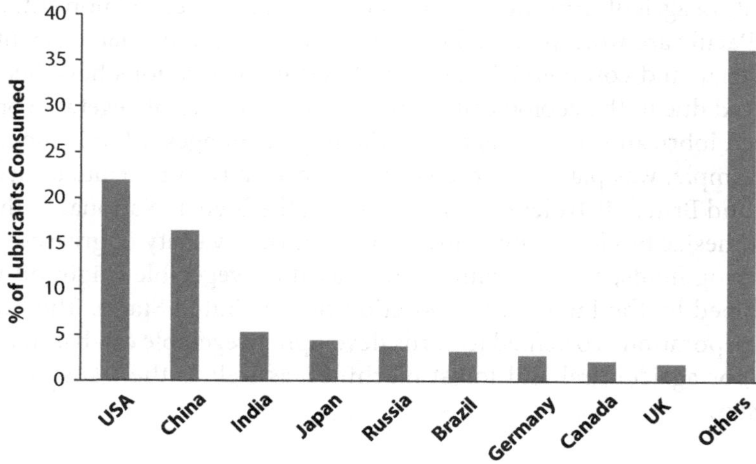

Figure 8.1 Global consumption of lubricants. Source: Global Industry Analyst Report (2010) [6].

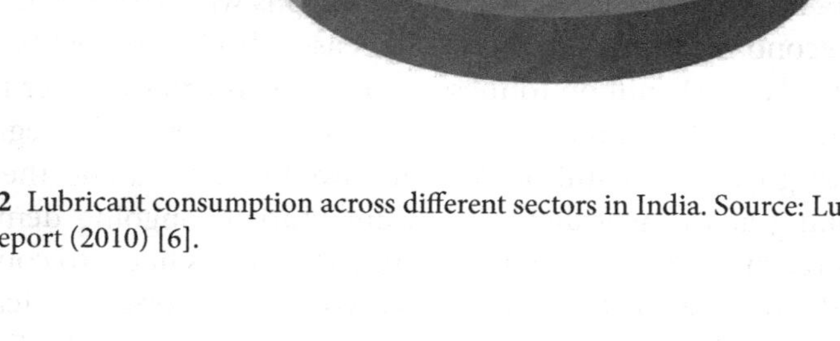

Figure 8.2 Lubricant consumption across different sectors in India. Source: Lubricant Market Report (2010) [6].

petroleum-based lubricants must be minimized since engine oil evaporation accounts for around one-third of diesel engine particulate emissions. The creation of ecologically friendly lubricants was motivated by the substantial lubricant losses into the environment.

8.1.2 Environmental Risks Due to Petrochemical-Based Lubricants

Environmental health issues like pollution have become more significant social challenges in recent years. With around 50 to 60% of petrochemical-based lubricants coming into direct contact with soil, water, and air, they pose a potential threat to ecosystems and presently one of the most concerned worries in the field of lubricants and their applications [2]. A recent study found that mineral oil and lubricants made of petrochemicals can harm ground water for up to 100 years, affecting tree growth and aquatic life span [7].

In industry, lubricants are mostly composed of hydrocarbons with minor amounts of metals, sulphides, and nitrogen compounds. Petrochemical-based lubricants constitute a long-term hazard to the environment and huge groundwater supplies because of its inherent toxicity and lack of biodegradability [8].

About 45-55% of all lubricants consumed worldwide were utilized in the automobile industry, which is the leading source of air pollution. The majority of hydrocarbons and a large quantity of sulfur are found in lubricants made from petrochemicals. Sulphur creates sulfur dioxide and sulfur trioxide as it burns, the majority of which is released into the environment. Acid rain is most likely caused by sulfur trioxide reacting with water vapor to generate sulphuric acid. The sulfur traces in the goods raises particulate matter emissions substantially. Particulate matter in the air is a crucial contributor to health issues such as asthma and respiratory disorders. Evaporation during operation emits a variety of pollutants, including smoke, carbon monoxide, nitrogen oxides, particulate matter, and carbon dioxide.

Government laws to mitigate the consequences of air pollution have been greatly enhanced in recent years [9]. Currently, petrochemical-based lubricants are extensively utilised for a variety of purposes, but rising environmental consciousness has prompted authorities to seek alternatives.

8.1.3 Biodegradable Lubricants

Biolubricants are bio-based products that must be made from renewable resources. Biodegradable lubricants are biodegradable and can be synthesized from various sources, which include non-renewable petroleum and renewable feedstock that can be broken down naturally into common components and returned to the environment. Biolubricants are environmentally friendly, quickly biodegradable, safer to the environment, economical and have improved lubricity and anticorrosion characteristics.

Readily biodegradable: A quickly biodegradable material will decompose 80% in a regulated standard test setting in 21 days. This is desirable since the material will decompose before causing environmental harm. Synthetic ester-based lubricants, certain polyalkylene glycols, and vegetable-based lubricants all share this property. There is evidence that some biodegradable substances don't break down in the environment if the conditions aren't right or the amount of bacteria isn't high enough.

Naturally biodegradable: Natural biodegradable compounds are only capable of degrading since there are no time or degree restrictions. Some of these products, like conventional fuels and lubricants, include some petroleum content. Products that are naturally biodegradable but not quickly biodegradable can linger in the environment for a long time. The sole thing that makes them more environmentally friendly than petroleum-based lubricants is that they are bio-based and do not include heavy metal additions [10].

The biodegradable gasoline engine oils that are used in Western Europe are an equal proportion mixture of synthetic esters, polyalpha olefins, and hydrocracked oils with traditional additives. By adding more ester, a product's biodegradability can be enhanced. In comparison to conventional products, using ester lubricants in engine oil compositions has a variety of advantages, such as reduced oil consumption, improved engine cleaning (low deposition), and biological degradability. Biodegradable lubricants can be produced from renewable feedstocks like vegetable oil and must adhere to specified biodegradation standards.

Ongoing research and development on bio-derived lubricants to reduce the effect of everyday tasks on the environment has demonstrated concern for environmental protection. The researchers' most difficult tasks are designing a comparable product in terms of performance, environmental sustainability, and affordability. Only cutting-edge research and products produced from renewable raw materials can accomplish this. Vegetable oils as a foundation stock for biolubricants have shown to be a viable solution to both past and future problems. They offer high performance, little environmental impact, and long-term viability.

Governments have enacted stringent regulations to facilitate the usage of biolubricants and to decrease the leakage of petroleum-based lubricants into the environment. Several countries, including Germany, Scandinavia, the United States, and India, have adopted eco-labeling and granted eco-friendly seals like the Blue Angel, White Swan, Green Cross, and Eco-mark [11].

Five main types of biodegradable lubricant base stock—synthetic esters, polyol esters, polyalkylene glycols, polyalphaolefins, and vegetable oil-based lubricants—are now very important, among which vegetable oil-derived lubrived are more important from an environmental point of view [2].

8.1.3.1 *Synthetic Esters*

Synthetic esters are a broad category of compounds that are derived from either petrochemical or oleochemical sources [12]. Mineral oils do not fulfill the toxicity and biodegradability standards when compared to synthetic esters. Synthetic esters offer greater biodegradability, flammability, lubricity, fluidity, volatility, and thermal and oxidative stability than natural esters. To fit the actual application, these qualities can be further increased by molecular design. As a result, when environmental protection is important, synthetic esters are potential choices.

Synthetic esters are appealing because of their excellent lubrication and biodegradability. Mineral oils, vegetable oils, and synthetic lubricants are the three types of base oils utilised in lubricant compositions. Synthetic esters in lubricants were synthesized as a response to the requirement for high-temperature lubricants, primarily for military and aircraft engine applications [13]. Synthetic esters compete with mineral origin oils in most cases and may be combined to improve performance.

8.1.3.2 *Polyol Esters*

High-performance lubricants necessitated the development of polyol esters. As a result, polyol esters have become increasingly important and popular in both military and industrial settings. Polyol esters are typically made by reacting fatty acids and alcohols with standard catalysts. The catalysts employed in the process have a problem with disposal, generate a base fluid with a high acidity, and are hazardous to the environment [14].

Saturated fatty acid polyol esters had remarkable oxidative stability, heat resistance, are biocompatible, and low temperature properties. Polyol esters have a worse oxidative stability when used with unsaturated fatty acids [15].

8.1.3.3 Polyalkylene Glycols (PAG)

Polyalkylene glycol lubricants can be utilized in various situations in cases where petrochemical-derived lubricants fail to meet expectations. These are prepared by combining ethylene oxide and propylene oxide monomers to form a polymer. PAG is a highly polar substance with ether bonds and a hydrocarbon structure. Polarity increases water solubility, which is beneficial for biodegradable fluids but detrimental for many lubricant applications since it quickly absorbs and contaminates moisture.

PAG has a high viscous index, is nontoxic, has a wide range of solubility qualities, and is lubricious. Some of the most often utilized applications are lubricants for gears and compressor engines, metalworking fluids, lubricating fluid for fibers, and hydraulic fluids for resisting fire [16].

8.1.3.4 Poly-Alpha-Olefins (PAO)

These are low molecular weight dimers to tetramers which have negligible side branching and short synthesized molecules. They were developed as high-performance base oils for industrial and automotive hydraulic, engine, and transmission fluids. PAO are oligomers of hydrogenated olefins (alkenes). PAO has a more uniform molecular structure and molecular mass distribution than mineral oils, resulting in more consistent lubricating qualities.

PAO's viscosity indices and vapor pressure are often higher than those of mineral oils. As there are no double bonds or other reactive functional groups in the structure, PAO has higher oxidative stability compared to goods made from petroleum. PAO is a desirable biodegradable lubricant substitute for water, especially used in different fields at less temperature [15].

8.1.3.5 Lubricants Made from Vegetable Oil and Its Derivatives

Recent research on the deteriorative implications of lubricants of mineral origin on ecology and the environment has confirmed groundwater contamination for more than 100 years, as well as the impact on the longevity of trees and aquatic life. Because natural esters have greater potential in lubricating and tribological processes, consumers are more aware of the need to utilise ecologically friendly products. Oils of vegetable origin, fats originating from animals, and synthetic esters are all potential possibilities for environmentally responsible lubricants [17]. Compared to synthetic esters, vegetable origin oils are more likely to be used as lubricants due to economic factors and price stability. Triglycerides, which have three hydroxyl groups and are esterified with fatty acid carboxyl groups, make up the majority of vegetable oils. A variety of fatty acids with different levels of unsaturation and similar chain lengths (14 to 22 carbons) make up triglycerides. A lubricant's triglyceride structure has beneficial properties. Triglycerides' lengthy polar fatty acid chains interact with metallic surfaces to generate robust lubricant coatings that lessen wear and friction. Triglycerides are resistant to temperature changes due to their strong intermolecular bonds, which maintain their viscosity and lubricity throughout time [18].

Even though they have worse oxidative, thermal, and cold flow qualities to mineral oils, vegetable oils have superior lubricity, biodegradability, nontoxicity, low volatility, and a high viscosity index [19, 20]. Vegetable origin oils outperformed mineral-derived oils in antiwear and friction tests without the use of additives, proving that they had superior lubricating qualities [21]. In comparison to mineral oil lubricants, vegetable oil lubricants have higher pitting tolerance, a lower coefficient of friction, and poorer heat and oxidative stability [22].

Vegetable oil advantages and limits are being used to produce lubricant compositions. In the bulk of published data, vegetable oils are used as hydraulic fluids, gear oil, and transmission fluids. In hydraulic fluid bench testing, the performance of soybean oil was studied at various stages of refinement, indicating that soy-based hydraulic fluids were feasible [23]. Field testing of vegetable oil–based hydraulic and transmission fluid formulations yielded promising results [24]. Vegetable oil lubricants performed along with or better than commercial petrochemical-based lubricants in two-stroke engines. When compared to commercial lubricants, Jojoba oil demonstrated a good compatibility with gasoline, showed enhanced wear performance, comparable scuffing and deposit-producing tendencies in two-stroke engines [25]. Vegetable oils' performance and stability were investigated using a customized engine test rig. Vegetable origin oils were found as effective lubricants with no defect; however, a modest rise in consistency occurred over a period of time, causing settling on the test rig's bore [26].

Vegetable oils serve as great fluid additives because they have many of the qualities that lubricants need, like high index viscosity, low volatility, and strong lubricity. The two most significant fatty acids in nature are oleic acid and ricinoleic acid, which can be isolated from plant oils and utilized as lubricants [27, 28]. A lubricant's pour and cloud points are reliable indications of its low-temperature fluidity. Vegetable oils have weak cold flow characteristics, limiting their application in sub-zero conditions. Vegetable oils have a tendency to develop macro crystalline structures by homogeneous stacking at low temperatures. Transesterification and epoxidation on the fatty acid chain can interfere with this stacking process, resulting in better low-temperature characteristics [29]. Environmentally friendly goods are attracting a lot of attention that are made from vegetable oils rich in oleic acid and ricinoleic acid are viable alternatives for mineral-derived products [30, 31].

Chemical modifications like alcoholysis in presence of catalysis (transesterification), selective hydrogenation, and epoxidation are employed to enhance the oxidation stability and cold flow characteristics of plant-based oils [8, 28].

8.1.3.6 Long-Chain Fatty Acid Esters

Esters of long carbon chain fatty acids, which are practical functional molecules, can be found in nutrients, cosmetics, metalworking fluids, rolling oils, petrochemicals, and synthetic lubricants. Over the past ten years, the demand for long-chain fatty acid esters has gradually grown [32, 33]. Researchers that looked into the lubricity and base fuel properties of alkyl esters found that all esters had higher kinematic viscosity as the carbon chain length increased and that ethyl esters had better lubricity than methyl esters of the same vegetable oils [34]. According to some researchers, long-chain fatty acids and alcohols produce monomolecular layers on metals, enhancing the lubricity of machine parts using a four-ball tribometer for testing as their molecules are adsorbed with their central axes perpendicular to the surface of metals [35, 36]. Maleque *et al.* (2003) had discovered that adding 5% palm oil methyl esters (POME) to base lubricating oil significantly reduces wear and friction and revealed that blended POME serves as a lubricating oil additive in a two-stroke engine [37]. According to the results of the tests, the palm oil–based lubricating oil performed better in terms of friction and wear. Masjuki *et al.* (1997), tested engines with palm oil and a mineral oil-based commercial lubricating oil, indicating that palm oil-based lubricants significantly reduced exhaust pollutants while improving lubricity [38].

According to Sharma *et al.* (2006) [39], it has been demonstrated that chemically altering the fatty chains of soybean oil can change both its lubricating and physical characteristics, such as pour point and cloud point. When greased, the lubricant's hydroxy and ester groups create active oxygen sites that attach to the metal surface. The performance of both methyl and ethyl esters of Canola oil was examined as a lubricity additive and it was found that these esters reduced wear scar area by 16% while improving the lubricity number of the basic fuel by 20% [40]. The lubricity of various fatty compounds was compared to that of hydrocarbon compounds present in petro-diesel. It was found that the lubricity of low-level (1-2%) blends of methyl esters with low-lubricity petro-diesel was primarily caused by free fatty acid and monoacylglycerol contaminants present in the methyl esters [41]. Esters of castor and Lesquerella oils increase the lubricity of low sulfur petrol-diesels at lower bending levels than esters of non-hydroxylated vegetable oils [34, 42]. According to Hincapie *et al.* (2011), castor oil methyl esters have an extraordinarily high viscosity compared to typical methyl esters [43]. Therefore, they may be employed in contemporary internal combustion (IC) engines in combination with fossil fuel and also as a biolubricant base stock. However, when compared with typical vegetable oils, the hydroxyl group (ricinoleic acid, which is a fatty acid that includes both a double bond and a hydroxyl group) boosts lubrication property, making it a great contender as a diesel fuel additive. Due to severe government regulations, the sulfur level of petro diesel must be less than 0.005%, according to Bharat IV guidelines. Minimal sulfur diesel fuel reduces lubricity, so blended alkyl esters mainly esters of hydroxy fatty acid plant oils such as castor oil at smaller amount into petro-diesel, maintains engine rheological properties [44].

8.2 Biolubricants from Vegetable Oil

Direct use of vegetable oils as lubricant formulations is constrained due to its inherent limitations, including poor thermal and oxidative stability, poor low-temperature fluidity and hydrolytic instability [45]. Low oxidation stability is due to the presence of bis allylic protons, which are easily attacked by radicals and then oxidized to generate polar-oxy compounds. This oxy polymerization eventually produces insoluble residues that raise the oil's acidity, and increase its viscosity [1]. Lubricants of vegetable oil provide inferior durability and presence of glyceride linkage renders those oils susceptible to hydrolytic breakdown [45, 46]. At subzero temperatures, most vegetable origin oils exhibit cloudiness, precipitation, poor flow, and solidification. Furthermore, the thermal and oxidative stability of the vegetable oil is inadequate and vegetable oil unsaturation can be altered, reduced, or eliminated entirely to prevent these negative effects. These drawbacks can be avoided by altering, lowering, or completely removing unsaturation from vegetable oils. The triglyceride structure in vegetable oil contains double bonds, allylic carbons, and ester groups, all of which are potential chemical modification sites. Among the procedures explored, epoxidation, hydrogenation, and oil transesterification generate a cleaner and more ecologically friendly lubricant.

8.2.1 Epoxidation

Epoxidation is a chemical process, which results in the formation of cyclic, three-membered ether which occurs due to action of oxygen. Epoxidation results in the formation of oxirane molecules, often known as epoxides. In case of epoxidation of vegetable oil, unsaturation present in fatty acids of the triglyceride can be oxidized by hydrogen peroxide in the presence of acetic or formic acid [12].

Due to the oxirane ring's high chemical reactivity, epoxides get wide range of chemical reactivity, which opens up a wide range of commercial uses. It has garnered a lot of interest because of the broad range of reactions that may be carried out under mild reaction conditions. Epoxide can, for example, combine with various nucleophiles to form hydroxyesters, diesters, and thioesters derivatives. High-temperature lubricants may be made from epoxidized vegetable oil, whereas low-temperature lubricants can be made from its ring-opening product [47]. Epoxide of superior grade has a lower iodine value and a better oxirane score. Epoxides are crucial to the polymer and lubricant industries, hence numerous vegetable oils were investigated on a massive level. The largest producer, with 200,000 tonnes produced annually globally, was soybean oil [48, 49].

8.2.2 *In Situ* Transesterification or Reactive Extraction

Traditional procedures for making alkyl esters from vegetable oil requires refined extracted oil from raw seed and then alcoholysis of refined oil by transesterification in the presence of catalyst [50, 51]. When purified oil is employed as raw material, the various processing stages account for more than 70% of the overall cost of manufacturing of alkyl esters [52]. Reactive extraction (RE) is an *in situ* method where extraction and transesterification reaction occurs simultaneously in a single process, where alcohol serves as both an extraction solvent and a transesterification reagent [53]. Unlike other alkyl ester synthesis processes, RE uses oil transporting material in alcohol instead of reacting pre-extracted oil. As a result, RE eliminates the time-consuming and costly oil extraction technique as well as the transesterification reaction, minimizing the cost, reaction time, and solvent [54–56]. In RE, more alcohol is used because it functions as both an extraction medium and solvent in alcoholysis. After the completion of reaction, the leftover alcohol in the product was recovered and utilized as a starting material for the reaction. Kaul *et al.* (2010) [57] explored reactive extraction and optimized parameters such as seed size, seed/solvent ratio, catalyst concentration, and reaction duration from *Jatropha curcas* and obtained ester conversion up to 98%. Haas *et al.* (2004) [54] improved the process parameters for *in situ* methanol-alkaline transesterification of soybean oil utilizing statistical experimental methodologies and response surface regression analysis. Some researchers also examined acid catalyzed *in situ* transesterification of sunflower oil and discovered that the fatty ester composition and characteristics of *in situ* products are virtually the same as those of traditional pre-extracted oil transesterification [58].

According to Hincapie *et al.* (2011), *in situ* transesterification of castor seed generated a maximum transformation of 95% when utilizing 60:1 molar ratio of ethyl alcohol to oil [43]. Some researchers used an alkaline catalyst in methanol to perform *in situ* transesterification of cotton seeds to produce cotton seed oil methyl ester. They discovered that the alkaline methanol aids lipids extraction from seeds and then functions as a reaction solvent [59]. *In situ* transesterification of sunflower seeds was studied by Zeng *et al.* (2009) and optimized reaction conditions such as moisture content, catalyst type, molar ratio of catalyst to oil, molar ratio of methanol to oil, molar ratio of co-solvent to oil, reaction time, reaction temperature, and agitation speed [52]. *In situ* methanolic esterification of rice bran oil using sulfuric acid catalyst was done and the effect of specific factors such as FFA levels, *in situ* esterification time, reaction temperature, catalyst amount, moisture content, and methanol amount on the yield and purity of methyl esters was investigated [60]. According to Haas and Scott [61], moisture reduction improves the efficiency of *in situ* methyl esters synthesis from soybeans. Some studies investigated the effect of the catalyst type on the transesterification of fats and oils of both animal and plant origins to produce fatty acid esters from triglycerides [62–64].

8.2.3 Transesterification

Transesterification is the alcoholysis of plant-based oil triglycerides in existence of a catalyst to create fatty acid alkyl esters (FAAE) and glycerol. Alcohol from one ester is replaced by another alcohol during the process of transesterification, which is also known as alcoholysis and works similarly to hydrolysis. Scheme 8.1 depicts the transesterification reaction.

According to stoichiometry, three molar proportions of alcohol are required to react with 1 mol of triglyceride to produce alkyl esters. Since transesterification is a reversible process, at least six times more alcohol is required for completion of reaction.

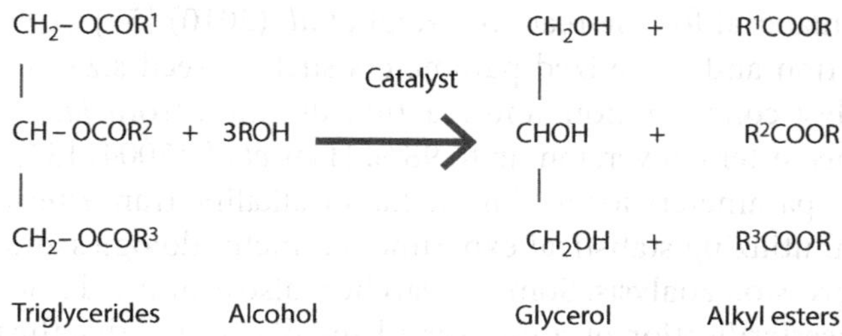

Scheme 8.1 Alcoholysis of triglyceride.

8.2.3.1 Transesterification Using Chemical Catalyst

Bronsted acids, primarily sulfonic and H_2SO_4, and strong bases such as NaOH and KOH can act as a catalyst in the transesterification process. Transesterification processes catalyzed by alkali are quick and the mechanism of reaction is depicted in Scheme 8.2. In the first step, the triglyceride molecule's carbonyl carbon is attacked by alkoxide ion, resulting in the formation of a tetrahedral intermediate. During the second-stage mechanism, the reaction of alcohol with intermediate formed alkoxide ions. The tetrahedral intermediate is rearranged in the final step to produce an ester and a diglyceride [9].

Acid-catalyzed processes are sluggish, need high temperatures, and take a longer time to complete [65]. Scheme 8.3 depicts the transesterification pathway catalyzed by acid. The carbonation of the ester's carbonyl group results from protonation, which generates the tetrahedral intermediate following a nucleophilic attack of the alcohol. To build a fresh ester and replenish the catalyst, this intermediate removes glycerol.

Because of the fast reaction time and low temperature requirement, alkali catalysts are often used to transesterify vegetable oils. There are other different catalysts, which include bronsted acids, heterogeneous catalysts such as basic zeolites, alkaline metal complexes, enzymes (lipase), and anion exchange resins [40, 66]. The transesterification reaction can be carried out in supercritical alcohols without the use of a catalyst [44].

Prior to transesterification, the rancid vegetable origin oils must be pretreated to eliminate FFA. FFA in the reaction medium causes soap to develop when it reacts with the catalyst and functions as a surface active agent, making glycerol separation harder. As a result, alkali soaps or acids

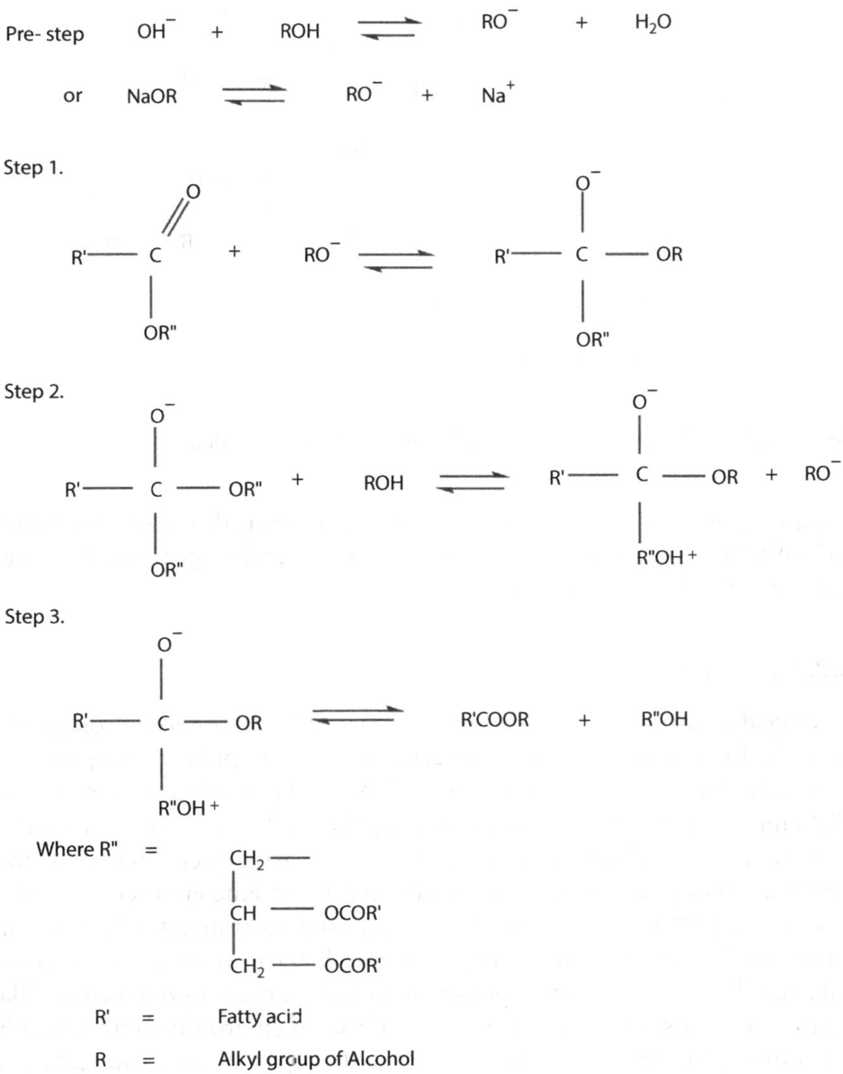

Scheme 8.2 Mechanism of the alkali-catalyzed transesterification process.

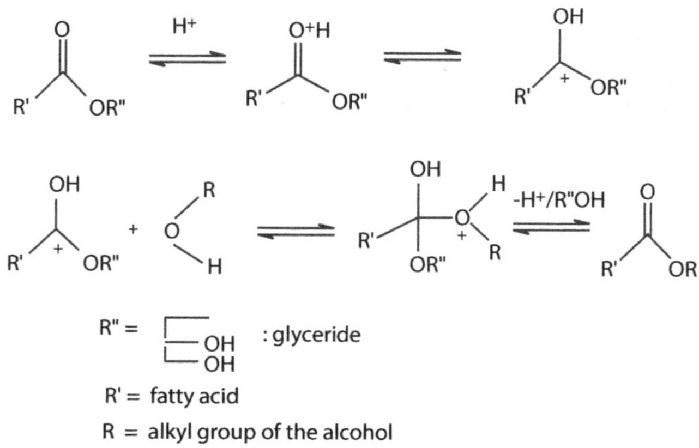

Scheme 8.3 Mechanism of acid-catalyzed transesterification reaction.

must be used to discard excess free fatty acids from this acidic feedstock [67, 68]. Water, which interacts with triacylglycerol to generate FFA, has the same effect as free fatty acids.

8.2.3.2 *Enzymatic Transesterification*

Chemical catalysts are replaced by enzymes, which are more ecologically friendly. Enzymatic alcoholysis is becoming more popular among the several catalysts used for transesterification. Enzymatic transesterification (Scheme 8.4) is becoming more appealing due to the availability of a range of feedstocks, moderate reaction parameters, easy glycerol recovery and purification of product, and the promise for a cost-efficient method [69].

Gog *et al.* (2012) stated that the catalyst used for transesterification can be grouped as alkaline, acid, enzymatic and heterogeneous. Among these catalysts, lipases are the most powerful tool that can catalyze reactions like hydrolysis, transesterification, and aminolysis in environmentally friendly conditions [70]. Paizs *et al.* (2003) reported that lipases have strong catalytic activity and non-aqueous medium stability, and their specificity, regioselectivity, and enantioselectivity can be used advantageously in a variety of organic synthesis applications [71]. Methanol has long been used to make enzymatic methyl esters, but it inhibits enzyme activity compared to other higher alcohols, resulting in reduced enzyme performance [72]. Different lipases are used to carry out enzymatic operations employing alcohol as an acyl group accepter in the presence of diverse solvents, such as hexane and tert-butanol as well as in absence of solvent [73]. However, in solventless

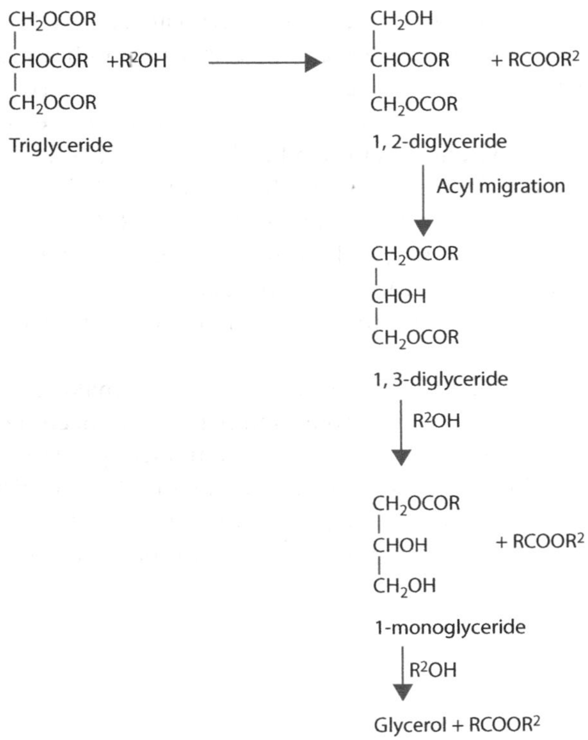

Scheme 8.4 Lipase-catalyzed transesterification of triglyceride.

systems, the substrates' mass transfer characteristics are weak, and lipase may be inactivated if there is too much alcohol in the solution. Organic solvent systems have further drawbacks, such as high solvent costs, environmental problems, and the need to separate the solvent from the reaction media [74–76]. As a result of the demand for maximum yields, good separation methods for high purity products, and ecologically safer methods, new technologies such as sub and supercritical CO_2 have been sought.

8.2.3.3 Enzymatic Transesterification in Sub and Supercritical CO_2

Biochemical reactions in supercritical fluids (SCF) are a new study topic in green and clean technology. Both subcritical and supercritical CO_2 has been hailed as an ecologically friendly reaction medium that has tremendous potential for the commercial evolution of enzyme synthesis in supercritical fluid [77, 78]. A state where a compound is above its critical points is known as the supercritical phase (critical pressure Pc, and critical temperature Tc). A chemical has characteristics of both a gas and a liquid at this phase. Carbon dioxide is the most frequently used supercritical fluid due to its benefits, which include lower critical parameters (Pc = 73.8 atm and Tc = 31.06°C), non-flammability, low toxicity, and great availability [79]. Figure 8.3 depicts the CO_2 phase diagram. When compared to organic solvents, mass transfer in supercritical CO_2 (SC-CO_2) is increased because of its low consistency, high diffusive property, and low surface tension [95]. Furthermore, because the products and enzyme are insoluble in CO_2, separating CO_2 from the product may be easily accomplished by lowering pressure [18, 80].

Exceptional solubilizing properties of SC-CO_2 make it possible for reactants which are insoluble at room temperature to become miscible at supercritical temperatures. The increased solubility speeds up the rate of reaction and product yield. As shown in Table 8.1, SCF has distinct transport qualities since it is thick like liquids yet "mobile" like gas (extremely low viscosity and intermediate diffusivity). Compared to liquid solvents,

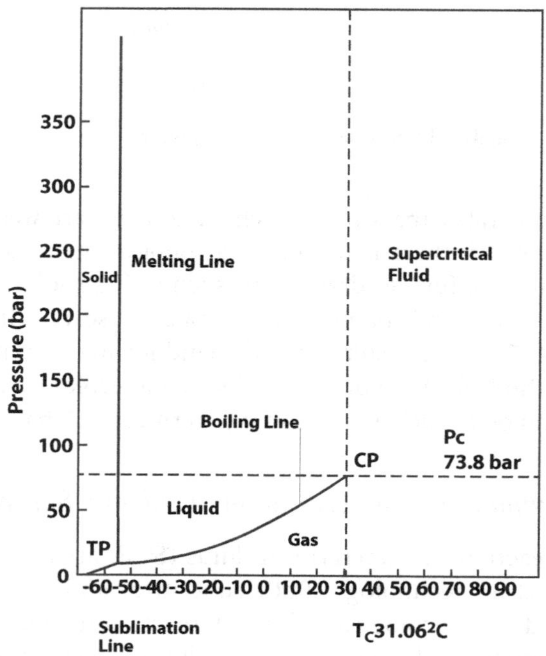

Figure 8.3 Phase (pressure-temperature) diagram for CO_2: CP=critical point, TP=triple point, Pc=critical pressure, Tc=critical temperature. (Source: Brogle, 1982).

Table 8.1 Correlation characteristics of Gas, SCF and liquid.

Property	Gas	SCF	Liquid
Density (g/ml)	0.001	0.1 - 0.5	1
Viscosity (Pa s)	1x 10^{-5}	1x10^{-4} – 1x10^{-5}	10^{-3}
Diffusivity (cm^2/s)	0.1	0.001	10^{-5}

mass transport characteristics in SCF are quick. SCF also diffuses quickly into reaction mediums because of its high diffusivity compared to liquids. The most significant attribute of SCF is its tunability, or the fluid's ability to alter density, viscosity, and diffusivity within a temperature range, because even tiny temperature changes produce pressure changes.

Figure 8.4 depicts a classic illustration showing variations in density and viscosity of $SCCO_2$ as a function of pressure. It demonstrates that pressure increases density and viscosity. The density and diffusivity of a supercritical fluid determine its solvent power, and solvent power increases as density increases. A small variation in pressure and/or temperature around a critical point of a supercritical fluid may modify the density and hence raise solubility dramatically. The physicochemical characteristics of

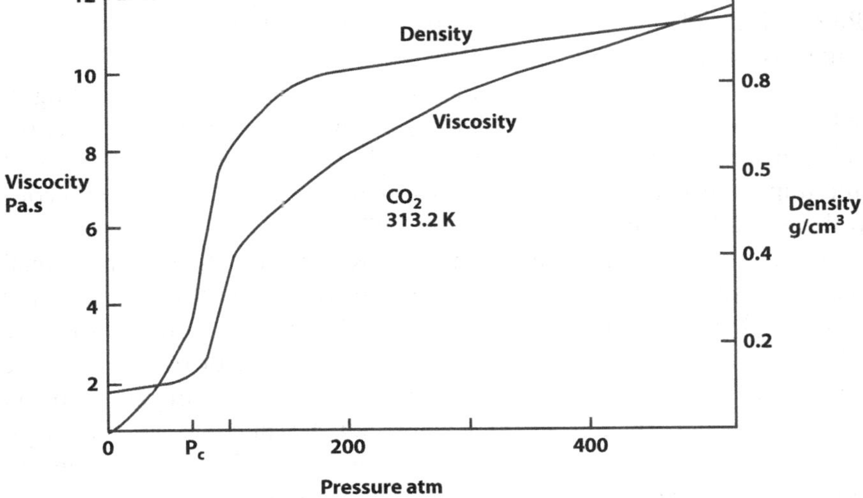

Figure 8.4 Variation of viscosity and density of CO_2 versus pressure at temperature of 313.2 K.

supercritical fluid may vary with no effect on the substance's molecular structure. $SCCO_2$ is a great solvent since it may operate as a replacement for a variety of different solvents.

Supercritical CO_2 has a number of benefits, including the ability to substitute organic solvents, better diffusivity, and reduced viscosity, all of which improve kinetics of reaction. At the end of the reaction, easily adjustable solvating power of supercritical CO_2 makes it simple to isolate reactants, products, and enzymes, whereas isolating product from the solvent is expensive, time-consuming, and unsuitable for commercial use. Because of very high volatility CO_2, it may be extracted from the product entirely and readily, resulting in a solventless reaction system [32, 33, 81], which is necessary for the commercial manufacturing of alkyl esters as biolubricants.

The biggest drawback for use lipase in the industrialization of the enzymatic processes is the increased economic rates of the enzyme; however, the enzyme cost has decreased recently and it is now accessible in an immobilized form that increases stability of operation, allowing it to be recycled and reused [82, 83].

The batch containers are ideal for screening enzymes in the early stages of product development. Continuous reactors, such as packed bed reactors and plug flow reactors can be helpful for obtaining sufficient reaction performance data for up-scaling. Packed bed reactors (PBR) are generally employed in industrial-level applications because they are ideal for solid–fluid interaction in heterogeneous catalysis processes. Benefits of PBR include: (1) it promotes the contact between the solid-liquid system and its isolation; (2) it allows for constant discard of inhibitory substances; (3) it permits the re-use of enzyme; (4) also permits for the handling of substrates which have less solubility by use of excess volumes having low substrate concentrations; (5) it leads to highly consistent quality of product and improved stability of enzyme because of ease of automation and control [84]; and (6) it is suitable [32, 33, 85–87].

The substrate volume handled in a PBR is higher than in traditional reactors systems, and the substrate to enzyme ratio in a PBR is significantly less than that of typical batch reactors, resulting in improved performance of reaction. The emptying and refilling of batch type of reactor must be done at the conclusion of every run; that is time demanding and ends up in a less productivity. It is also vulnerable to batch fluctuation. As a result, with a set quantity of enzyme in PBR, a more clarified and repeatable end product may be generated than in batch procedures.

Some researchers adjusted several reaction conditions such as enzyme loading, molar ratio of methanol to oil, reaction temperature, and reaction duration for the synthesizing methyl esters using castor and linseed oils in supercritical condition of alcohols with no catalyst and in supercritical CO_2 using lipase [88]. Oliveira and Oliveira (2000) [75] used Lipozyme IM and Novozyme 435 to investigate batch ethanolysis of palm kernel oil in the n-hexane and $SCCO_2$, concluding that $SCCO_2$ has good transport qualities that is able to speed mass-transfer-limited enzymatic processes. Madras *et al.* (2004) [89] studied the influence of enzyme charging, oil to alcohol ratio, time period, and temperature on yield in synthesis of methyl esters of sunflower oil in $SCCO_2$. Some researchers observed the impact of a variety of parameters influencing transesterification of non-edible oils such as Pongamia pinnata and Jatropha curcas in $SCCO_2$ and found that 70% of the ethyl esters were converted [90]. Lee *et al.* (2009) studied batch enzymatic methanolysis of a variety of edible and non-edible oils in $SCCO_2$ and reported the conversion yield of 92% and also investigated the reusability of immobilized enzymes up to eight times and confirmed that yield of methyl esters was unchanged [91]. Laudani *et al.* (2007) [33] carried out lipase-catalysed production of *n*-octyl oleate by esterification of oleic acid with 1-octanol in $SCCO_2$ in packed-bed bioreactor and elucidated the impact of many process parameters, like pressure, temperature, flow rates of substrate on esterification. They investigated the influence of many process factors on esterification, including pressure, temperature, and substrate flow rates, in lipase-catalyzed synthesis of n-octyl oleate by esterification of oleic acid with 1-octanol in $SCCO_2$ in packed-bed bioreactors. Fomuso *et al.* (2002) worked on *Rhizomucor miehei* lipase-catalyzed acidolysis of olive oil and caprylic acid in a bench-scale packed bed bioreactor to yield structured lipids [92]. Researchers explored batch enzymatic methanolysis of several edible and non-edible oils in $SCCO_2$ and found a conversion rate of 92% [91]. They also looked into the reusability of immobilized enzymes up to eight times and confirmed that the output of methyl esters remained the same. To create structured lipids, Fomuso *et al.* (2002) used a bench-scale packed bed bioreactor to perform Rhizomucor miehei lipase-catalyzed acidolysis of olive oil and caprylic acid [92].

Only a few researches have looked at the transesterification process kinetics of vegetable origin oil in $SCCO_2$. Diasakou *et al.* (1998) studied kinetics of non-catalytic transesterification process of soybean oil, assuming irreversible stepwise processes [93]. The kinetic parameters of the rapeseed oil transesterification process using supercritical methanol were investigated by [44]. As a function of triglyceride content and reaction temperature, the reaction was supposed to follow a first order process. Total transesterification reaction's rate constants were established. With the addition of the solvent tetrahydrofuran, Ataya *et al.* (2006) investigated the kinetic parameters of canola oil transesterification process in two-phase as well as in single-phase systems and observed a first order reaction mechanism [94]. Laudani *et al.* (2007) [32] studied the kinetics of oleic acid esterification in $SCCO_2$ and estimated activation energies at various carbon dioxide pressures. A few studies also reported lipase-catalyzed fatty ester synthesis from higher chain alcohols in $SCCO_2$ medium.

8.3 Renewable Feedstocks Available in India

The always increasing demand for oil, decreasing oil supplies and noxious environmental effects of petrochemical-derived product use have regenerated curiosity in biologically derived goods as a means of achieving long-term sustainability [95].

Renewable raw resources are crucial in the development of ecologically friendly goods. Over the last six decades, research on lubricating oil has enhanced the qualities of the oil and broadened its uses in all disciplines of lubrication [37]. The increased curiosity in tribology of using vegetable origin oil and esters in lubricating operations is due to increased environmental consciousness and the utilization of renewable resources and eco-friendly feedstock. This worry has arisen because of a mix of municipal and national rules, and also due to consumer pressure. It provides wide range of application options that are rarely satisfied by petrochemical resources [96]. In some lubricant uses where instant contact with the environment is expected, renewable resources like vegetable origin oil and its derivatives are currently being studied as viable substitutes for mineral origin oil. Sunflower, soybean, rapeseed, linseed, cottonseed, canola, and other key oilseeds are utilized over the world, with the majority of them being edible. According to Masjuki *et al.* (2003), vegetable oils provide considerable environmental benefits in aspects of resource renewability and biodegradability, as well as good properties in number of lubricant applications [97]. Vegetable origin oil-based lubricants, according to [1], offer several benefits over mineral oil–based lubricants, including greater innate lubricity qualities, higher viscosity index, improved bulk modulus, flash and fire point, and improved additive solvency property. Due to the constraints of using edible oils like soybean, rapeseed, sunflower, palm, peanut, etc., in developing nations like India, non-edible oil crops like *Jatropha Curcus*, *Pongamia Pinnata*, and *Castor* can be utilized for the synthesis of alkyl esters among which castor alkyl esters is promising bio additive to diesel fuels [98]. With an output of roughly 2,339,000 MT in 2011 [99], India is a major producing and exporting country of castor seeds. The cultivation of several non-edible oil seeds in India is mentioned in Table 8.2. Currently more than 72% of India's crude oil needs are met by imports, and presently India is the second-largest consumer of oils of vegetable origin after China [100]. About 10 million tonnes of edible oil are consumed in India each year, of which 5 million tonnes are imported. Therefore, numerous non-edible oils from the seeds of domesticated and untamed plants

Table 8.2 Non-edible oilseed plants in India.

Botanical (common) name	Fatty acid composition	Potential metric ton (MT)	Oil (%)	Use
Azadirachta indica (Neem)	C16:0 acid (19.4%), C18:0 acid (21.2%), C18:1 acid (42.1%), C18:2 acid (14.9%), and C20:0 acid (1.4%)	5,00,000	35-40	Medicinal, biopesticides
Pongamia pinnata (Karanja)	C18:1 acid 71.3%; C18:2 acid 10.8%; C20:1 acid 9.5%; C16:0 acid 8%.	2,00,000	27-39	Soaps, lubricants, ill-umination, industrial
Calophyllum inophyllum (Undi)	C18:1 acid 53.1%; C18:2 acid 28.8%; C18:0 acid 13.2%, C16:0 acid 14.1%	11,467	50-73	Soaps, illumination,
Jatropha curcas (Jatropha)	C18:1 acid 44.7%; C18:2 acid 32.8%; C16:0 acid 14.2%		30-40	Industrial, medicinal soap
Simamondisa chinensis (Jojoba)	C20:1 acid 30.3%; C22:1 acid 14.2%; C22:2 acid 33.7%; Godoelic acid (C20:1) 14.6%		50-55	Lubricants, adhesives, pharmaceuticals
Vernicia fordii (Tung)	α - C18:3 acid 82%; C18:2 acid 8.5%; C18:1 acid 4%; C16:0 acid 5.5%		40-50	Wood finishing, Paints, adhesives

including Castor, Karanja, Jojoba, Tung, Neem, and Jatropha can be used to meet the demand for feedstock [60].

8.3.1 Castor as a Feedstock for Biolubricants

India has a lot of potential for creating biolubricants from sources of non-edible oil because of its extensive biodiversity. There are thousands of acres of waste land accessible in India that are home to wild castor plants. Contrary to popular belief, castor biolubricants are not well researched in India.

Castor, Ricinus communis L. (castor bean), is an oleaginous species that belongs to the Euphorbiaceae family of plants. It may currently be seen and farmed in temperate, tropical, and subtropical zones across the world. Following China and Brazil as the world's leading producers of castor oilseeds, India is the world's largest producer. India accounts for around 80% of world output, but just 15% of local consumption. In 2011, India produced roughly 2,339,000 MT of castor seed, with an average yield of 1-1.2 tonnes/ha [99].

Castor seed contains around 45–55% oil by weight on average. Because ricin is present in castor seeds, ricinine and specific allergens that are harmful to people and animals, also harmful to both humans and animals [100, 101]. Ricinoleic acid accounts for 80–90% of the oil, followed by 3–6% linoleic acid, 2–4% oleic acid, and 1–5% saturated fatty acids. Ricinoleic acid is the single unsaturated fatty acid found in natural vegetable origin oils having a hydroxyl group at the 12th position in the carbon chain [102]. The presence of this hydroxyl group, which is uncommon in vegetable oils, is responsible for castor oil's very high viscosity. Such a functional group lends added stability to the oil as well as to the oil derivatives by inhibiting the production of hydroperoxides [101, 103]. As compared with conventional vegetable origin oils, this hydroxyl group boosts lubricity, making it a good option for lubricants and as an additive in diesel fuel [43, 104–106].

Pharmaceutical and chemical industries have long used castor oil as a feedstock [107]. Castor oil is thick, pale yellow, non-drying oil having a bland flavor. It has a long shelf life as compared with other vegetable origin oils, and doesn't get rancid until exposed to extreme heat.

In most tropical and subtropical areas, castor crop thrives in huge quantities in the wild and is frequently found on wasteland. It is inexpensive, and the plant can withstand a variety of weather conditions. To produce optimal yields, castor plants require a temperature of 25 to 30 degrees Celsius with low humidity throughout the growing season. Castor seed contains ricin, a highly toxic polar chemical that causes stomach upset.

Ricin can cause discomfort, vomiting, and diarrhoea if consumed unintentionally, and as little as 1 mg of ricin can be lethal to adult humans. On the other hand, 100% refined oil does not contain ricin and can be used as a laxative in appropriate dosages. Likewise, the seed cake is toxic and unfit for animal consumption. People who work with castor cake and meal may get hypersensitivity responses including asthma and respiratory problems [108].

8.4 Ester-Based Lubricants from Vegetable Origin Oils (Edible and Non-Edible Oil)

Transesterification is the process of exchange of the alkyl group of one ester for that of another (Figure 8.5). Biolubricants derived from non-edible vegetable origin oil are environmentally beneficial lubricants because they are easily biodegradable, non-toxic, and emit no greenhouse gases [109]. Transesterification of vegetable origin oils (both edible and non-edible) results in producing various fatty acids. Transesterification of vegetable origin oils produces a variety of FAAE by reaction with low and high homologous alcohols in presence of both acid and base catalyst [20, 110]. Then the product FAAE after separation and purification is used as a biolubricant and also bio additive to low sulfur diesel fuels. These reactions require 1 mol of triglyceride and 3 mol of alcohol, according to stoichiometry. However, an excessive amount of alcohol is utilized to enhance alkyl ester quantity and facilitate phase separation from produced glycerol. The alcohol to oil molar ratio, reaction temperature, catalyst type (alkaline or

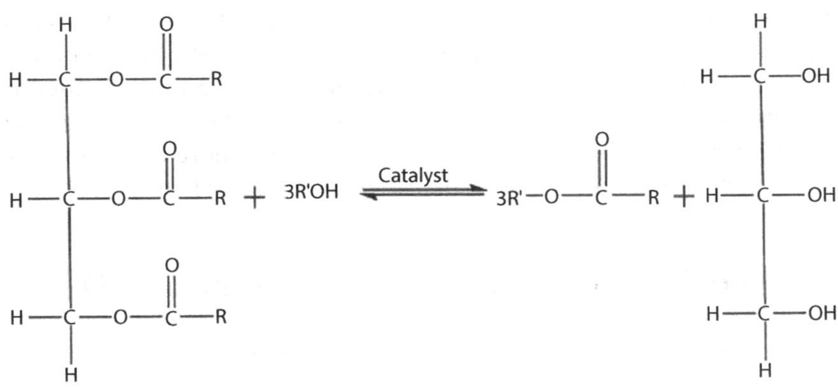

Figure 8.5 Transesterification of TAG to produce esters and glycerol.

acid), catalyst concentration, reactant purity, and free fatty acid content are all important factors to consider during transesterification.

8.4.1 Acid-Catalyzed Transesterification

In industrial settings, base-catalyzed transesterification is more frequently used than acid-catalyzed processes. An important factor to take into account is that the reaction catalyzed by the base is around 4,000 times faster than the reaction catalyzed by the acid. In this approach, acids, particularly RSO_3H, H_2SO_4, HCl, and H_3PO_4 catalyze the transesterification process, which may cause reactor corrosion. Although the product yield is high when using an acid catalyst, the reaction rate is slow, at temperatures above 100 °C and greater than 3 hours to complete the reaction. Excess amount of alcohol favors alkyl esters formation but too much alcohol makes glycerol recovery difficult, so it is critical to utilize the best possible alcohol to oil ratio [111–113].

8.4.2 Base-Catalyzed Transesterification

In presence of bases, esters generate an anionic intermediate that may dissociate into the original ester or produce a new ester, according to a proposed process for base-catalysed transesterification. Oils of vegetable origin undergo base-catalysed transesterification more frequently than acid-catalysed transesterification [111]. Alkaline metal alkoxides and hydroxides are used in the majority of commercial transesterification because of this and the fact that base catalysts damage industrial equipment less than acidic catalysts [111–113]. The most common types of alkalis are NaOH, KOH, carbonates, and alkoxides like methoxide, ethoxide, propoxide, and butoxide. Alkali metal alkoxides are the most effective catalysts because they can produce extremely high yields (> 98%) in a shorter amount of time (30 minutes), even at low molar concentrations (0.5 mol %). They are not suited for the majority of industrial procedures, nevertheless, as they do require a lack of water [53]. Metal alkoxides are more expensive than alkaline metal hydroxides despite being less active (KOH and NaOH).

8.5 Epoxide-Based Lubricants from Vegetable Oil

The term epoxide is used to describe cyclic ethers with three components in the epoxide ring. Unsaturated fatty acid undergoes epoxidation frequently by *in situ* reaction in presence of acid catalysts such as nitric acid, sulphuric acid, phosphoric acid, hydrochloric acid and strongly acidic ion exchange resins. Vegetable oil double bonds are utilized as reactive sites in coating purposes, and they may also be functionalized via epoxidation. The unsaturation of vegetable oil is a disadvantage, limiting its use as a lubricant in high-temperature applications. The use of epoxidized vegetable origin oil has grown in popularity in recent times. Furthermore, it has been demonstrated that plasticizers and additives for polymer PVC derived from vegetable origin oil improve heat and light resistance [114]. Epoxidized oil have epoxide or oxirane rings. An epoxidation reaction is a chemical reaction, a broad term for the reaction that occurs when an alkene is reacted with an organic peroxy acid to produce epoxide groups. In India, rice bran oil, cottonseed oil, groundnut oil, sunflower oil, rapeseed oil, sesame oil, palm kernel oil, coconut oil, linseed oil, castor oil, sal, neem oil, and karanja oil are all widely available. These oils by epoxidation can produce epoxide-based lubricants. Epoxidation of soyabean oil was done in presence of acidic ion exchange catalyst by *in situ* formed peracetic acid. The reaction was carried out with 1.1 mol of H_2O_2 as per unsaturated acid present in soyabean oil, 0.5 mol of CH_3COOH, 5% of ion exchange catalyst at 75°C and the optimum conversion was 91% with 5.99% epoxide content after 8 hrs [115]. Several researchers also synthesized epoxide-based lubricant from Jatropha, Karanja and castor oil using different catalyst and optimized reaction conditions.

8.5.1 Epoxide-Based Lubricants from Waste Cooking Oil

Renewable feedstocks are expected to gradually replace fossil fuels (oil, gas, and coal) as a form of energy and raw material resource for the chemical industry (51). To reduce environmental impact, the chemical industry made significant efforts throughout the twentieth century to replace petroleum-based feedstock with renewable resources. Because of their potential to synthesize various high-value domestic and industrial end products, waste cooking oils (WCO) have received a lot of attention among the various renewable sources [51]. WCO epoxidation was also carried out in a baffled flask with a plain bottom. WCO was first heated to 60°C before adding glacial acetic acid according to mole calculations, H_2SO_4 was added by weight, and then H_2O_2 was added into the reaction mixture drop by drop. The amount of H_2O_2 was calculated using the molecular weight of the WCO. A stirring speed of 1100–1500 rpm was used to facilitate uniform mixing. Following the complete addition of H_2O_2, the reaction was allowed to run for the desired amount of period. The optimized conditions for synthesis of epoxide-based lubricant from WCO was 2 moles H_2O_2,

1.5% catalyst loading, in which about 4.7% epoxide content was found after 6 h reaction time [78].

8.6 Conclusion

The world's limited crude oil supply, rising crude oil prices, and environmental challenges seem to be a grim reality check for the problems that the next generation will have to deal with. The current article focuses on vegetable oil–based lubricants. Plant oil–based lubricants are a desirable alternative to conventional petro-lubricants because of their physical properties like renewable nature, easily biodegradable, high lubricating action, and improved flash points. Esterification, transesterification and epoxidation are the better methods adopted for chemical modification of vegetable oils into biolubricants. After chemical modification the physico-chemical properties of vegetable oils change and that can be used as biolubricant to low sulfur diesel fuels. The global lubricant market will see a considerable replacement of the current products during the next 10 to 15 years, and it will unquestionably remain an interesting field of development for lubricant makers. The literature review in this article indicates that it is feasible to investigate vegetable oil–based resources for the production of lubricants and greases with performance that is comparable to petroleum-based lubricants.

References

1. Chatra, S. K. R., Kailas, S. V. 2012. Natural oil-based lubricants, Green Tribology, 287-329.
2. Nagendramma, P., Kaul, S. 2012. Development of ecofriendly/biodegradable lubricants: an overview. Renewable Sustainable Energy Reviews, 16, 764-774.
3. Mang T, Dresel W. Lubricants and lubrication. Second completely revised and extended edition Wiley-VCH; 2007.
4. Fuchs, M. 2000. The world lubricants market, current situation and outlook. In: 12th International colloquium on tribology.
5. Erhan, S. Z., Sharma, B. K., Doll, K. M. 2009. Opportunities for industrial uses of agricultural products Chimica Oggi, 27(1), 38-41.
6. Lubricant Market Report-2010; https://web.pucsl.gov.lk/english/wp-content/themes/pucsl/pdfs/lubricant_market_report_2010.
7. Bartz, W. J. 1998. Lubricants and the environment. Tribology International, 31, 35-47.

8. Adhvaryu, A., Liu, Z., Erhan, S. Z. 2005. Synthesis of novel alkoxylated triacylglycerols and their lubricant base oil properties. Industrial Crops & Products, 21, 113–119.
9. Ma, F., Hanna, M. A. 1999. Biodiesel production: a review. Bioresource Technology, 70, 1-15.
10. Rensselar, J. V., 2011. The perfect Biodegradable lubricant. Tribology and Lubrication Technology, June, 44-55.
11. Vander Waal, G., Kenbeek D. 1993. Testing application and future development of environmentally friendly ester base fluids. Journal of Synthetic Lubrication, 10(1), 67–83.
12. Lathi, P., Mattiasson, B. 2007. Green approach for the preparation of biodegradable lubricant base stock from epoxidized vegetable oil. Applied Catalysis B: Environmental, 69, 207–212.
13. Kasim, F. H., Harvey, A. P. 2011. Influence of various parameters on reactive extraction of Jatropha curcas L. for biodiesel production. Chemical Engineering Journal, 171(3), 1373-1378.
14. Kodali Dharma R. 2002. Process for modifying unsaturated triacylglycerol oils: resulting products and uses thereof. US 6,420,322.
15. Smith T.G. In: Gunderson RC, Hart WA, editors. 1962. Neopentyl polyol esters. New York: Reinhold Publishing Corporation.
16. Brown, M., Fotheringham, J., D., Hoyes, T., J. *et al.* 1997. Synthetic base fluids, Chemistry and technology of lubricants. 61.
17. Goyan, R. L., Melley R. E., Wissner P. A., Ong W. C. 1998. Biodegradable lubricants. Lubrication Engineering, 54, 10-17.
18. Fox, N. J., Stachowiak, G. W. 2007. Vegetable oil-based lubricants-A review of oxidation. Tribology International, 40, 1035-1046.
19. Battersby, N. S., Pack, S. E., Watkinson, R. J. 1992. A correlation between the biodegradability of oil products in the CEC L-33-T-82 and modified sturm tests. Chemosphere 24, 1998–2000.
20. Becker, R., and Knorr, A. 1996. An Evaluation of Antioxidants for Vegetable Oils at Elevated Temperatures, Lubrication Science 8(2), 96-117.
21. Asadauskas, S., Perez, J. M. and Duda, J. L. 1997, Lubrication Properties of Castor Oil- Potential Base Stock for Biodegradable Lubricants. Lubrication Engineering, 53(12), 35-40.
22. Hohn, B. R., Michaelis, K., Dobereiner, R. 1999. Load Carrying Capacity of Fast Biodegradable Gear Lubricants. Lubrication Engineering, 55(11), 15-38.
23. Honary, L. A. T. 1996. An investigation of the use of soybean oil in hydraulic systems. Bioresource Technology, 56, 41-47.
24. Adamczewska, J. Z., Wilson, D. 1997. Development of ecologically responsive lubricants. Journal of Synthetic Lubrication, 14, 129-142
25. Sivasankaran, G. A., Bisht, R. P. S., Jain, V. K., Gupta, M., Sethuramiah, A., Bhatia V. K. 1988. Jojoba-oil-based two-stroke gasoline engine lubricant. Tribology International, 21(6), 327-333.

26. Permsuwan, A., Picken, D. J., Seare K. D. R., Fox M. F. 1996. Engine wear monitoring vegetable oil lubricant. International Journal of Ambient Energy, 17, 157-161.
27. Beilafi-Bakoi, K., Dombi, A., Szaboi, L., Nagy, E. 1994. Triacylglycerol hydrolysis by lipase in a flat membrane bioreactor. Biotechnology Techniques, 8 (9), 671-674.
28. Madankar, C. S., Pradhan, S., Naik, S. N. 2013. Parametric study of reactive extraction of castor seed (*Ricinus communis* L.) for methyl ester production and its potential use as bio lubricant. Industrial Crops and Products, 43, 283-290.
29. Sharma, B., Doll, K. M., Erhan, S. Z. 2008. Ester hydroxy derivatives of methyl oleate: tribological, oxidation and low temperature properties. Bioresource Technology, 99, 7333-7340.
30. Randles, S. J., Wright, M. 1992. Environmentally considerate ester lubricants for the automotive and engineering industries. Journal of Synthetic Lubrication, 9, 145–161.
31. Singh, A. K. 2011. Castor oil-based lubricant reduces smoke emission in two-stroke engines. Industrial Crops and Products, 33, 287-295.
32. Laudani, C. G., Habulin, M., Knez, Z., Della Porta, G., Reverchon, E. 2007. Lipase catalyzed long chain fatty ester synthesis in dense carbon dioxide: kinetics and thermodynamics. The Journal of Supercritical Fluids, 41, 92-101.
33. Laudani, C. G., Habulin, M., Knez, Z., Porta, G. D., Reverchon, E. 2007. Immobilized lipase-mediated long-chain fatty acid esterification in dense carbon dioxide: bench-scale packed-bed reactor study. The Journal of Supercritical Fluids, 41, 74-81.
34. Drown, D. C., Harper, K., Frame, E., 2001. Screening vegetable oil alcohol esters as fuel lubricity enhancers. Journal of the American Oil Chemists' Society, 78, 579-584.
35. Daniel, S. G. 1951. The adsorption on metal surfaces of long chain polar compounds from hydrocarbon solutions. Transactions of the Faraday Society, 47, 1345-1359.
36. Kipling, J. J. 1965. Adsorption from Solutions of Non-Electrolytes, Academic Press, London, 55.
37. Maleque, M. A., Masjuki, H. H., Sapuan, S. M. 2003. Vegetable-based biodegradable lubricating oil additives. Industrial Lubrication and Tribology, 55(3), 137-143.
38. Masjuki, H. H., Maleque, M. A. 1997b. Investigation of the anti-wear characteristics of palm oil methyl ester using a four-ball tribometer test. Wear, 206, 179-186.
39. Sharma, B. K., Adhavaryu, A., Liu, Z., Erhan S. Z. 2006. Chemical modifications of vegetable oil for lubricant application, Journal of the American Oil Chemists' Society, 83(2), 129-136.

40. Dalai, A. K., Kulkarni, M. G., Meher, L. C. 2006. Biodiesel Productions from Vegetable Oils using Heterogeneous Catalysts and Their Applications as Lubricity Additives. IEEE, 1-8.
41. Knothe, G., Steidley, K. R., 2005. Lubricity of Components of Biodiesel and Petrodiesel. The Origin of Biodiesel Lubricity. Energy & Fuels, 19, 1192-1200.
42. Goodrum, J. W., Geller, D. P. 2005. Influence of Fatty Acid Methyl Ester from Hydroxylated Vegetable Oils on Diesel Fuel Lubricity. Bioresource Technology, 96(7), 851-855.
43. Hincapie, G., Mondragon, F., Lopez, D. 2011. Conventional and *in situ* transesterification of castor seed oil for biodiesel production. Fuel, 90, 1618-1623.
44. Kusdiana, D., Saka, S. 2001. Kinetics of transesterification in rapeseed oil to biodiesel fuel as treated in supercritical methanol. Fuel, 80(5), 693-698.
45. Onkawa, S., Konishi, A., Hanato, H., Ishihama, K., Tanaka, K., Awamura, M. 1995. Oxidation and corrosion characteristics of vegetable base biodegradable hydraulic oils. SAE Technical Paper, 951038, 55-63.
46. Rhodes, B., N., Mammel, W., Landis, P., Erickson, F., L. 1995. Water rejection of vegetable oil base stocks for tractor/hydraulic fluids. SAE Technican Paper, 952073, 1-4.
47. Hang, X., Yang, H., 1999. Model for a cascade continuous epoxidation process. Journal of the American Oil Chemists' Society, 76, 89-92.
48. Goud, V. V., Patwardhan, A. V., Dinda, S., Pradhan, N. C. 2007. Epoxidation of karanja (Pongamia glabra) oil catalysed by acidic ion exchange resin. European Journal of Lipid Science and Technology, 109, 575-584
49. Klaas, M. R., Warwel, S., 1999. Complete and partial epoxidation of plant oils by lipase catalyzed perhydrolysis. Industrial Crops and Products, 9, 125-132.
50. Balabin, R. M., Syunyaev, R. Z., Karpov, S. A. 2007. Quantitative measurement of ethanol distribution over fractions of ethanol-gasoline fuel. Energy Fuels, 21, 2460–2465.
51. Wang, R., Zhou, W. W., Hanna, M. A., Zhang, Y. P., Bhadury, P. S., Wanga, Y. *et al.* 2012. Biodiesel preparation, optimization, and fuel properties from non-edible feedstock, Datura stramonium L. Fuel, 91, 182-186.
52. Zeng, J., Wang, X., Zhao, B., Sun, J., Wang, Y. 2009. Rapid *in situ* transesterification of sunflower oil. Industrial & Engineering Chemistry Research, 48, 850-856
53. Georgogianni, K. G., Kontominas, M. G., Pomonis, P. J., Avlonitis, D., Gergis, V. 2008. Conventional and *in situ* transesterification of sunflower seed oil for the production of biodiesel. Fuel Processing Technology, 89, 503-509
54. Haas, M. J., Scott, K. M., Marmer, W. N., Foglia, T. A. 2004. *In situ* alkaline transesterification: an effective method for the production of fatty acid esters from vegetable oils. Journal of the American Oil Chemists' Society, 81(1), 83-89.
55. Nagendramma, P., Kaul S. 2008. Study of complex esters as automotive gear lubricants. Journal of Synthetic Lubrication, 25, 131-136.

56. Shuit, S. H., Lee, K. T., Kamaruddin, A. H., Yusup, S. 2010. Reactive extraction and *in situ* esterification of Jatropha curcas L. seeds for the production of biodiesel. Fuel, 89, 527-530.
57. Kaul, S., Porwal, J., Garg, M. O. 2010. Parametric study of jatropha seeds for biodiesel production by reactive extraction. Journal of the American Oil Chemists' Society, 87, 903-908.
58. Siler-Marinkovic, S., Tomasevic, A. 1998. Transesterification of sunflower oil *in situ*. Fuel, 77(12), 1389-1391.
59. Qian, J., Wang, F., Liu, S., Yun Z. 2008. *In situ* alkaline transesterification of cottonseed oil for production of biodiesel and nontoxic cottonseed meal. Bioresource Technology, 99, 9009-9012.
60. Ozgul Yucel, S., Turkays S. 2002. Variables Affecting the Yields of Methyl Esters derived from *in situ* Esterification of Rice Bran Oil. Journal of the American Oil Chemists' Society, 79(6), 611-614.
61. Haas, M. J., Scott, K. M. 2007. Moisture removal substantially improves the efficiency of *in situ* biodiesel production from soybeans. Journal of the American Oil Chemists' Society, 84(2), 197-204.
62. Vicente, G., Martınez, M., Aracil, J., 2004. Integrated biodiesel production: a comparison of different homogeneous catalysts systems. Bioresour. Technol. 92, 297–305.
63. Syama, A. M., Yunusb, R., Ghazia, T. I. M., Choong, T. S. Y., 2012. Synthesis of Jatropha curcas oil-based biodiesel in a pulsed loop reactor. Ind. Crops Prod. 37, 514–519.
64. Uosukainen, E., Lamsa, M., Linko, Y. Y., Linko, P., Leisola, M., 1999. Optimization of enzymatic transesterification of rapeseed oil ester using response surface and principal component methodology. Enzyme Microb. Technol. 25 (3–5), 236–243.
65. Dalai, A. K., Kulkarni, M. G., Meher, L. C. 2006. Biodiesel Productions from Vegetable Oils using Heterogeneous Catalysts and Their Applications as Lubricity Additives. IEEE, 1-8.
66. Meher, L. C., Sagar, V., Naik, S. N. 2006. Technical aspects of biodiesel production by transesterification: A review. Renewable Sustainable Energy Reviews, 10(3), 248-268.
67. Achten, W. M. J., Verchot, L., Franken, Y. J., Mathijs, E., Singh, V. P., Aerts, R. *et al.* 2008. Jatropha biodiesel production and use. Biomass and Bioenergy, 32(12), 1063-1084.
68. Tiwari, A. K., Kumar, A., Raheman, H., 2007. Biodiesel production from jatropha oil (Jatropha curcas) with high free fatty acids: An optimized process. Biomass and Bioenergy, 31, 569-575.
69. Chang, H. M., Liao, H. F., Lee, C. C., Shieh, C. J. 2005. Optimized synthesis of lipase catalyzed biodiesel by Novozym 435. Journal of Chemical Technology and Biotechnology, 80, 307-312.

70. Gog, A., Roman, M., Tosa, M., Paizs, C., Irimie, F. D. 2012. Biodiesel production using enzymatic transesterification-Current state and perspectives. Renewable Energy 39, 10-16.
71. Paizs, C., Tosa, M., Bodai, V., Szakacs, G., Kmecz, I., Simandi, B., *et al.* 2003. Kinetic resolution of 1-(benzofuran-2-yl) ethanols by lipase-catalyzed enantiomer selective reactions. Tetrahedron: Asymmetry, 14, 1943-1949.
72. Lu, J., Nie, K., Xie, F., Wang, F., Tan, T. 2007. Enzymatic synthesis of fatty acid methyl esters from lard with immobilized Candida sp. 99-125. Process Biochemistry, 42, 1367-1370.
73. Nie, K., Xie, F., Wang, F., Tan, T., 2006. Lipase catalyzed methanolysis to produce biodiesel: optimization of the biodiesel production. Journal of Molecular Catalysis B: Enzymatic, 43, 142-147.
74. Mittelbach, M. 1990. Lipase catalyzed alcoholysis of sunflower oil. Journal of the American Oil Chemists' Society, 67, 168-170.
75. Oliveira, J. V., Oliveira, D. 2000. Kinetics of the enzymatic alcoholysis of palm kernel oil in supercritical CO_2. Industrial and Engineering Chemistry Research, 39, 4450-4454.
76. Ciftci, O. N., Temelli, F. 2011. Continuous production of fatty acid methyl esters from corn oil in a supercritical carbon dioxide bioreactor. The Journal of Supercritical Fluids, 58, 79- 87.
77. Mesiano, A. J., Beckman, E. J., Russell, A. J. 1999. Supercritical biocatalysis. Chemical Reviews, 99, 623-634
78. Beckman, E. J. 2004. Supercritical and near-critical CO2 in green chemical synthesis and processing. The Journal of Supercritical Fluids, 28, 121-191.
79. Black, H. 1996. Supercritical carbon dioxide: the "greener" solvent. Environmental Science and Technology, 30(3), 124A-128A.
80. Ikushima, Y. 1997. Supercritical fluids: an interesting medium for chemical and biochemical processes. Advances in Colloid and Interface Science, 71–72, 259-280.
81. Wen, D., Jiang, H., Zhang, K. 2009. Supercritical fluids technology for clean biofuel production. Progress in Natural Science, 19, 273-284.
82. Wei, D., Wei, L., Ting, S., Xin, C., Dehua, L. 2008. Perspectives for biotechnological production of biodiesel and impacts. Applied Microbiology and Biotechnology, 79, 331-337.
83. Kaewthong, W., Sirisansaneeyakul, S., Prasertsan, P., H-Kittikun, A. 2005. Continuous production of monoacylglycerols by glicerolysis of palm olein with immobilized lipase. Process Biochemistry, 40, 1525-1530.
84. Roca, E., Meinander, N., Hahn-Hagerdal, B. 1996. Xylitol production by immobilized recombinant Saccharomyces cerevisiae in a continuous packed-bed bioreactor. Biotechnology and Bioengineering, 51(3), 317-326.
85. Mu, H., Xu, X., Hoy, C. E. 1998. Production of specific-structured triacyglycerols by lipase-catalysed interesterification in a laboratory scale continuous reactor. Journal of the American Oil Chemists' Society, 75(9), 1187-1193.

86. Xu, X., Balchen, S., Hoy, C. E., Adler-Nissen, J. 1998. Production of specificstructured lipids by enzymatic interesterification in a pilot continuous enzyme bed reactor. Journal of the American Oil Chemists' Society, 75(11), 1573-1579.
87. Shimada, Y., Suenaga, M., Sugihara, A., Nakai, S., Tominaga, Y. 1999. Continuous production of structured lipid containing γ-linolenic and caprylic acids by immobilized Rhizopus delemar lipase. Journal of the American Oil Chemists' Society, 76(2), 189-193.
88. Varma, M. N., Madras, G. 2007. Synthesis of biodiesel from castor oil and linseed oil in supercritical fluids. Industrial and Engineering Chemistry Research, 46, 1-6
89. Madras, G., Kolluru, C., Kumar, R. 2004. Synthesis of biodiesel in supercritical fluids. Fuel, 83, 2029-2033.
90. Rathore, V., Madras, G. 2007. Synthesis of biodiesel from edible and non-edible oils in supercritical alcohols and enzymatic synthesis in supercritical carbon dioxide. Fuel, 86(17-18), 2650-2659.
91. Lee, J. H., Kwon, C. H., Kang, J. W., Park, C., Tae, B., Kim, S. W. 2009. Biodiesel production from various oils under supercritical fluid conditions by Candida Antartica lipase B using a stepwise reaction method. Applied Biochemistry and Biotechnology, 156, 454-464.
92. Fomuso, L. B., Akoh, C. C. 2002. Lipase-catalysed acidolysis of olive oil and caprylic acid in a bench-scale packed bed bioreactor. Food Research International, 35(1), 15-21.
93. Diasakou, M., Louloudi, A., Papayiannakos, N. 1998. Kinetics of the non-catalytic transesterification of soybean oil. Fuel, 77(12), 1297-1302.
94. Ataya, F., Dube, M. A., Ternan, M. 2006. Single-phase and two-phase base-catalyzed transesterification of canola oil to fatty acid methyl esters at ambient conditions. Industrial and Engineering Chemistry Research, 45, 5411-5417.
95. Azam, M. M., Waris, A., Nahar, N. M. 2005. Prospects and potential of fatty acid methyl esters of some non-traditional seed oils for use as biodiesel in India. Biomass and Bioenergy, 29, 293-302.
96. Biermann, U., Friedt, W., Lang, S., Lühs, W., Machmüller, G., Metzger, J. O., Klaas, M. R., Schäfer, H. J., Schneider, M. P. 2000. New synthesis with oils and fats as renewable raw materials for the chemical industry. Angewandte Chemie International Edition, 39, 2206–2224.
97. Masjuki, H. H., Sapuan, S. M., 2003. Vegetable based biodegradable lubricating oil additives, Industrial Lubrication and Tribology, 55(3), 137-143.
98. Om Tapanes, N., Gomes Aranda, D., de Mesquita Carneiro, J., Ceva Antunes, O. 2008. Transesterification of Jatropha curcas oil glycerides: theoretical and experimental studies of biodiesel reaction. Fuel, 87, 2286-2295.
99. http://faostat.fao.org, 2011.
100. Economic Editors' Conference 19-20 Oct, 2011

101. Ogunniyi, D. S. 2006. Castor oil: A vital industrial raw material. Bioresource Technology, 97, 1086-1091.
102. Hayes, D. G., Carlson, K. D., Kleiman, R. 1996. The isolation of hydroxyl acids from lesquerella oil lipolysate by a saponification/extraction technique. Journal of the American Oil Chemists' Society, 73(9), 1113-1119.
103. Encinar, J. M., González, J. F., Pardal, A., 2012. Transesterification of castor oil under ultrasonic irradiation conditions. Preliminary results, Fuel Processing Technology, 103, 9-15.
104. Goodrum, J. W., Geller, D. P. 2005. Influence of Fatty Acid Methyl Ester from Hydroxylated Vegetable Oils on Diesel Fuel Lubricity. Bioresource Technology, 96(7), 851-855.
105. Pena, R., Romero, R., Martinez, S., Ramos, M., Martinez, A., Natividad, R. 2009. Transesterification of castor oil: effect of catalyst and co-solvent. Industrial and Engineering Chemistry Research, 48, 1186-1189.
106. Madankar, C. S., Pradhan, S., Naik, S. N. 2013. Green synthesis of biolubricant base stock from canola oil. Industrial Crops and Products, 44, 139-144.
107. Scholz, V., Nogueira da Silva, J. 2008. Prospects and risks of the use of castor oil as a fuel. Biomass and Bioenergy, 32, 95-100.
108. http://www.aseanfood.info
109. Singh, Y., Farooq, A., Raza, A., Mahmood, M. A., & Jain, S. (2017). Sustainability of a non edible vegetable oil based bio-lubricant for automotive applications: A review. *Process Safety and Environmental Protection, 111*, 701–713.
110. Zaher, F.A. and Nomany, H.M. (1988) Vegetable Oils as Lubricants. Grasas y Aceites, 39, 235-238
111. Freedman, B., Pryde, E. H., Mounts, T. L. 1984. Variables affecting the yields of fatty esters from transesterified vegetable oils. Journal of American Oil Chemists Society, 61(10), 1638-1643.
112. McHugh, M., Krukonis, V. J. 1986. Supercritical Fluid Extraction, Butterworth, MA, 3, 118-230.
113. Schuchardt, U., Sercheli, R., Vargas, R. M. 1998. Transesterification of Vegetable Oils: a Review, Journal of the Brazilian Chemical Society, 9(1), 199-210.
114. Saurabh, T. & Patnaik, M. & Bhagt, S.L. & Renge, Vilas. 2011. Epoxidation of Vegetable oils: A review. International Journal of Advanced Engineering Technology. 2. 491-501.
115. Fiser, S.S., Jankovic. M., Petrovic, Z., 2001Kinetics of *in situ* epoxidation of soybean oil in bulk catalyzed by ion exchange resin. JAOCS, 78, 725-732.

9

Vegetable Oils in Pharmaceutical Industry

Shruti Mishra[1]*, Shubhankar Anand[2] and Achyut Pandey[3]

[1]*School of Basic and Applied Sciences, Galgotias University, Uttar Pradesh, India*
[2]*School of Biochemical Engineering, Indian Institute of Technology (BHU) Varanasi, Uttar Pradesh, India*
[3]*School of Biotechnology, Jawaharlal Nehru University, New Delhi, India*

Abstract

Since the dawn of time, several uses for vegetable oils have been known to exist such as in the nutritional, industrial, biofuel and pharmaceutical industries. Vegetable oils have significant pharmacological properties as antioxidant, anti-inflammatory, antifungal, antipyretic, analgesic, antibacterial, and anti-hyperglycemia that have gained popularity in the pharmaceutical industry. These oils are incredibly popular for treatment of many illnesses among the general public in nations like India due to their significant efficacy, manageable cost, low adverse effect profile, and favourable efficacy. Vegetable oils like Sunflower, sweet almond, olive and coconut oils, etc., were successfully used in the improvement of liver functioning, boosting immunity, encouraging hair development, and treating skin and scalp conditions, superficial wounds, sore throats, obesity, cancer and many other pathological conditions.

Numerous different oils have been employed, and newer ones are always being developed. Vegetable oils have been incorporated into nanoemulsions which is thought to be a viable method for making these natural substances more usable and for amplifying their effects. A promising area for the treatment of several illnesses is the creation of lipid nanoparticles using vegetable oils.

The purpose of this review is to report on some properties and uses and assess current studies that have been published in the literature about the use of vegetable oils with therapeutic properties.

Keywords: Vegetable oil, lipid carrier, pharmaceutical industry, antioxidant, anticancerous

**Corresponding author*: shruti.mishra@galgotiasuniversity.edu.in

Divya Bajpai Tripathy, Anjali Gupta, Pooja Agarwal, Anuradha Mishra, and Arvind Kumar Jain (eds.) Oils and Fats as Raw Materials for Industry, (231–266) © 2024 Scrivener Publishing LLC

9.1 Introduction

Vegetable oils (VOs) are important biorenewable resources that may be derived from a variety of plants. They are typically identified by the biological source of their production, such as oil from soybean and palm plants. These are the fats typically obtained from seeds of the plants like soybean oil, sesame oil, rapeseed oil, etc., using the process of simple pressing and/or solvent extraction or less frequently from other parts of plant fruits (such as olive oil, palm oil, etc.) [1]. Their composition is extremely diverse and complex and depends on the place of origin, the crop properties, the type of growing season, and the techniques of production [2]. Vegetable oils are mostly made up of triglycerides, which are made when glycerol and other fatty acids react. Vegetable oil triglycerides include fatty acids with carbon chains between 14 and 22 carbons long and various degrees of unsaturation [3, 4]. Triglycerides are classified based on the level of fatty acid saturation into mono-, and poly-unsaturated and saturated fatty acids. The length of the fatty acid chain, the degree of unsaturation, and the stereochemistry of the double bonds are three factors that affect the properties of VO [5]. Vegetable oil's physical state is dependent on the distribution and composition of its fatty acids. At room temperature, the majority of vegetable oils are liquid. Vegetable oils with higher melting points typically have a longer fatty acid chain, fewer carbon-carbon double bonds, an E (trans) structure, and conjugated carbon-carbon double bonds [6].

The demand for vegetable oils in various areas such as for industrial and domestic uses has significantly expanded due to the growing global population. The use of herbal medicines, phytonutrients, and nutraceuticals continues to gain widespread recognition and public attention. Vegetable oils made from plants have been recognized as one of the oils with a high nutritional value. The use of natural products in pharmaceuticals and cosmetics industries has become increasingly popular in recent years. Because of this, the market for oils derived from indigenous plants at the national and international levels have expanded quickly and significantly. In addition to being used for edible products in both virgin and refined forms, VOs are essential for the treatment and prevention of illnesses like diabetes, inflammation, atherosclerosis and cancer [7, 8].

9.1.1 Pharmaceutical Uses of Vegetable Oil

Vegetable oils have a unique chemical composition and a variety of pharmacological qualities that make it possible to investigate their potential as a treatment for several diseases. Numerous studies have been done on their qualities, offering up novel therapeutic possibilities for a variety of ailments [9]. They are known to be used as anti-cancer, anti-fungal, antibacterial, anti-aggregation, anti-inflammatory, anti-fibrillation, and antioxidant agents due to their pharmacological properties [10–14]. According to a study spanning over the last 10 years the main therapeutic uses of products containing vegetable oils in nanostructured forms are anti-inflammatory, antibacterial, anticancer, and antioxidant agents.

Vegetable oils can also be employed in lipid compositions as a carrier. Vegetable oil has a biological impact, thus when mixed with the active ingredient, the product can have synergistic effects. Additionally, even in the presence of component incompatibility, the use of vegetable oil as a lipid carrier can lower the risk of irritations and allergies [15]. In a study by Arbain *et al.*, palm oil was employed as a lipid carrier to operate in combination with quercetin, due to its anti-cancer action, and a synergistic effect was discovered [16]. Similarly, oil from Fructus bruceae plant (as a lipid carrier) enhanced the anticancer efficacy of evodiamine in research by Zhao *et al.* [17]. Due to its anti-inflammatory properties, seed oil from a pomegranate plant was employed as a lipid carrier to enhance the effects of ketoprofen [18]. Organic UV filters octocrylene and avobenzone photoprotective properties were enhanced by bocaiuva oil [19]. Due to their anticancer qualities, ginger and frankincense essential oils were utilised to boost the activity of mitomycin C, the drug used in chemotherapy [20]. In a study by Kumar *et al.*, a derivative of colchicoside, thiocolchicoside (a natural glycoside from the plant Gloriosa superba), was used as a lipid carrier in a formulation that contained oil from linseed plant, fostering a synergistic effect, for treating rheumatologic illnesses [21]. Additionally, when combined with amaranth oil, nanocarriers containing UV-A and UV-B filters demonstrated a greater UV absorption impact [22]. Vegetable oils have so far demonstrated a good method for enhancing the therapeutic effects.

In this chapter we have tried to elaborately present various uses of oils from plants especially in the pharmaceutical industry.

9.2 Olive Oil

Traditional medicine has given olives a lot of attention, including the use of several parts of olive plants such as leaves to cure malaria and fever [23, 24]. Virgin olive oil, the primary source of fat in the Mediterranean diet, contains mono-unsaturated fatty acids and phenolic compounds that may help to reduce the risk of obesity, metabolic syndrome, and chronic illnesses, including type II diabetes, hypertension, and high cholesterol [25, 26]. Olive oil consumption has been suggested to have preventive benefits against the development of hypertension and hyperlipidemia [27]. In this context, consuming oleic acid has been linked to lower levels of platelet aggregation, inflammation, and low-density lipoprotein (LDL) cholesterol oxidation. Additionally, it appears that virgin olive phenolic chemicals are highly bioavailable, which could account for their favorable impact on plasma lipoprotein profiles [28]. Their intake is associated with greater plasma high-density lipoprotein (HDL) and reduced triglyceride levels, total cholesterol, and LDL cholesterol levels [29, 30]. Inhibition of platelet aggregation, a decline in pro clotting factors connected to coronary heart disease, and a drop in biomarkers of inflammation linked to cardiovascular disease are other effects of olive oil phenolics [31]. It has also been suggested that olive oil has a sound impact on immune system health, including defense against autoimmune conditions like rheumatoid arthritis [32]. Consuming a Mediterranean diet has also been linked to a lower risk of developing colon, prostate, and breast cancers [33–35]. Phenolic compounds and Oleic acid, which may lessen the oxidation of protein, deoxyribonucleic acid (DNA), and lipids that contribute to tumour growth, have been linked to the anticarcinogenic action of olive oil [36, 37].

9.2.1 Antioxidant Property

The term "oxidative" refers to an imbalance between the body's antioxidant defenses and reactive oxygen species (ROS) generation [38]. Numerous illnesses, including cancer, cardiovascular disease, atherosclerosis, diabetes, and neurodegenerative illnesses like Parkinson's and Alzheimer's disease, have been linked to oxidative stress [39]. The phenolic acids and polyphenols in virgin olive oil have significant antioxidant activity, which contributes to both the oil's stability and possible health benefits. Because phenolic substances have an aromatic ring with an attached hydroxyl group that scavenges free radicals, they have potent antioxidant properties. Hydrophilic phenolic compounds exhibit stronger antioxidant activity in oil bulk media than hydrophobic antioxidants. The phenolic compounds that have been isolated from virgin olive oil have a variety of antioxidant effects depending on their chemical composition and the quantity of hydroxyl groups they contain [40]. The three phenolic compounds with the greatest significant antioxidant activity are hydroxytyrosol, decarboxymethyl oleuropein aglycone, and oleuropein aglycone. Using the 2,2-diphenyl-1-picrylhydrazyl (DPPH) technique, virgin olive oil's extracted oleuropein and hydroxytyrosol showed substantial free radical scavenging activity. However, hydroxytyrosol had more potent antioxidant activity since it contains the o-dihydroxy phenyl moiety. Consuming phenolic compounds from olive oil has shown promise for improving health by lowering oxidative stress and raising plasma's overall antioxidant capacity [41]. Olive oil high in phenols has also been associated with lower levels of LDL oxidation, oxidised glutathione, lipid peroxides, and F2-isoprostanes, suggesting that it may help prevent cardiovascular disease [42].

Deoxyribonucleic acid oxidative damage is a crucial contributor to the development and spread of malignancies. Studies using extract of phenolic compounds from extra virgin olive oil on humans, animals, and *in vitro* cells have shown a 30% reduction in DNA oxidation. In this line, it has been demonstrated that the hydroxytyrosol phenols extracted from olive oil and caffeic acid shield DNA against oxidation [43].

9.2.2 Antidiabetic Properties

The most widespread public health issue in the world is diabetes mellitus. According to estimates, 347 million people worldwide suffer from diabetes, with type II diabetes accounting for 95% of all cases [44]. Hyperglycemia brought on by insulin resistance or inadequate insulin production characterizes type II diabetes, a disease in the metabolism of lipids and carbohydrates [45]. Due to the hypothesis that oxidative stress contributes to the onset of both type I and type II diabetes, interest in the use of natural antioxidants in the treatment of diabetes has grown [46]. As a result, there is increasing interest in creating biological therapeutic agents for type II diabetes that focus on using phytochemicals and medicinal plants to prevent and treat this condition [47]. By inhibiting carbohydrate-digesting enzymes like amylase and glucosidase, treatments for type II diabetes and its complications can manage postprandial hyperglycemia, which can be a significant factor in the disease's consequences [48]. The salivary amylase enzyme starts the breaking down of carbohydrates in the mouth, but the pancreatic amylase enzyme is crucial for the breakdown of carbohydrates in the small intestine lumen. Maltose, dextrins, and maltotriose are released when alpha-amylase hydrolyzes endo 1,4-glucosidic connections. Brush boundary glucosidase enzymes like maltase, sucrase, and lactase hydrolyze these disaccharides, releasing the monosaccharides glucose, galactose, and fructose in the process. Treatment for type II diabetes has been suggested for alpha-amylase inhibitors [49]. Numerous *in vitro* studies have looked at the significance of various natural plant phenolic extracts in the prevention and treatment of type II diabetes as well as their ability to inhibit the enzyme amylase [50]. However, the chemical makeup and structural characteristics of phenolic compounds affect their affinity for and capacity to bind to amylase. This capacity of phenolic compounds to conjugate with amylase may be used to explain how phenolic compounds slow down amylase action [51]. The phenolic compounds with the highest-amylase inhibitory actions include flavonoids, oleuropein aglycone, and phenolic extracts from eight virgin olive oils [52]. Oleuropein aglycone was shown to have the strongest amylase inhibitory activity [53]. Due to its positive impact on glucose metabolism, the Mediterranean diet has been linked to decreased type II diabetes incidences and the avoidance of diabetic complications. When compared to low fat diets, a Mediterranean diet high in virgin olive oil has been shown to cut the risk of type II diabetes in half [54]. In the past, olive leaves were used to treat type II diabetes. Extracts from olive leaves improved glucose homeostasis in people with type II diabetes in a well-conducted study [55]. It has been shown that extracts from olive leaves rich in hydroxytyrosol and oleuropein have antidiabetic properties and reduce plasma glucose levels in diabetic rats [56].

9.2.3 Antihypertensive Properties

Hypertension, also referred to as high systolic and diastolic blood pressure, is a major risk factor for cardiovascular disease [57]. Moreover, two-thirds of people with type II diabetes develop hypertension, making it one of the chronic consequences of diabetes [58].

The control of blood pressure is dependent mainly on the renin-angiotensin-aldosterone system. Angiotensinogen is transformed into angiotensin I by the kidney enzyme rennin, which is then changed into angiotensin II by the angiotensin-converting enzyme (ACE) [59]. Blood arteries constrict as a result of angiotensin II vasoconstriction impact, raising blood pressure [60]. Additionally, angiotensin II boosts the kidneys' aldosterone hormone production, enhancing salt and water absorption and raising blood pressure [61]. The main antihypertensive drugs are ACE inhibitors, which have considerably decreased mortality in hypertension patients [62].

Modifying one's lifestyle to include a healthy diet, exercise, and weight control helps reduce hypertension. Numerous studies have demonstrated the potential of dietary phenolic compounds to function as ACE inhibitors and regulate blood pressure [63]. Flavonoid-rich plant diets have effectively inhibited ACE both *in vivo* and *in vitro*. A diet high in virgin olive oil from the Mediterranean region has been demonstrated to lower arterial blood pressure. Virgin olive oil consumption has been demonstrated to lower systolic and diastolic blood pressure in both hypertensive elderly patients and in healthy people, which is helpful for hypertension [64]. It has been established that the antihypertensive properties of olive oil are a result of both the substance's high oleic acid content and the presence of polyphenolic compounds. When compared to patients who ate olive oil free of polyphenols, hypertensive individuals who consumed virgin olive oil rich in polyphenol components experienced considerably lower blood pressure and enhanced endothelial function. The significant *in vitro* ACE inhibitory action was found to be associated with olive oil with the highest phenol concentration [65]. Eight different types of virgin olive oil's phenolic components were used to demonstrate how ACE is inhibited [66].

It has also been demonstrated that consuming olive leaf extract lowers blood pressure, and oleacein, a secoiridoid derived from olive leaves, exhibits strong *in vitro* ACE inhibitory action [67]. It has been demonstrated that consuming olive leaf extract can help hypertensive individuals lower their blood pressure [68].

9.2.4 Anticarcinogenic Properties

Cancer is characterized by aberrant and unregulated cell division that results in the development of cancerous cells [69]. According to estimates, cancer accounts for 25% of fatalities in the United States. It is suggested that a healthy body weight, a well-balanced diet full of fruits and vegetables, and regular exercise are critical factors in cancer prevention [70]. A third of malignancies are thought to be preventable through modifying dietary practices. It has been assumed that dietary antioxidants help prevent cancer by lowering DNA damage and ROS production and decreasing the peroxidation of membrane lipids [71].

Due to their capacity to shield DNA from damage, phenolic compounds may help prevent cancer [72]. DNA damage is a significant factor in the development and start of several tumour types. Olive oil consumption has been related to the Mediterranean diet's reduced risk of developing a number of malignancies [73]. Olive oil phenolic compounds have anticarcinogenic and antiproliferative characteristics because they promote tissue free radical defence mechanisms that scavenge free radicals created either endogenously or as a result of exposure to UV light and radiation [74]. The phenolic compounds in olive oil have been demonstrated to have anticarcinogenic properties in numerous *in vivo* and *in vitro* studies [75].

In this context, it has been demonstrated that the phenolics in virgin olive oil can inhibit the growth of human leukemia cells and cause them to undergo programmed cell death (apoptosis) [76]. Hydroxytyrosol has demonstrated anticarcinogenic properties against leukemia and breast tumours in animal models [77]. Flavonoids, lignans, and secoiridoids from virgin olive oil have demonstrated antitumorigenic effects in human mammary cancer cells [78]. Similarly, against human breast carcinoma cells, Lutolein, apigenin, and oleuropein have all shown antiproliferative effects [79].

As one of the leading causes of cancer mortality globally, colorectal cancer poses a serious public health concern and claims the lives of half a million individuals yearly [80]. The primary risk factors for colorectal cancer include diets high in fat and poor in fibre, sedentary lifestyles, and reduced consumption of fruits and vegetables [81]. The phenolics found in virgin olive oil have been shown in numerous studies to have antiproliferative effects on human colon cancer cells. For instance, Gill *et al.* (2005) discovered that olive oil phenolic extracts inhibit the onset, progression, and metastatic phases of colorectal cancer [82]. Similar to this, colorectal cancer cell lines' viability was reduced by extra virgin olive oil pinoresinol extracts. In the HT-29 colon cancer cell line, hydroxytyrosol from virgin olive oil has been demonstrated to produce cell cycle arrest and encourage apoptosis.

9.3 Rice Bran Oil

After removing the chaff or rice husk, rice bran oil is produced from the tough outer brown layer of rice [83]. It is renowned for having a mild flavour and a high smoke point of 232 °C, or 450 °F, which makes it suitable for high-temperature cooking techniques like stir-frying and deep-frying [84]. For a variety of reasons, rice bran oil is perhaps the healthiest option among cooking oils [85].

The ratio of mono-unsaturated fats (MUFA) to polyunsaturated fats (PUFA) is perfect [86]. Nearly a 1:1 ratio of 37% polyunsaturated to 45% mono-unsaturated fats may be found in rice bran oil. Since rice bran oil is derived from bran, it contains a lot of the antioxidant Vitamin E [87]. It may be manufactured in India and is less expensive than other cooking oils. Most notably in Bangladesh, Japan, India, and China, it is a widely used cooking oil in Asian nations [88].

9.3.1 Helps in Relieving Menopause Symptoms

According to research, using a rice bran oil extract during menopause might have a good impact [89]. One study showed that more than 90% of the women who took the supplement experienced a decrease in hot flashes and other uncomfortable menopausal symptoms, which all women eventually experience [90].

Women who use rice bran oil are advised to get the same advantages. Hot flashes that might happen during menopause can be controlled using y-oryzanol found in rice bran oil.

9.3.2 Good for Heart Patients

Compared to other vegetable oils, rice bran oil contains the best possible composition of mono-unsaturated, polyunsaturated, and saturated fats, according to the World Health Organization and the American Heart Association [91]. Due to the proper balance of fatty acids, the body's harmful cholesterol levels can be kept from sticking together [86].

Additionally, rice bran oil can reduce diseases like atherosclerosis and its associated disorders, such as heart attacks and strokes. Since Indians are genetically predisposed to heart disease, consuming rice bran oil will support heart health [92].

9.3.3 Rice Bran Oil Prevents Cancer

Antioxidants are abundant in rice bran oil, which contains oryzanol and other forms of vitamin E [93]. The rice bran oil contains significant amounts of tocopherol and tocotrienol, two forms of Vitamin E [94]. Free radicals, which are harmful byproducts of cellular metabolism that can cause healthy cells to change into malignant cells, are claimed to be neutralized by vitamin E, which is thought to operate as an antioxidant. As a result, eating this rice bran lowers the risk of getting several forms of cancer [95].

9.3.4 Rice Bran Oil for Acne Prone Skin and Skin Whitening

The substance squalene may be found in rice bran oil. It maintains the skin smooth, supple, and soft by being quickly absorbed by the skin [96].

Increased levels of vitamin E in the body can hasten wound healing, boost cellular regeneration, smooth skin, minimize wrinkles, and protect against sunburn. Levels of vitamin E in the body can be raised by using rice bran oil [97].

Vitamin E-rich rice bran oil aids in strengthening the barrier function of skin cells, keeping infections and other toxins at bay and preventing them from penetrating the skin. Tocopherol, often known as vitamin E, has a direct impact on the well-being of the skin [98].

9.3.5 Rice Bran Oil Prevents Allergies

Typically, rice bran oil is hypoallergenic [99]. As a result, when used in cooking, rice bran oil won't trigger allergic responses and can even soothe the body's immune system. As a result, it guards against hypersensitivity to other uncommon allergens [100].

9.3.6 Benefits of Rice Bran Oil for Weight Loss

Although officially high-calorie content, rice bran oil only absorbs roughly 20% of the oil due to its thickness [101]. The meal keeps its natural flavour when less oil is absorbed, and you may still take advantage of rice bran oil's health advantages. Compared to other vegetable oils, rice bran oil has fewer calories overall [102]. Cook with rice bran oil if you want to manage your weight or are seeking ways to lose weight [91].

9.3.7 Rice Bran Oil as Health Guard Immunity Booster

The antioxidant vitamin E found in rice bran oil is abundant [103]. People benefit from having a more robust immune system. Trocotrienol rich fraction's antioxidant qualities can have a variety of positive benefits on cellular health [104].

9.4 Soybean Oil

The Soya bean is a kind of legume that has a lot to offer in terms of health [105]. One of the most popular dishes worldwide, it is the most crucial nutrient that the body needs. In addition to providing high-quality protein, it is also incredibly rich in antioxidants, fibre, phytoestrogens, Omega-3 fatty acids, saturated fats, and fibre [106].

Due to the massive demand from the vegan community, there are many soy-based goods on the market that have become quite popular [107]. According to health experts, Soya bean oil is crucial for sustaining a healthy lifestyle [108].

Soybeans have long been a staple of the human diet. It is a vegan-specific cuisine that is very nutrient-dense. A few fantastic health advantages of soy oil are described below [109].

9.4.1 Treats Sleep Disorders

This unique health advantage of soya bean oil may not be well known [110]. Soya bean oil may be used to treat the signs and symptoms of sleep disorders. Soybeans should be a regular part of the diet for those who have insomnia [111].

Soya beans have a lot of magnesium, which is directly related to bettering the duration and quality of sleep [112].

9.4.2 Manages Diabetes

If you already have diabetes, it efficiently regulates your condition and stops any future harm the disease may bring [113].

Soybeans contain relatively little in the way of carbs. Because of this, health professionals frequently advise consumers to include this perfect meal in their regular diet. The anti-diabetic benefits of soy oil [114].

9.4.3 Improves Blood Circulation

Recent research indicates that soya bean oil includes iron and copper, two essential nutrients for the production of red blood cells (RBCs) [115]. It efficiently increases blood circulation.

Soybeans and meals prepared with soybean oil are good choices for postmenopausal women. It aids in symptom management and increases the body's general performance efficiency [116].

9.4.4 Essential for Pregnancy

Folic acid and vitamin B complexes are abundant in soybeans [117]. For women who are currently pregnant, these elements are crucial. According to experts, iron and vitamin B are essential for the growth of the fetus [118].

Babies born to mothers who don't meet their dietary needs frequently have birth abnormalities. Pregnant women should be careful to consume enough soybeans [119].

9.4.5 Anticancer Properties

The soya bean oil is a perfect meal to prevent numerous forms of cancer since it contains antioxidants [120]. It operates by eliminating free radicals from the body and eliminating the cancer-causing cells [121].

The body makes healthy cells from soybeans. Soya beans can help people with cancer who are already ill by reducing the side effects of their therapy [122].

Soybean oil can provide a wealth of advantages. The only thing you need to be careful about is making sure it is a scheduled component of your diet [123]. It not only guarantees your general health but also shields you against risky conditions like cancer, diabetes, and birth deformities [124].

9.5 Walnut Oil

Dried nuts are used to make unrefined walnut oil, which is then cold-pressed. The colour of high-quality walnut oil is topaz, and it has a deep, nutty flavour [125].

About 6–12 months is the maximum shelf life for walnut oil [126]. After being opened, all nut oils should be refrigerated or stored in a cold, dark place to prevent them from getting rancid. According to research from several medical studies, walnut oil is a substantial source of antioxidants, especially ellagic acid [127].

It has been discovered that this antioxidant detoxifies several chemicals connected to certain malignancies' emergence. Additionally, ellagic acid aids in preventing these cancer cells from proliferating [128].

Along with melatonin, a hormone that helps to regulate the body's circadian clock, walnut oil is also high in manganese, copper, and these minerals. The following are outstanding health advantages of walnut oil [129].

9.5.1 Improves Blood Circulation

Omega 9 and other mono-unsaturated fatty acids found in walnut oil aid to keep the arteries flexible [130]. The arteries' elasticity encourages blood flow without any hardening that might cause hypertension or heart disease [131].

9.5.2 Lowers Heart Disease Risk

The main components of walnut oil's cardio-protective qualities are alpha-linolenic acid (ALA), an omega-3 fatty acid that is ultimately converted to eicosapentaenoic acid (EPA), and docosahexaenoic acid (DHA) [132, 133]. According to a 2009 Yale University research, daily eating of walnuts may enhance blood vessel health, lowering the risk of heart disease [134].

9.5.3 Cuts Inflammation

Additionally, walnut oil contains a lot of polyunsaturated fatty acids like omega 3 and omega 6, which lower inflammation and help ward against conditions like arthritis [135, 136].

9.5.4 Maintains Hormone Levels

Walnuts are a great source of selenium, phosphorus, magnesium, zinc, iron, and calcium, all of which are essential for maintaining hormone balance in the body. They are also high in phytonutrients [137, 138].

9.5.5 Improves Skin

Vitamins B1, B2, and B3 are abundant in walnuts, walnut oil, vitamin E, and niacin [139]. Walnut oil contains fatty acids and phytonutrients that help improve the skin's quality and texture [140].

9.5.6 Prevents Eczema

Eczema patients may benefit from taking a few tablespoons of walnut oil daily to treat their skin rashes [141].

9.5.7 Anti-Aging

Among tree nuts, walnuts and walnut oil are the finest providers of antioxidants [142]. Antioxidants block the effects of free radicals, which harm cells and hasten aging [143].

9.6 Sesame Oil

An edible vegetable oil made from sesame seeds is called sesame oil [144]. In addition to being used as cooking oil in South India, it also gives Middle Eastern, African, and Southeast Asian dishes more flavour. It tastes and smells very nutty [145].

Sesame oil is produced from sesame seeds and is widely used in alternative medicine, including both traditional and modern massages and therapies [146]. Although sesame oil is a staple in Asia and one of the earliest known crop-based oils, its mass production is still restricted globally due to the labor-intensive manual harvesting procedure needed to extract the oil [147].

9.6.1 Treats Premature Graying Hair

Sesame oil massages on the hair and scalp can delay the onset of grey hairs and keep hair coloured naturally for longer [148]. In actuality, sesame oil has the ability to darken hair. Sesame oil can maintain healthy black hair with regular application [149].

9.6.2 Helps in Rheumatoid Arthritis

Minerals and vitamins abound in sesame seeds [150]. They are abundant in minerals including calcium, magnesium, iron, zinc, and copper. Sesame oil still retains the majority of sesame oil's health benefits despite the fact that it may not contain as many nutrients as sesame seeds because some are lost during the extraction process [151]. It is notably renowned for its zinc and copper compositions, which support metabolism, blood circulation, and the synthesis of red blood cells [152]. Sesame oil contains copper, which is very well recognized for its anti-inflammatory qualities. Copper helps to strengthen bones and reduces arthritic pain and joint swelling [153].

9.6.3 Lowering Blood Pressure

Since ancient times, cooking using sesame oil has been a widespread practice [154]. According to a study by an Indian researcher from Annamalai University, utilizing sesame oil as an edible oil improves the antioxidant state of hypertension patients, lowers blood pressure, and lowers lipid peroxidation [155].

9.6.4 Used in Stress and Depression Condition

Tyrosine, an amino acid found in sesame oil, directly impacts serotonin function [156]. A chemical that affects mood is serotonin. Its imbalance may result in stress and despair. Experts claim that including sesame oil in your diet helps to increase serotonin synthesis, which allows you to feel good and avoid chronic stress [157].

9.6.5 Sesame Oil Pulling for Improving Oral Health

A traditional Ayurvedic method for enhancing dental health and eliminating plaque is called sesame oil pulling [158]. On an empty stomach, a teaspoon of sesame oil is consumed, swished around the mouth for 20 minutes, and then spat out. It is said to help the body get rid of toxins [159]. Because of its therapeutic qualities, sesame oil is frequently utilized in this procedure [160].

9.6.6 Sesame Oil Skincare Benefits

Sesame oil is prized in Ayurveda for its antimicrobial and anti-inflammatory properties [161]. Sesame oil is commonly used in skincare products because it is a great moisturiser, promotes the regeneration of healthy skin, has anti-aging properties, and is thought to be a natural SPF. It is also frequently used as massage oil due to its warming effect and capability to permeate the skin deeply [162].

Sesame oil's antioxidant content aids in skin detoxification. All those water-soluble poisons are absorbed by the antioxidants, facilitating detoxification [163]. Regular face washing with a solution of 1/4 cup water, 1/2 cup apple cider vinegar, and 1/2 cup sesame oil helps to cleanse the skin and promotes healthy, radiant skin [164].

9.6.7 Natural Anti-Inflammatory Agent

Due to its anti-inflammatory characteristics, sesame seed oil is a natural healer [165]. Athlete's foot fungus, streptococci, and staphylococcus are a few examples of the microorganisms that help the antibacterial characteristic to fight against when they infect the skin. For vaginal yeast infections, a solution of lukewarm water and sesame seed oil works well [166, 167].

9.6.8 Benefits for Diabetes Patients

Sesame seed oil is a rich source of magnesium and many other minerals [168]. Sesame can prevent diabetes by lowering blood glucose levels, thanks to all of these factors working together. Sesame seed oil is a good option for people with diabetes who want to cook [169].

9.6.9 Sesame Oil for Anemia

The amount of iron in sesame oil is very high [170]. They are, therefore, among the most frequently advised natural treatments for anemia and other iron-deficiency issues [171].

9.6.10 Anti-Cancerous Properties

Magnesium, a mineral with a strong anti-cancerous reputation, is a component of sesame oil [172]. These also include phytate, a cancer-fighting substance. Due to the combined effects of these components, sesame oil can lower the risk of colorectal cancers and even stop them from developing [173].

9.6.11 Helps in Improving Eye Health

Traditional Chinese medicines said that the link between the liver and eyes was particularly friendly [174]. It was believed that the liver could store blood and that the blood it produced was distributed through its branches to the eye to maintain appropriate function. The natural liver tonic sesame oil increases blood flow, which is said to nourish the eyes [175]. These beneficial effects can alleviate eye pain, eye fatigue, and blurry vision. Sesame oil, often used on the eyelids, might also aid in reducing wrinkles and dark under-eye circles [176].

9.7 Peanut Oil

The low-growing, annual plant known as the peanut (Arachis hypogea), which has a characteristic peanut flavour and aroma, is the source of peanut oil, a sweet edible vegetable oil (Leguminosae) [177].

Peanut oil is commonly used in the preparation of dishes from China, South Asia, and Southeast Asia, both for basic cooking and, in the case of roasted oil, for added flavour. Peanut oil is widely used for frying food due to its high smoke point when compared to other cooking oils.

The majority of peanut oil is composed of fatty acids, including oleic acid (46.8% as olein), linoleic acid (33.4% as linolein), and palmitic acid (10.0% as palmitin) in different concentrations. A small amount of stearic acid, arachidic acid, behenic acid, lignoceric acid, and other fatty acids are also present in the oil. Vitamin E and occasionally additional antioxidants are added to the oil's shelf life [178].

9.7.1 Used for Body Massage Therapy

A massage with peanut oil is energizing and revitalizing and can aid with joint issues [178]. If applied regularly, this provides excellent relief for sore joints and muscles. Vitamin E is included in peanut oil and is perfect for protecting the skin and improving health. Peanut oil may be used to hydrate the skin of an infant [179].

9.7.2 Helps in Improving Heart Health

Peanut oil has a comparatively high calorie content as a result of the fatty acid makeup [180]. This oil contains a lot of mono-unsaturated fatty acids (MUFA), which help to raise good cholesterol and lower bad cholesterol levels in the blood. Peanut oil assists heart patients by preventing coronary artery disease and heart attacks and maintaining a healthy blood lipid profile [181].

Linoleic acid, a precursor to prostaglandins, is present in peanut oil [182]. The body needs prostaglandin to perform numerous crucial processes, including the contraction and dilation of blood vessels and other muscles [183].

9.7.3 Peanut Oil Maintains Cholesterol Level

To keep cholesterol levels in check and promote heart health, reasonable amounts of peanut cooking oil should be consumed daily [184].

Mono-unsaturated fats found in peanut oil can help decrease high blood pressure. This indirectly lowers the chance of developing cardiac problems [185].

9.7.4 Skin Benefits

Taking a few drops of peanut oil and two to three drops of lime juice is the standard treatment for acne [186]. For dry skin and natural skincare, peanut oil is frequently utilized. This has miraculous effects in preventing blackheads on your skin as well [187].

9.7.5 Peanut Oil Helps to Prevent Stomach Problems

Numerous gastrointestinal diseases can be avoided with the aid of peanut oil [188]. As a result, it is a great treatment for conditions including constipation, digestive issues, diarrhoea, etc. [189].

9.7.6 Aromatherapy by Peanut Oil

Due to its mild and nutty scent, peanut oil is frequently used in aromatherapy [190]. It may be used to massage the body for a wonderfully soothing and revitalizing experience [191].

9.7.7 Provides Skincare

Peanut oil is found in a variety of creams and lotions [192]. It is most often found in bath products and lotions for diaper rash. For added moisturization, certain nipple creams may also include peanut oil [161].

9.7.8 For Hair Growth

Reducing protein loss, thickening hair, moisturizing split ends, and repairing damaged hair are benefits of using peanut oil. It lessens scalp flaking as well [193, 194].

9.7.9 Used as an Antispasmodic

It is advised that diabetic people regularly consume peanut oil to raise their body's insulin level [195]. The oil balances blood sugar levels and aids in bringing them down to normal [196].

9.7.10 Peanut Oil Helps Strengthen the Body

A highly abundant source of protein is peanut oil. It should be ingested in moderation daily for body strengthening and muscular growth [181, 197].

9.8 Sunflower Oil

The non-volatile oil extracted from sunflower seeds is called sunflower oil (*Helianthus annuus*). The majority of the oils in this mono-unsaturated (MUFA)/polyunsaturated (PUFA) mixture belong to the linoleic acid (omega-6) and oleic acid (omega-9) categories [198]. The refined oil is a bright yellow tint, while the crude oil has a mellow, agreeable flavour [199]. Low refining losses have been noted, and the oil has good storage properties with a lesser propensity to change in flavour [200].

Sunflower oil contains significant levels of sterols, vitamin E, squalene, and other aliphatic hydrocarbons. Sunflower oil is frequently used as frying oil in food and as an emollient in cosmetic products [201]. In 2014, almost 16 million tonnes of sunflower oil were produced worldwide, with Ukraine and Russia being the two biggest producers [202].

9.8.1 Cardiovascular Benefits

Low in saturated fat and high in vitamin E is sunflower oil. It also contains a lot of heart-healthy phytochemicals including choline and phenolic acid [203]. It is one of the healthiest oils to consume since it also includes vitamins and mono-unsaturated and polyunsaturated fats [204]. At least 80% of the lipids in high oleic sunflower oils are mono-unsaturated. So, sunflower oil aids in reducing the risk of heart attack and cardiovascular disorders [205].

9.8.2 Helps in Preventing Arthritis

For individuals who are concerned about getting arthritis, sunflower oil is a beneficial remedy [206]. Rheumatoid arthritis can be prevented with sunflower oil [207].

9.8.3 Prevention of Colon Cancer and Other Cancers

No other cooking oil has a higher level of vitamin E than sunflower oil [208]. Therefore, using this oil in our diet offers a defense against colon cancer. Sunflower oil contains carotenoids that aid in the prevention of skin, lung, and uterine malignancies [209]. The tocopherols in sunflower oil, which include vitamin E, are a resource in scavenging cancer-causing free radicals. The immune system and cells are both harmed by free radicals [210].

9.8.4 Lowering Cholesterol

Other oils include saturated fats that are harder for the body to break down entirely during metabolism [211]. As a result, they are transported by circulation to various areas of the body where they are deposited, raising LDL cholesterol levels (bad cholesterol). As a result, people have birth abnormalities, atherosclerosis, joint discomfort, and embolism [212].

The mono-unsaturated and polyunsaturated fats in sunflower oil help the body maintain a healthy LDL/HDL (good cholesterol) ratio while also supplying energy [213]. Lecithin, another component of sunflower oil, aids in decreasing cholesterol levels. This is because it adheres to the cell and nerve membranes, preventing cholesterol and fat from sticking to the cell [214].

9.8.5 Body Repairing

Sunflower oil contains proteins that are necessary for the creation of hormones, enzymes, and tissue growth and repair [215]. The average person needs a lot of protein. Sunflower oil fills the need for proteins as the body cannot store them [216].

9.8.6 Sunflower Oil Helps in Preventing Infant Infection

Low birth weight preterm infants have a decreased risk of infection because of sunflower oil [217]. Because their immune systems and other organs, like their skin, are still developing, newborns are more vulnerable to contracting diseases. Such an infection is stopped by the protective barrier provided by sunflower oil [218].

9.8.7 Healthy Nervous System

Sunflower oil's vitamin B level aids normal digestion and a healthy neurological system and is a fantastic energy source [219].

9.8.8 Helps in Reducing Cardiac Problems

Selenium, which is included in sunflower oil and helps lower the risk of cardiac issues and hepatic deterioration, is also present. Reduced skin and lung cancer risk is another benefit of high blood selenium levels [220].

9.8.9 Sunflower Oil for Skin Benefits

Vitamin E-rich sunflower oil is directly linked to enhancing skin health and cell regeneration. As a result, the skin is better guarded against UV ray damage and the aging process that results from the presence of free radicals in the body. Sunflower oil is beneficial for skin lightening, allergies, and skin irritability [221].

Free radicals are neutralized by antioxidants like vitamin E, preventing them from killing or harming healthy cells. Scarring is reduced more, wounds heal more quickly, and the skin is more beneficial and more radiant. This is a critical factor in the widespread usage of sunflower oil in cosmetic applications [222].

9.8.10 Anti-Aging Property

Sunflower oil's antioxidant qualities aid in reducing early aging symptoms [223]. Free radicals and sunlight accelerate the aging process of the skin, resulting in the development of wrinkles and fine lines at a young age. Sunflower oil's antioxidants reduce the likelihood of acquiring early aging symptoms [224].

9.8.11 Controlling Frizzy Hair

Sunflower oil softens hair and gives it a lovely sheen because of its thin texture and mild flavour [225]. Sunflower seed oil makes the hair manageable, combats dryness and damage, and controls frizz. This adaptable oil may be used as a natural conditioner and preserves the hair's shine and structure. For hair loss, rub sunflower oil into the scalp each week before taking a bath [226].

9.8.12 Anti-Inflammatory Properties

Millions of people throughout the world suffer from asthma, a respiratory ailment that can be minor or fatal [227]. Sunflower oil has been positively associated with a reduction in the frequency and intensity of asthma attacks due to its anti-inflammatory effects, which are derived from its vitamin content as well as the advantageous fatty acids it contains [228].

9.9 Conclusions

Vegetable oils, both essential and non-essential, have been successfully incorporated into various compositions to increase their advantages (antibacterial, antifungal, and antiviral properties) in the pharmaceutical industry. The majority of the vegetable oils examined are by-products of tropical regions, where they are plentiful and simple to grow. Considering their role, in the ensuing decades, its use in the pharmaceutical and cosmetic industries will further increase their demand. Therefore, research into additional plant sources, novel methods that ensure quality and yield, and most importantly, the creation of commercial-scale crops that ensure world demand are all vital.

The role of vegetable oils especially in the pharmaceutical industry touches different dimensions. Several examples are the recently studied use of vegetable oils as an efficient alternative treatment for wounds with the help of balanced composition of fatty acids obtained from natural sources.

From the future perspectives, vegetable oil-derived bioactives that are used to increase the potency of the same medications, as well as nanoformulations made to boost the nutraceutical qualities of compounds based on vegetable oil are gaining attention. Advantages of the use of vegetable oils in enhancing the preservation effect in food, medicinal, and cosmetic applications that the bioactive compounds from the vegetable oils is also being explored. This sheds light on the importance of bioactive substances from the viewpoint of human health and the sustainability of the world's natural resources. The potential of bioactive chemicals in numerous fields is sure to increase as technology advances, opening up new opportunities. Although encapsulation technology appears promising, formulation development still needs work. Due to the numerous uses of vegetable oils in the cosmetic and pharmaceutical industries, this finding is of significant commercial significance.

There are still many difficult issues to be solved, but it seems inevitable that better materials based on vegetable oil will be created.

References

1. Tripathy, D. B., & Mishra, A. (2016). Renewable plant-based raw materials for industry. *Sustainable Inorganic Chemistry*, 353.
2. Lligadas, G., *et al.*, Plant oils as platform chemicals for polyurethane synthesis: current state-of-the-art. 2010. **11**(11): pp. 2825-2835.
3. Folayan, A.J., *et al.*, Experimental investigation of the effect of fatty acids configuration, chain length, branching and degree of unsaturation on biodiesel fuel properties obtained from lauric oils, high-oleic and high-linoleic vegetable oil biomass. 2019. **5**: pp. 793-806.
4. Fox, N. and G.W. Stachowiak, Vegetable oil-based lubricants—a review of oxidation. 2007. *Tribology International* **40**(7): pp. 1035-1046.
5. Riezman, H.J.C., The long and short of fatty acid synthesis. 2007. **130**(4): pp. 587-588.
6. Rustan, A.C. and C.A. Drevon, Fatty acids: structures and properties. 2001.
7. Ganesan, K., *et al.*, Impact of consumption and cooking manners of vegetable oils on cardiovascular diseases - A critical review. 2018. **71**: pp. 132-154.
8. Kochikuzhyil, B.M., K. Devi, and S.R. Fattepur, Effect of saturated fatty acid-rich dietary vegetable oils on lipid profile, antioxidant enzymes and glucose tolerance in diabetic rats. *Indian J Pharmacol.* 2010. **42**(3): p. 142.

9. Lubbe, A., R.J. Verpoorte, Cultivation of medicinal and aromatic plants for specialty industrial materials. *Industrial Crops and Products,* 2011. **34**(1): p. 785-801.
10. Ghiasi, F., *et al.*, Development of highly stable colloidal dispersions of gelled-oil nanoparticles loaded with cuminaldehyde. 2019. **541**: pp. 65-74.
11. Petropoulos, S.A., *et al.*, Nutritional value and bioactive compounds characterization of plant parts from Cynara cardunculus L.(Asteraceae) cultivated in central Greece. 2018. **9**: p. 459.
12. Xing, Y. and Z.J.C. Shi, Nucleocapsid protein VP15 of White spot syndrome virus colocalizes with the nucleolar proteins nucleolin and fibrillarin. *Can. J Microbiol.* 2011. **57**(9): pp. 759-764.
13. da Silva, M.J.F., *et al.*, Development and characterization of a babassu nut oil-based moisturizing cosmetic emulsion with a high sun protection factor. 2020. **10**(44): pp. 26268-26276.
14. de Carvalho, F.A.L., *et al.*, Effect of replacing backfat with vegetable oils during the shelf-life of cooked lamb sausages. 2020. **122**: p. 109052.
15. Rai, S., D. Charak, and R.J.P.A. Bharat, Scenario of oilseed crops across the globe. 2016. **16**(1): pp. 125-132.
16. Arbain, N.H., *et al.*, Optimization of quercetin loaded palm oil ester based nanoemulsion formulation for pulmonary delivery. 2018. **67**(8): pp. 933-940.
17. Zhou, X., *et al.*, Stability and *in vitro* digestibility of beta-carotene in nanoemulsions fabricated with different carrier oils. 2018. **6**(8): p. 2537-2544.
18. Ferreira, C.D. and I.L. Nunes, Oil nanoencapsulation: development, application, and incorporation into the food market. *Nanoscale Research Letters.* 2019. **14**(1): pp. 1-13.
19. Dario, M.F., *et al.*, Synergistic photoprotective activity of nanocarrier containing oil of Acrocomia aculeata (Jacq.) Lodd. Ex. Martius—Arecaceae. 2018. **112**: pp. 305-312.
20. Ansari, M.J., *et al.*, Formulation and evaluation of self-nanoemulsifying drug delivery system of brigatinib: Improvement of solubility, *in vitro* release, *ex-vivo* permeation and anticancer activity. 2021. **61**: p. 102204.
21. Kumar, A., A. Sharma, and K.C. Upadhyaya, Vegetable oil: nutritional and industrial perspective. *Current Genomics*, 2016. **17**(3): pp. 230-240.
22. Badea, G., *et al.*, Use of various vegetable oils in designing photoprotective nanostructured formulations for UV protection and antioxidant activity. 2015. **67**: pp. 18-24.
23. Covas, M.-I. Olive oil and the cardiovascular system. *Pharmcol Res.* 2007. **55**(3): p. 175-186.
24. Tuck, K.L. and P.J. Hayball, Major phenolic compounds in olive oil: metabolism and health effects. *J. Nutr Biochem.* 2002. **13**(11): pp. 636-644.
25. Linos, A., *et al.*, The effect of olive oil and fish consumption on rheumatoid arthritis-a case control study. 1991. **20**(6): pp. 419-426.
26. Hodge, A.M., *et al.*, Foods, nutrients and prostate cancer. 2004. **15**(1): pp. 11-20.

27. Kremer, J.M., *et al.*, Dietary fish oil and olive oil supplementation in patients with rheumatoid arthritis clinical and immunologic effects. 1990. **33**(6): pp. 810-820.
28. Jeffery, N., *et al.*, The effects of olive oil upon rat serum lipid levels and lymphocyte functions appear to be due to oleic acid. 1996. **40**(2): pp. 71-80.
29. Vissers, M.N., *et al.*, Olive oil phenols are absorbed in humans. 2002. **132**(3): pp. 409-417.
30. López-Miranda, J., *et al.*, Olive oil and health: summary of the II international conference on olive oil and health consensus report, Jaén and Córdoba (Spain) 2008. 2010. **20**(4): pp. 284-294.
31. Berbert, A.A., *et al.*, Supplementation of fish oil and olive oil in patients with rheumatoid arthritis. 2005. **21**(2): pp. 131-136.
32. Fitó, M., *et al.*, Antioxidant effect of virgin olive oil in patients with stable coronary heart disease: a randomized, crossover, controlled, clinical trial. 2005. **181**(1): pp. 149-158.
33. Mataix, J., Barbancho, F.I, Olive oil in Mediterranean food. In *Olive Oil and Health*, 2006: pp. 1-44.
34. Gorinstein, S., *et al.*, Olive oils improve lipid metabolism and increase antioxidant potential in rats fed diets containing cholesterol. 2002. **50**(21): pp. 6102-6108.
35. Martin-Moreno, J.M., *et al.*, Dietary fat, olive oil intake and breast cancer risk. 1994. **58**(6): pp. 774-780.
36. Petroni, A., *et al.*, Inhibition of platelet aggregation and eicosanoid production by phenolic components of olive oil. 1995. **78**(2): pp. 151-160.
37. Ruano, J., *et al.*, Intake of phenol-rich virgin olive oil improves the postprandial prothrombotic profile in hypercholesterolemic patients. 2007. **86**(2): pp. 341-346.
38. Birben, E., *et al.*, Oxidative stress and antioxidant defense. 2012. **5**(1): pp. 9-19.
39. Hutcheson, R. and P. Rocic, The metabolic syndrome, oxidative stress, environment, and cardiovascular disease: the great exploration. *Exp Diabetes Res.* 2012. **2012**.
40. Porter, W.L. Health, Paradoxical behavior of antioxidants in food and biological systems. *Toxicol Ind Health*, 1993. **9**(1-2): pp. 93-122.
41. Servili, M., *et al.*, Health and sensory properties of virgin olive oil hydrophilic phenols: agronomic and technological aspects of production that affect their occurrence in the oil. 2004. **1054**(1-2): pp. 113-127.
42. Decker, E.A. The role of phenolics, conjugated linoleic acid, carnosine, and pyrroloquinoline quinone as nonessential dietary antioxidants. *Nutr Rev.* 1995. **53**(3): p. 49-58.
43. Pandey, K.B., Rizvi, S.I. Plant polyphenols as dietary antioxidants in human health and disease. *Oxid Med Cell Longev.* 2009. **2**(5): pp. 270-278.
44. Danaei, G., *et al.*, National, regional, and global trends in fasting plasma glucose and diabetes prevalence since 1980: systematic analysis of health

examination surveys and epidemiological studies with 370 country-years and 2· 7 million participants. 2011. **378**(9785): pp. 31-40.
45. Alberti, K.G. and P.Z. Zimmet, Definition, diagnosis and classification of diabetes mellitus and its complications. Part 1: diagnosis and classification of diabetes mellitus. Provisional report of a WHO consultation. *Diabet Med* 1998. **15**(7): pp. 539-553.
46. Martínez-González, M.Á., *et al.*, Adherence to Mediterranean diet and risk of developing diabetes: prospective cohort study. 2008. **336**(7657): pp. 1348-1351.
47. Loizzo, M., *et al.*, Inhibitory activity of phenolic compounds from extra virgin olive oils on the enzymes involved in diabetes, obesity and hypertension. 2011. **35**(2): pp. 381-399.
48. Naziroğlu, M. and Butterworth, P. J. Protective effects of moderate exercise with dietary vitamin C and E on blood antioxidative defense mechanism in rats with streptozotocin-induced diabetes. *Can J Appl Physiol.* 2005. **30**(2): p. 172-185.
49. Sales, P.M., *et al.*, α-Amylase inhibitors: a review of raw material and isolated compounds from plant source. 2012. **15**(1): pp. 141-183.
50. Xiao, J., *et al.*, Interaction of natural polyphenols with α-amylase *in vitro*: molecular property–affinity relationship aspect. 2011. **7**(6): pp. 1883-1890.
51. Rahimi, R., *et al.*, A review on the role of antioxidants in the management of diabetes and its complications. 2005. **59**(7): pp. 365-373.
52. Tiwari, A.K. and Rao, J.M. Diabetes mellitus and multiple therapeutic approaches of phytochemicals: Present status and future prospects. *Current Science*, 2002: pp. 30-38.
53. Melzig, M.F. and I. Funke, Inhibitors of alpha-amylase from plants--a possibility to treat diabetes mellitus type II by phytotherapy? *Wiener Medizinische Wochenshrift*, 2007. **157**(13-14): pp. 320-324.
54. Mccue, P., Y.I. Kwon, and K. Shetty, Anti-amylase, anti-glucosidase and anti-angiotensin i-converting enzyme potential of selected foods. *Journal of Food Biochemistry*, 2005. **29**(3): pp. 278-294.
55. Oboh, G., A.J. Akinyemi, and A.O. Ademiluyi, Inhibitory effect of phenolic extract from garlic on angiotensin-1 converting enzyme and cisplatin induced lipid peroxidation–*in vitro*. *Int J Biomed Sci.* 2013. **9**(2): p. 98.
56. Ranilla, L.G., *et al.*, Phenolic compounds, antioxidant activity and *in vitro* inhibitory potential against key enzymes relevant for hyperglycemia and hypertension of commonly used medicinal plants, herbs and spices in Latin America. 2010. **101**(12): pp. 4676-4689.
57. Fitzgerald, R.J. and B.A. Murray, Bioactive peptides and lactic fermentations. *Int J of Dairy Technology*, 2006. **59**(2): pp. 118-125.
58. Ferrannini, E. and W.C. Cushman, Diabetes and hypertension: the bad companions. *Lancet* 2012. **380**(9841): pp. 601-610.

59. Atlas, S.A., The renin-angiotensin aldosterone system: pathophysiological role and pharmacologic inhibition. *J Manag Care Pharm,* 2007. **13**(8 Supp B): pp. 9-20.
60. Van Vark, L.C., *et al.*, Angiotensin-converting enzyme inhibitors reduce mortality in hypertension: a meta-analysis of randomized clinical trials of renin–angiotensin–aldosterone system inhibitors involving 158 998 patients. 2012. **33**(16): pp. 2088-2097.
61. Ahmed, F., *et al.*, Radical scavenging and angiotensin converting enzyme inhibitory activities of standardized extracts of Ficus racemosa stem bark. 2010. **24**(12): pp. 1839-1843.
62. Edgar, L., *et al.*, ACE inhibitors for the treatment of hypertension drug selection by means of the SOJA method. 2011. **6**(1): pp. 70-94.
63. Bacon, S.L., *et al.*, Effects of exercise, diet and weight loss on high blood pressure. 2004. **34**(5): pp. 307-316.
64. Perona, J.S., *et al.*, Plasma lipid modifications in elderly people after administration of two virgin olive oils of the same variety (Olea europaea var. hojiblanca) with different triacylglycerol composition. 2003. **89**(6): pp. 819-826.
65. Dong, J., *et al.*, Inhibition of angiotensin converting enzyme (ACE) activity by polyphenols from tea (Camellia sinensis) and links to processing method. 2011. **2**(6): pp. 310-319.
66. Actis-Goretta, L., *et al.*, Inhibition of angiotensin converting enzyme activity by flavanol-rich foods. 2006. **54**(1): pp. 229-234.
67. Balasuriya, B.N., H.P.V. Rupasinghe, Plant flavonoids as angiotensin converting enzyme inhibitors in regulation of hypertension. In *Functional Foods in Health and Disease*, 2011. **1**(5): pp. 172-188.
68. Kwon, E.-K., *et al.*, Flavonoids from the buds of Rosa damascena inhibit the activity of 3-hydroxy-3-methylglutaryl-coenzyme a reductase and angiotensin I-converting enzyme. 2010. **58**(2): pp. 882-886.
69. Klaunig, J.E., L.M. Kamendulis, and toxicology, The role of oxidative stress in carcinogenesis. *Annu Rev Pharmacol Toxicol.* 2004. **44**: p. 239.
70. Jemal, A., *et al.*, Cancer statistics, 2009. 2009. **59**(4): pp. 225-249.
71. Bennett, L.L., *et al.*, Role of antioxidants in the prevention of cancer. 2012. **4**(4): pp. 215-222.
72. Hillestrøm, P.R., *et al.*, Effect of dietary virgin olive oil on urinary excretion of etheno–DNA adducts. 2006. **41**(7): pp. 1133-1138.
73. Cooke, M.S., *et al.*, Oxidative DNA damage: mechanisms, mutation, and disease. 2003. **17**(10): pp. 1195-1214.
74. Manna, C., *et al.*, Protective effects of synthetic hydroxytyrosol acetyl derivatives against oxidative stress in human cells. 2005. **53**(24): pp. 9602-9607.
75. Corona, G., *et al.*, Hydroxytyrosol inhibits the proliferation of human colon adenocarcinoma cells through inhibition of ERK1/2 and cyclin D1. 2009. **53**(7): pp. 897-903.

76. Covas, M.-I., Bioactive effects of olive oil phenolic compounds in humans: reduction of heart disease factors and oxidative damage. *Inflammopharmacology*, 2008. **16**(5): p. 216-218.
77. Escrich, E., *et al.*, Olive oil and other dietary lipids in breast cancer. 2014: pp. 289-309.
78. Evans, M.D., M. Dizdaroglu, and M.S. Cooke, Oxidative DNA damage and disease: induction, repair and significance. *Mutat Res*. 2004. **567**(1): pp. 1-61.
79. Omar, S.H., Cardioprotective and neuroprotective roles of oleuropein in olive. *Saudi Pharm J*. 2010. **18**(3): pp. 111-121.
80. Fabiani, R., *et al.*, Oxidative DNA damage is prevented by extracts of olive oil, hydroxytyrosol, and other olive phenolic compounds in human blood mononuclear cells and HL60 cells. 2008. **138**(8): pp. 1411-1416.
81. Awad, A.B., H. Williams, and C.S. Fink, Effect of phytosterols on cholesterol metabolism and MAP kinase in MDA-MB-231 human breast cancer cells. *J Nutr Biochem*. 2003. **14**(2): p. 111-119.
82. Gill, C.I., *et al.*, Potential anti-cancer effects of virgin olive oil phenolson colorectal carcinogenesis models *in vitro*. 2005. **117**(1): pp. 1-7.
83. Orthoefer, F.T., J. Eastman, Rice bran oil. 2005. **2**: pp. 465-489.
84. Ghosh, M., Review on recent trends in rice bran oil processing. *Journal of the American Oil Chemists' Society*, 2007. **84**(4): p. 315-324.
85. Lakkakula, N.R., M. Lima, and T. Walker, Rice bran stabilization and rice bran oil extraction using ohmic heating. 2004. *Bioresource Technology* **92**(2): pp. 157-161.
86. Sugano, M. and E. Tsuji, Rice bran oil and cholesterol metabolism. *J. Nutr.* 1997. **127**(3): pp. 521S-524S.
87. Cicero, A. and A. Gaddi, Rice bran oil and γ-oryzanol in the treatment of hyperlipoproteinaemias and other conditions. 2001. *Phytother Res*. **15**(4): pp. 277-289.
88. Rukmini, C. and T.C. Raghuram, Nutritional and biochemical aspects of the hypolipidemic action of rice bran oil: a review. *J Am Coll Nutr.* 1991. **10**(6): pp. 593-601.
89. Sawada, K., *et al.*, Absorption and metabolism of γ-oryzanol, a characteristic functional ingredient in rice bran. 2019. **65** (Supplement): pp. S180-S184.
90. Morelli, V. and C. Naquin, Alternative therapies for traditional disease states: menopause. 2002. *Am Fam Physician*. **66**(1): p. 129.
91. Kuriyan, R., *et al.*, Use of rice bran oil in patients with hyperlipidaemia. 2005. **18**(6): p. 292.
92. Most, M.M., *et al.*, Rice bran oil, not fiber, lowers cholesterol in humans. 2005. **81**(1): pp. 64-68.
93. Nayik, G., *et al.*, Rice bran oil, the future edible oil of India: A mini Review. 2015. **3**(151): p. 2.
94. Yu, Y., *et al.*, The anti-cancer activity and potential clinical application of rice bran extracts and fermentation products. 2019. **9**(31): pp. 18060-18069.

95. Law, B.M., *et al.*, Hypotheses on the potential of rice bran intake to prevent gastrointestinal cancer through the modulation of oxidative stress. 2017. **18**(7): p. 1352.
96. Nilforoushzadeh, M.A., *et al.*, Skin care and rejuvenation by cosmeceutical facial mask. 2018. **17**(5): pp. 693-702.
97. Lacatusu, I., *et al.*, Advanced bioactive lipid nanocarriers loaded with natural and synthetic anti-inflammatory actives. 2019. **200**: pp. 113-126.
98. Phasha, V., *et al.*, Review on the Use of Kojic Acid—A Skin-Lightening Ingredient. 2022. **9**(3): pp. 64.
99. Punia, S., *et al.*, Rice bran oil: Emerging trends in extraction, health benefit, and its industrial application. 2021. **28**(3): pp. 217-232.
100. Friedman, M., Rice brans, rice bran oils, and rice hulls: composition, food and industrial uses, and bioactivities in humans, animals, and cells. 2013. *J Agric Food Chem.* **61**(45): pp. 10626-10641.
101. Lai, O.-M., *et al.*, Nutritional studies of rice bran oil. 2019: pp. 19-54.
102. Ali, A. and S. Devarajan, Nutritional and health benefits of rice bran oil, in *Brown Rice*. 2017, Springer. pp. 135-158.
103. Dutta, S., *et al.*, Role of nutraceuticals on health promotion and disease prevention: A review. 2018. **8**(4): pp. 42-47.
104. Mostafa, M.E., M.S. Abbady, Secondary metabolites and bioactivity of the Monascus pigments review article. *Global Journal of Biotechnology & Biochemistry*, 2014. **9**(1): pp. 1-13.
105. Hammond, E.G., *et al.*, Soybean oil. 2005.
106. Clemente, T.E. and E.B. Cahoon, Soybean oil: genetic approaches for modification of functionality and total content. 2009. *Plant Physiology,* **151**(3): pp. 1030-1040.
107. Noureddini, H. and D. Zhu, Kinetics of transesterification of soybean oil. *Journal of the American Oil Chemists' Society*, 1997. **74**(11): pp. 1457-1463.
108. Jung, M., S. Yoon, and D. Min, Effects of processing steps on the contents of minor compounds and oxidation of soybean oil. *Journal of the American Oil Chemists' Society*, 1989. **66**(1): pp. 118-120.
109. List, G., *et al.*, Oxidation and quality of soybean oil: a preliminary study of the anisidine test. 1974. **51**(2): p. 17-21.
110. Zhao, M., *et al.*, The effects of dietary nutrition on sleep and sleep disorders. 2020. **2020**.
111. Lee, J., *et al.*, Effects of Passiflora incarnata Linnaeus on polysomnographic sleep parameters in subjects with insomnia disorder: a double-blind randomized placebo-controlled study. 2020. **35**(1): pp. 29-35.
112. Yi, G., *et al.*, A study of the mechanism of small-molecule soybean-protein-derived peptide supplement to promote sleep in a mouse model. 2020. **10**(19): pp. 11264-11273.
113. Colles, S.L., *et al.*, Dietary beliefs and eating patterns influence metabolic health in type 2 diabetes: A clinic-based study in urban North India. 2013. **17**(6): pp. 1066.

114. Visser, J., *et al.*, Prevention of diabetes by a hydrolysed casein-based diet in diabetes-prone BioBreeding rats does not involve restoration of the defective natural regulatory T cell function. 2009. **52**(7): pp. 1445-1447.
115. Pscheidl, E., *et al.*, Fish oil-supplemented parenteral diets normalize splanchnic blood flow and improve killing of translocated bacteria in a low-dose endotoxin rat model. 2000. **28**(5): pp. 1489-1496.
116. Siqueira, J., *et al.*, Substitution of standard soybean oil with olive oil-based lipid emulsion in parenteral nutrition: comparison of vascular, metabolic, and inflammatory effects. 2011. **96**(10): pp. 3207-3216.
117. Rooke, J., I. Bland, and S. Edwards, Relationships between fatty acid status of sow plasma and that of umbilical cord, plasma and tissues of newborn piglets when sows were fed on diets containing tuna oil or soyabean oil in late pregnancy. *Br J Nutr,* 1999. **82**(3): pp. 213-221.
118. Rooke, J., I. Bland, and S. Edwards, Effect of feeding tuna oil or soyabean oil as supplements to sows in late pregnancy on piglet tissue composition and viability. *Br J Nutr,* 1998. **80**(3): pp. 273-280.
119. de Velasco, P.C., *et al.*, Maternal intake of trans-unsaturated or interesterified fatty acids during pregnancy and lactation modifies mitochondrial bioenergetics in the liver of adult offspring in mice. 2017. **118**(1): pp. 41-52.
120. Timsina, B., M. Shukla, and V. Nadumane, A review of few essential oils and their anticancer property. *Journal of Natural Pharmaceuticals*, 2012. **3**(1): pp. 1-8.
121. Rashad, M.M., *et al.*, Production, characterization and anticancer activity of Candida bombicola sophorolipids by means of solid state fermentation of sunflower oil cake and soybean oil. 2014. **65**(2): p. e017-e017.
122. Awad, A.B., *et al.*, Peanuts as a source of β-sitosterol, a sterol with anticancer properties. 2000. **36**(2): pp. 238-241.
123. Rani, P., *et al.*, Material, antibacterial and anticancer properties of natural polyphenols incorporated soy protein isolate: A review. 2021. **152**: p. 110494.
124. Yang, H., *et al.*, Recovery of phytosterols from waste residue of soybean oil deodorizer distillate. 2010. **101**(5): pp. 1471-1476.
125. Martínez, M.L., *et al.*, Walnut (Juglans regia L.): genetic resources, chemistry, by-products. 2010. **90**(12): pp. 1959-1967.
126. Jokić, S., *et al.*, Optimization of walnut oil production. 2014. **6**(1): pp. 27-35.
127. Yang, R., *et al.*, A review of chemical composition and nutritional properties of minor vegetable oils in China. 2018. **74**: pp. 26-32.
128. Nael, A.T. and M. Wadaan, Utility and importance of walnut, Juglans regia Linn: A review. *African Journal of Microbiology Research*, 2011. **5**(32): pp. 5796-5805.
129. Liu, M., *et al.*, Walnut fruit processing equipment: academic insights and perspectives. 2021. **13**(4): pp. 822-857.
130. West, S.G., *et al.*, Effects of diets high in walnuts and flax oil on hemodynamic responses to stress and vascular endothelial function. 2010. **29**(6): pp. 595-603.

131. Zibaeenezhad, M., *et al.*, The effect of walnut oil consumption on blood sugar in patients with diabetes mellitus type 2. 2016. **14**(3).
132. Zibaeenezhad, M., *et al.*, Antihypertriglyceridemic effect of walnut oil. 2003. **54**(4): pp. 411-414.
133. Zibaeenezhad, M.J., *et al.*, Effects of walnut oil on lipid profiles in hyperlipidemic type 2 diabetic patients: a randomized, double-blind, placebo-controlled trial. 2017. **7**(4): pp. e259-e259.
134. Ma, Y., *et al.*, Effects of walnut consumption on endothelial function in type 2 diabetic subjects: a randomized controlled crossover trial. 2010. **33**(2): pp. 227-232.
135. Zhang, J., *et al.*, Walnut oil increases cholesterol efflux through inhibition of stearoyl CoA desaturase 1 in THP-1 macrophage-derived foam cells. 2011. **8**(1): pp. 1-13.
136. Miao, F., *et al.*, Effect of walnut (Juglans sigillata) oil on intestinal antioxidant, anti-inflammatory, immunity, and gut microbiota modulation in mice. 2021. **45**(1): pp. e13567.
137. Adelakun, S.A., *et al.*, Nitrite-induced testicular toxicity in rats: therapeutic potential of walnut oil. 2019. **23**(1): p. 15.
138. Hassan, M.Z., *et al.*, Study of walnut oil supplementation on serum biochemical parameters and histopathology of male rats. 2021. **19**(6): p. 8.
139. Radu, O., A. Fuior, and T. Capcanari, The study of biological and nutritional potential of walnut oil. 2013.
140. Mir, G., N. Owais, and I. Uzma, Scientific Processing of Walnuts Necessary for Amazing Health Benefits. *Journal of Chemistry and Chemical Sciences*, 2016. **6**: pp. 783-793.
141. Boroujeni, H.R., *et al.*, Formulation and clinical trial study of AJMT cream in treatment of eczema. 2017. **8**: pp. 46-50.
142. Milind, P. and K. Deepa, Walnut: not a hard nut to crack. *International Research Journal of Pharmacy*, 2011. **2**(5): pp. 8-17.
143. Zhao, H., *et al.*, Antioxidant effects of compound walnut oil capsule in mice aging model induced by D-galactose. 2018. **62**.
144. Hwang, L.S., Sesame oil. *Bailey's Industrial Oil and Fat Products*, 2005. **2**: pp. 547-552.
145. Budowski, P. and K.S. Markley, The chemical and physiological properties of sesame oil. *Chem Rev.*, 1951. **48**(1): pp. 125-151.
146. Mohamed, H. and I. Awatif, The use of sesame oil unsaponifiable matter as a natural antioxidant. *Food Chemistry*, 1998. **62**(3): pp. 269-276.
147. Abou-Gharbia, H.A., A.A.Y. Shehata, and F. Shahidi, Effect of processing on oxidative stability and lipid classes of sesame oil. *Food Research International*, 2000. **33**(5): pp. 331-340.
148. Anilakumar, K.R., *et al.*, Nutritional, medicinal and industrial uses of sesame (Sesamum indicum L.) seeds-an overview. 2010. **75**(4): pp. 159-168.
149. Gupta, A., *et al.*, Indian medicinal plants used in hair care cosmetics: a short review. 2010. **2**(10): pp. 361-364.

150. Musaeva, N., *et al.*, Influence of Sesame oil immunogenesis and hemopoiesis in imuran inducedimmunodeficiency. 2021. **3**: pp. 1544-1550.
151. Koopman, M., *et al.*, Bodybuilding, sesame oil and vasculitis. 2005. **44**(9): pp. 1135-1135.
152. Sharma, D., P. Chaubey, and V. Suvarna, Role of natural products in alleviation of rheumatoid arthritis—A review. *J Food Biochem.*, 2021. **45**(4): p. e13673.
153. Khattab, A., D.M.N. Abouhussein, Development of injectable tenoxicam *in situ* forming microparticles based on sesame oil and poly-DL-lactide: Characterization, efficacy and acute toxicity. *Journal of Drug Delivery, Science and Technology*, 2019. **51**: pp. 682-694.
154. Devarajan, S., *et al.*, A blend of sesame oil and rice bran oil lowers blood pressure and improves the lipid profile in mild-to-moderate hypertensive patients. 2016. **10**(2): pp. 339-349.
155. Sankar, D., *et al.*, Effect of sesame oil on diuretics or ß-blockers in the modulation of blood pressure, anthropometry, lipid profile, and redox status. 2006. **79**(1): p. 19.
156. Takemoto, H., *et al.*, Effects of sesame oil aroma on mice after exposure to water immersion stress: Analysis of behavior and gene expression in the brain. 2020. **25**(24): p. 5915.
157. Kesmati, M., M. Mard-Soltani, and L. Khajehpour, Anxiogenic effects of acute injection of sesame oil may be mediated by β-1 adrenoceptors in the basolateral amygdala. *Adv Pharm Bull*, 2014. **4**(1): p. 35.
158. Asokan, S., Oil pulling therapy. *Indian J Dent. Res.*, 2008. **19**(2): p. 169.
159. Naseem, M., *et al.*, Oil pulling and importance of traditional medicine in oral health maintenance. 2017. **11**(4): p. 65.
160. Woolley, J., *et al.*, The effect of oil pulling with coconut oil to improve dental hygiene and oral health: A systematic review. 2020. **6**(8): p. e04789.
161. Athar, M. and S.M. Nasir, Taxonomic perspective of plant species yielding vegetable oils used in cosmetics and skin care products. *African Journal of Biotechnology*, 2005. **4**(1): pp. 36-44.
162. Chen, J., *et al.*, Development and evaluation of resveratrol, Vitamin E, and epigallocatechin gallate loaded lipid nanoparticles for skin care applications. 2017. **117**: pp. 286-291.
163. Jondhale, R.S., *et al.*, Potential of Herbs in Skin-Care. 2013. **4**(2): pp. 67-72.
164. Djiobie Tchienou, G.E., *et al.*, Multi-response optimization in the formulation of a topical cream from natural ingredients. 2018. **5**(1): p. 7.
165. Narasimhulu, C.A., *et al.*, Anti-atherosclerotic and anti-inflammatory actions of sesame oil. 2015. **18**(1): pp. 11-20.
166. Henriques Monteiro, E.M., *et al.*, Antinociceptive and anti-inflammatory activities of the sesame oil and sesamin. 2014. **6**(5): pp. 1931-1944.
167. Ismail, M., *et al.*, Anti-inflammatory, antioxidative, and hepatoprotective effects of trans 9-tetrahydrocannabinol/sesame oil on adjuvant-induced arthritis in rats. 2018. **2018**.

168. Abbasi, Z., *et al.*, Effects of sesame oil on the reproductive parameters of diabetes mellitus-induced male rats. 2013. **31**(2): pp. 141-149.
169. Mitra, A., Study on the benefits of sesame oil over coconut oil in patients of insulin resistance syndrome, notably type 2 diabetes and dyslipidaemia. *Journal of Human Ecology*, 2007. **22**(1): pp. 61-66.
170. Soltan, S.S.A., The protective effect of soybean, sesame, lentils, pumpkin seeds and molasses on iron deficiency anemia in rats. *World Applied Sciences Journal*, 2013. **23**(6): pp. 795-807.
171. Hall, B.E., F.H. Krusen, and H.W. Woltman, Vitamin B12 and coordination exercises for combined degeneration of the spinal cord in pernicious anemia. *J Am Med Assoc.*, 1949. **141**(4): pp. 257-260.
172. Pathak, N., *et al.*, Value addition in sesame: A perspective on bioactive components for enhancing utility and profitability. 2014. **8**(16): pp. 147.
173. Yashaswini, P.S., *et al.*, *In vivo* modulation of LPS induced leukotrienes generation and oxidative stress by sesame lignans. 2017. **41**: pp. 151-157.
174. Johnsen, J., *et al.*, Pure sesame oil vs isotonic sodium chloride solution as treatment for dry nasal mucosa. 2001. **127**(11): pp. 1353-1356.
175. Askari, A., *et al.*, The efficacy of topical sesame oil in patients with knee osteoarthritis: a randomized double-blinded active-controlled non-inferiority clinical trial. 2019. **47**: p. 102183.
176. Ji, J., *et al.*, Effect of roasting treatment on the chemical composition of sesame oil. 2019. **101**: pp. 191-200.
177. Carrín, M.E., A.A. Carelli, Peanut oil: Compositional data. *European Journal of Lipid Science and Technology*, 2010. **112**(7): pp. 697-707.
178. Akhtar, S., *et al.*, Physicochemical characteristics, functional properties, and nutritional benefits of peanut oil: a review. 2014. **54**(12): pp. 1562-1575.
179. Kirby, W.M., *et al.*, Intramuscular and subcutaneous administration of penicillin in beeswax-peanut oil. 1945. **129**(14): pp. 940-944.
180. Suchoszek-Łukaniuk, K., *et al.*, Health benefits of peanut (Arachis hypogaea L.) seeds and peanut oil consumption, in *Nuts and Seeds in Health and Disease Prevention*. 2011, Elsevier. pp. 873-880.
181. Coelho, S.B., *et al.*, Effects of peanut oil load on energy expenditure, body composition, lipid profile, and appetite in lean and overweight adults. 2006. **22**(6): pp. 585-592.
182. Arya, S.S., *et al.*, Peanuts as functional food: a review. 2016. **53**(1): pp. 31-41.
183. Jiang, L., *et al.*, Aqueous enzymatic extraction of peanut oil and protein hydrolysates. 2010. **88**(2-3): pp. 233-238.
184. Kritchevsky, D., S.A. Tepper, and J.A. Story, Cholesterol vehicle in experimental atherosclerosis: Part 16. Effect of peanut oil on pre-established lesions. *Atherosclerosis*, 1978. **31**(3): pp. 365-370.
185. Korrapati, D., *et al.*, Coconut oil consumption improves fat-free mass, plasma HDL-cholesterol and insulin sensitivity in healthy men with normal BMI compared to peanut oil. 2019. **38**(6): pp. 2889-2899.

186. Ring, J., M. Möhrenschlager, Allergy to peanut oil–clinically relevant? *J Eur Acad Dermatol Venereol.*, 2007. **21**(4): pp. 452-455.
187. Taylor, S.L., *et al.*, Peanut oil is not allergenic to peanut-sensitive individuals. 1981. **68**(5): pp. 372-375.
188. Bock, S.A., F.M. Atkins, The natural history of peanut allergy. *J Allergy Clin Immunol.*, 1989. **83**(5): pp. 900-904.
189. Bernard, H., *et al.*, Serological characteristics of peanut allergy in children. 2003. **58**(12): pp. 1285-1292.
190. Herz, R.S., Aromatherapy facts and fictions: a scientific analysis of olfactory effects on mood, physiology and behavior. *Int J Neurosci.*, 2009. **119**(2): pp. 263-290.
191. Sugumaran, M. and T.J.E.L. Vetrichelvan, Aromatherapy: The Power of Scent. 2008. **2008**(1): p. 78.
192. Cooke, A., *et al.*, Use of oil for baby skincare: a survey of UK maternity and neonatal units. 2011. **19**(6): pp. 354-362.
193. Zhou, Y., *et al.*, Study on the chemical constituents of nut oil from Prunus mira Koehne and the mechanism of promoting hair growth. 2020. **258**: p. 112831.
194. Kaszynski, E., The stimulation of hair growth in the flank organs of female hamsters by subcutaneous testosterone propionate and its inhibition by topical cyproterone acetate: dose-response studies. *Br J Dermatol.*, 1983. **109**(5): pp. 565-569.
195. Ramos, S., *et al.*, Antispasmodic Effects of Bauhinia microstachya. on Isolated Smooth Muscle. 2005. **43**(5): pp. 467-470.
196. Fraser, H., Oxyphenonium bromide as a mydriatic. *Br J Ophthalmol.*, 1956. **40**(12): p. 751.
197. Sukhotu, R., *et al.*, Changes in physiochemical properties and stability of peanut oil body emulsions by applying gum arabic. 2016. **68**: pp. 432-438.
198. Grompone, M.A., Sunflower oil. In *Bailey's Industrial Oil and Fat Products*, 2005.
199. Vicente, G., *et al.*, Kinetics of sunflower oil methanolysis. 2005. **44**(15): pp. 5447-5454.
200. Campbell, E.J., Sunflower oil. *Journal of the American Oil Chemists' Society*, 1983. **60**(2 Part 2): pp. 387-392.
201. Souza, A., *et al.*, A thermoanalytic and kinetic study of sunflower oil. 2004. **21**: pp. 265-273.
202. Garcés, R., *et al.*, Current advances in sunflower oil and its applications. 2009. **21**(4): pp. 79-82.
203. Bester, D., *et al.*, Cardiovascular effects of edible oils: a comparison between four popular edible oils. 2010. **23**(2): pp. 334-348.
204. Nandha, R., *et al.*, Therapeutic potential of sunflower seeds: An overview. 2014. **3**(3): pp. 967-972.

205. Adeleke, B.S., O.O. Babalola, Oilseed crop sunflower (Helianthus annuus) as a source of food: Nutritional and health benefits. *Food Science and Nutrition,* 2020. **8**(9): pp. 4666-4684.
206. Rosillo, M.A., *et al.*, Dietary extra-virgin olive oil prevents inflammatory response and cartilage matrix degradation in murine collagen-induced arthritis. 2016. **55**(1): pp. 315-325.
207. Aryaeian, N., F. Shahram, and M.J.L. Djalali, CLA has a useful effect on bone markers in patients with rheumatoid arthritis. 2016. **51**(12): pp. 1397-1405.
208. Braga, C., *et al.*, Olive oil, other seasoning fats, and the risk of colorectal carcinoma. 1998. **82**(3): pp. 448-453.
209. Tuyns, A.J., R. Kaaks, and M. Haelterman, Colorectal cancer and the consumption of foods: a case-control study in Belgium. 1988.
210. Galeone, C., *et al.*, Fried foods, olive oil and colorectal cancer. 2007. **18**(1): pp. 36-39.
211. Binkoski, A.E., *et al.*, Balance of unsaturated fatty acids is important to a cholesterol-lowering diet: comparison of mid-oleic sunflower oil and olive oil on cardiovascular disease risk factors. 2005. **105**(7): pp. 1080-1086.
212. Murphy, K., D. Saint, and P.R.C. Howe, Lack of effect of sugar cane and sunflower seed policosanols on plasma cholesterol in rabbits. *J Am Coll Nutr.* 2004. **13.**
213. Wardlaw, G.M. and J.T. Snook, Effect of diets high in butter, corn oil, or high-oleic acid sunflower oil on serum lipids and apolipoproteins in men. *Am J Clin Nutr.* 1990. **51**(5): pp. 815-821.
214. Katsarou, A.I., *et al.*, Serum lipid profile and inflammatory markers in the aorta of cholesterol-fed rats supplemented with extra virgin olive oil, sunflower oils and oil-products. 2015. **66**(7): pp. 766-773.
215. López-Varela, S., *et al.*, Decreased food efficiency ratio, growth retardation and changes in liver fatty acid composition in rats consuming thermally oxidized and polymerized sunflower oil used for frying. 1995. **33**(3): pp. 181-189.
216. Vaughn, A.R., *et al.*, Natural oils for skin-barrier repair: ancient compounds now backed by modern science. 2018. **19**(1): pp. 103-117.
217. Darmstadt, G.L., *et al.*, Topically applied sunflower seed oil prevents invasive bacterial infections in preterm infants in Egypt: a randomized, controlled clinical trial. 2004. **23**(8): pp. 719-725.
218. Darmstadt, G.L., *et al.*, Effect of topical treatment with skin barrier-enhancing emollients on nosocomial infections in preterm infants in Bangladesh: a randomised controlled trial. 2005. **365**(9464): pp. 1039-1045.
219. Kostik, V., *et al.*, Fatty acid composition of edible oils and fats. 2013. **4**: pp. 112-116.
220. Vijayakumar, M., *et al.*, A randomized study of coconut oil versus sunflower oil on cardiovascular risk factors in patients with stable coronary heart disease. 2016. **68**(4): pp. 498-506.

221. Eichenfield, L.F., A. McCollum, and P. Msika, The benefits of sunflower oleo-distillate (SOD) in pediatric dermatology. *Pediatr Dermatol.* 2009. **26**(6): pp. 669-675.
222. Stoia, M. and S. Oancea, Selected evidence-based health benefits of topically applied sunflower oil. *Applied Science Reports* 2015. **10**: pp. 45-49.
223. Chojnowski, M., F. Corbineau, and D. Côme, Physiological and biochemical changes induced in sunflower seeds by osmopriming and subsequent drying, storage and aging. *Seed Science Research*, 1997. 7(4): pp. 323-332.
224. Taheri, P.A., *et al.*, The effect of a short course of moderate pressure sunflower oil massage on the weight gain velocity and length of NICU stay in preterm infants. 2018. **50**: pp. 22-27.
225. Dias, M.F.R.G. Hair cosmetics: an overview. *Int J Trichology*, 2015. **7**(1): p. 2.
226. Gite, S.S., *et al.*, A Review on oils used in herbal cosmetics. 2013. **4**(2): pp. 61-64.
227. Odabasoglu, F., *et al.*, Beneficial effects of vegetable oils (corn, olive and sunflower oils) and α-tocopherol on anti-inflammatory and gastrointestinal profiles of indomethacin in rats. 2008. **591**(1-3): pp. 300-306.
228. Rodrigues, K.L., *et al.*, Cicatrizing and antimicrobial properties of an ozonised oil from sunflower seeds. 2004. **12**(3): pp. 261-270.

10

Non-Edible Oils as Biodiesel

Shilpi Khurana* **and Shilpi Bhatnagar**

Department of Chemistry, Deshbandhu College, University of Delhi, Kalkaji, New Delhi, India

Abstract
With the growing concern related to alleviating world energy demand, which is rising due to increasing urbanisation and population growth, we need to determine unconventional ways to replace fuels obtained from petroleum. Biodiesel has been an attractive and potential option to these petroleum-based fuels as they are ecologically friendly. The raw material used for the manufacturing of biodiesel can be obtained from non-edible as well as from edible oils. At present, the primary source of production of biodiesel are edible oils. Biodiesel can be produced mainly from edible oils, approximately 95%, but the expansive utilization of these has started causing a large imbalance in the human nutrition chain versus fuel. It has also caused food supply depletion as well as economic imbalance. So, to overcome the shortcomings of these it is necessary to take advantage of non-edible oils as primary material for commercial production of biodiesel. The potential source for the mass production of biodiesel from these non-edible oils can be secured from microalgae, jatropha, rubber seed, neem, karanja, mahua, silk cotton tree, etc. In a developing country like India, these are easily available in semi-wild conditions in the vicinity of villages and are quite economical as compared to edible oils.

***Keywords*:** Fossil fuels, non-renewable, urbanisation, economic imbalance, Jatropha

10.1 Introduction

Today the main sources of energy consumed by the world are from fossil fuels which are non-renewable, present in less amounts and going to be exhausted in the near future [1, 2]. The most used fossil fuel is diesel which

**Corresponding author*: khuranashilpi1@gmail.com

Divya Bajpai Tripathy, Anjali Gupta, Pooja Agarwal, Anuradha Mishra, and Arvind Kumar Jain (eds.) Oils and Fats as Raw Materials for Industry, (267–284) © 2024 Scrivener Publishing LLC

is employed in transportation and machinery due to its combustion properties [3]. Regarding diesel-powered vehicles, 33% are used in the US and Europe and consumption and transportation will rise to 75% by 2050 as reported by the International Energy Agency (IEA). This will result in an immense increase in CO_2 emissions [4, 5] which is a greenhouse gas and responsible for global warming. In order to replace these traditional fossil fuels, alternative resources of energy are now taken care of by the World Energy Council. Also, there is dire need to protect the environment from pollution, global warming and many social problems. Increasing population and increase in the number of vehicles is putting lot of pressure on use of these petroleum products. And it is a proven fact that these reserves will last only for a few years. It has been reported that 15 billion tons of CO_2 are getting disseminated every year. The term biodiesel was promoted in 1992 by the National Soy Development Board in USA [6]. It is reported that the production of biodiesel has risen from 9.2 Mtoe to 95 Mtoe from 2000 to 2018 [7]. Its production and consumption are leading in Europe and the US. The world biodiesel consumption by 56 countries was 26.8MT, out of which 58% is consumed by five countries as listed below in Table 10.1.

Countries like India and China have targeted to utilize 15% biodiesel blended fuel by 2020. The Malaysian government started B10 blended biodiesel in 2019 and it will increase B20 by 2020 [8]. Biodiesel is one of the most upcoming and promising alternatives to these fossil fuels; it is biodegradable, environmentally friendly and produces less carbon dioxide. It can minimise the dependency on fossil fuels and most importantly can be easily prepared using vegetable oils and non-edible oils [9–11]. Production of biodiesel is achieved by various agricultural and edible crops [12], non-edible oils [13] and algal biomass [14]. The total number of edible and non-edible crops known worldwide is 350 [15]. Table 10.2 encapsulates the presumptive and future feeding stuff for the global production of biodiesel.

Table 10.1 Biodiesel consumption of five countries.

USA	22%
Brazil	10%
France	10%
Indonesia	9%
Germany	7%
Others	42%

Table 10.2 Potential feedstock for the production of biodiesel.

Biofuels	Source	Examples	Advantages	Disadvantages
First generation	Edible oil	Sunflower, soybean, palm, rapeseed	Conversion process is simple	Causes deforestation and destruction of ecosystem Oil yield is low Debate on food vs. fuel economy
Second generation	Non-edible oil	Jatropha, candlenut, rubber neem	Large variety of non-edible available globally Dilemma over food vs. fuel economy	Intransigent structure of feedstock
Third generation	Algal biomass	Bacteria, fungi, microalgae	High lipid content Growth rate is high	Technology required for conversion is quite advanced

10.2 Tussle Between Food and Fuel

The use of vegetable oils and non-edible oils have a fair advantage since they do not add carbon dioxide gas in the atmosphere. These biofuels are not only saviours of our environment but also provide employment to rural people through plantations. At present, the primary source of production of biodiesel is edible oils. The biggest issue in using these edible oils is that they are causing destruction of the ecosystem due to urbanisation and expansion of plantation. Mainly, manufacturing biodiesel is done utilising the edible oils and that amounts to approximately 95% [16] but the enormous utilization of these has started causing a large imbalance in human nutrition chain versus fuel. It has also caused food supply depletion as well as economic imbalance. Many developing countries are still facing scarcity of food and conversion of these edible oils could lead to acute food shortage. These edible oils, for example, are palm oil, soya bean oil, and coconut oil. In a developing country like India, the price of unrivaled quality refined edible oil is quite high, almost four times the value of diesel. So, in order to curb the shortcomings of these it is necessary to utilize non-edible oils as feeding stock for economic productivity of biodiesel without tampering with the food supply. Non-edible oils plants can be easily cultivated in waste land and the cost of cultivation is quite low. A few examples of non-edible oils are jatropha (*Jatropha curcas*), neem (*Azadirachta indica*), karanja (*Pongamia pinnata*), rubber seed (*Hevea brasiliensis*), mahua (Madhuca indica), and polanga (*Calophyllum inophyllum*).

10.3 Non-Edible Oils as Potential Feedstock

Animal fats such as pork lard and beef tallow can also be considered as probable feedstock for the commercial manufacturing of biodiesel [17]. Algal biomass possess high lipid content and a small quantity gives a high yield of biodiesel. It can be a potential source for the production of biodiesel but it requires quite advanced technology [18]. Non-edible oils produce fuels which are considered as second-generation biofuels. The non-edible oils can be obtained from rubber seeds, neem, karanja, and yellow oleander, which make up the second-generation feedstock for the production of biodiesel. The compositions of edible and nonedible oils are the same with respect to fatty acids, saturated and unsaturated fats but edible oils have more nutrient value and contain antioxidants. However, toxic substances present in non-edible oils such as Jatropha and rubber seed makes them unfit for human consumption [19–21]. For example, Cyanogenic glucoside in rubber seed oil and purgative present in Jatropha seed oil makes these oil unsuitable for human consumption. However, these plants can be grown anywhere as compared to edible oil plants because they do not require much soil nutrition or an extensive irrigation system. The above-mentioned advantages make these non-edible oils potential feedstock for the commercial production of biodiesel and also overcome the issues related to food versus fuel. Another main advantage of using non-edible plants is that they can be easily grown in wastelands all over the world which will prevent deforestation also. Further, they can be transferred in liquid state and are biodegradable with low content of sulphur [22]. Most importantly, the oil content present in these plants makes them valuable feedstock for production, say for instance, jatropha seed contain 40-60 w% [23], candlenut 60-65wt% [24] and mango seed 40-50 wt% [25]. Thus, to decrease the

Table 10.3 Production of non-edible crops annually.

Crops that yield non-edible oils	Yield (Kg per ha per year)
Jatropha	2500
Mahua	0-200
Candlenut	16000
Rubber	100-150
Jojoba	500-5000
Karanja	900-9000
Sea mango	1900-2500
Neem	2670
Tobacco	1170

burden on land area these non-edible crops have unique features that they can be grown in saline, sandy and gravel soils that are unfit for production of food. Table 10.3 summarises the annual production of some of the non-edible crops. The total biodiesel production cost depends upon the raw material. Non-edible crops namely *jatropha, candle nut, neem,* and *karanja* grow in abundance. The utility of non-edible oil as a feedstock will reduce the cost for the production of biodiesel due to cheaper raw material, and they require least cultivation labour to grow anywhere as well as in the soils which are not good enough for the growth of edible oils.

10.4 Non-Edible Plants as Raw Material

Edible crops as biodiesel feedstock will be a burden on land area and it will decrease production of food, creating competition between food and fuel. Since non-edible crops possess unique botanical characteristics that they can be easily grown in lands which are no longer fertile, such as saline, gravel and sandy soils, makes the potential basic unprocessed material for the commercial manufacturing of biodiesel. Over the past few years, the synthesis from these non-edible crops has been greatly and broadly researched [26, 27]. The amount of oil content present in the various non-edible crops is summarised in Table 10.3.

Jatropha (Jatropha curcas)

Jatropha is the part of the family *Euphorbiaceae* which is a tropical tree; it is drought resistant having a height of 5-7 metres and can survive in abandoned agricultural land [28]. This tree is found in areas of Peru, Brazil Mexico, Argentina, Bolivia and tropics including Africa and Asia [29]. Its seed contain approximately 30-40% oil content and it is the leading plant towards the commercialization of non-edible vegetable oil. Jatropha is a very important non-edible crop as one hectare of Jatropha can absorb 25 tons of CO_2 from atmosphere per annum, which is very important for climate change concerns. Palmitic acid, stearic acid, oleic acid, linoleic acid are a few of the saturated and unsaturated fatty acids present in the oil (Table 10.4).

Mahua (Madhuca indica)

Mahua (*Madhuca indica*) is member of the family *Sapotaceae,* which is a very fast-growing medium-size evergreen tree. It is also known as butter tree found in India, Pakistan and Malaysia having medium height approximately 20 metres [30–32]. This tree is unique as the production of seeds starts after 10 years and advances up to 60 years. Mahua grows in hot and wet regions and each plant yield 20-40 kg of seeds per year. The percentage of oil and protein present in seed is 35% and 16%, respectively (Table 10.4).

Ricinus communis (Castor)

Castor (*Ricinus communis*) grows quite well within temperature range 20-25°C; it is a member of a plant species the family *Euphorbiaceae* [33]. This tree has an advantage over Jatropha and Pongamia in that its growing period is small and farmers have more experience regarding its cultivation. Its height is about 6 metres and it is considered as a small wooden tree found in wild states in all tropical and subtropical countries. Since the oil content is extremely high, i.e., 48–60%, it is ideally suitable as potential raw material for the commercial manufacturing of biodiesel. Since it is highly viscous, it causes choking of the fuel injectors and incomplete combustion [34]. The seeds are poisonous as they contain ricin and other toxic compounds that make them unfit for human and animal consumption.

Hevea brasiliensis (Rubber)

The rubber tree (Hevea brasiliensis) is associated with the family *Euphorbiaceae* which is 34 metres high and is native to the Amazon Forest in Brazil [35]. The annual plantation yield is approximately 100 to 150Kgha-1 rubber seeds annually. Hevea brasiliensis is found in Sri Lanka, Indonesia, Malaysia and India. The colour of rubber seed is dark and the oil content present is 40-60%. The composition is of saturated and unsaturated fatty acids such as stearic acid, palmitic acid, linoleic acid and linolenic acid (Table 10.4). Seed production yields approximately 70-500 Kg per hectare.

Sapindus mukorossi (Soap nut)

Soap nut (*Sapindus mukorossi*) is a perennial tree that grows well in deep loamy soils and leached soils to prevent soil erosion [36]; it belongs to the family *Sapindaceae*. It is found in tropical and sub-tropical areas. The seeds of Sapindus mukorossi have oil content 23-51%, out of which 93% are triglycerides [37].

Table 10.4 Composition of fatty acids present in non-edible oil for the production of biodiesel.

Non-edible oil	Palmitic acid	Stearic acid	Oleic acid	Linoleic acid	Linolenic acid
Jatropha curcas	14.6	7.6	44.6	31.9	0.3
Madhuca indica	21.36	18.97	38.98	19.47	0.16
Moringa oleifera	6.9	8.3	67.7	10.7	
Azadirachta indica	14.9	14.4	61.9	7.5	0
Simmondesia chinesis	1.59	4.14	42.84	31.52	NA
Aleurites moluccana L	6.23	2.23	26.26	39.71	24.86
Nicotiana tabacum	10.96	3.34	15.54	69.49	0.69
Pongamia pinnata	9.8	6.2	72.2	11.8	NA
Hevea brasiliensis	10.2	8.7	24.6	39.6	16.3

Pongamia pinnata (karanja)

The karanja is also known as *Pongamia pinnata*; it belongs to the family *Fabaceae* and has a height of 12-15 metres with branches that spread like a crown [38]. It can be easily grown in clayey soil, sandy soil and stony soil [39]. It is a perennial fas- growing leguminous tree found in coastal areas of Southeastern Asia and Australia. 30-40% oil is present in seeds and is cheaper and less toxic than Jatropha oil. Pongamia pinnata oil has similar physical and chemical properties as diesel but produces a large amount of carbon residue and is more viscous. The oil is non-edible as it contains Karajin and toxic ketone pongamol [40]. The various saturated and unsaturated acids present are linoleic acid, palmitic acid and stearic acids [41] (Table 10.4). Though the physical and chemical properties resemble quite closely diesel oil, it has high viscosity and produces a greater amount of carbon residue.

Nicotiana tabacum (Tobacca)

Nicotiana tabacum (Tobacca) generally called as tobacco, belongs to the family *Solanaceae* grown in South and North America, Russia and India. It is a herbaceous plant and its seed oil content varies from 33 to 40% [42]. The major fatty acids are oleic acid, linoleic acid palmitic acid and stearic acid [43].

Candle nut (*Aleurites moluccana L.*)

Aleurites moluccana L. is a flowering tree which is also known as Candle nut, having a height of approximately 20 metres. It is known by different names such as Kemiri in Indonesia and Kukui nut in Hawaii [44]. The gross annual production of candle nut seed is 16 ton per ha and oil yield is approximately 3200 kgha. Seeds of candle nut contain 30-60% candlenut oil which are utilised for the manufacture of biodiesel.

Simmondesia chinesis (jojoba)

Simmondesia chinesis (jojoba) is a part of the family *Simmondsiaceae* [45]. This tree is found in areas of the Sonoran deserts and Mexico; its height is 0.7-1.0 metres [46]. The high oil content i.e., 45-55% [47] and its bio degradable, non-toxic nature makes it a best choice as non-edible crop for the production of biodiesel. It has a high flash point, low volatility, high viscosity and high dielectric constant. It is quite unique as its seed contains 50% oil by weight which is more than any oil seed crops.

Azadirachta indica (Neem)

Azadirachta indica (Neem) is associated with the family Meliaceae. It is a portentous, fast-growing, tropical tree 25 metres in height [48].

Azadirachta indica is found in Asia, Africa, Central and South America, Pakistan, Burma and India. This tree has a unique feature; not only can it be easily grown in non-fertile and degraded soils but it can also countenance high temperatures. Though it starts bearing fruit between 3 to 5 years it becomes completely gratifying in 10 years. Azadirachta indica seeds contain oil content 45% and the various saturated and unsaturated fatty acids are present such as oleic acid, palmitic acid and stearic acid.

Vernica fordii (Tung)

Tung belongs to the family Euphorbecaeae; it is an oil-bearing woody plant. It is native to China, Burma, and Vietnam having average height of 20 m [49]. It is a deciduous tree which is small to medium in size and produces 450-600 Kg per ha oil per annum [50].

Moringa oleifera (drumstick tree)

Moringa oleifera, commonly known as drumstick tree, belongs to the family Moringaceae. It is a fast-growing softwood perennial tree and is cultivated in the regions of the Philippines, Sudan, and Latin America. Its average height is 10 m; it is drought resistant and can endure poor soil with outspread rainfall ranging from 25-300 cm per year [51]. 3000 Kg *Moringa oleifera* seeds could be obtained from one hectare and oil content varies from 30-40% depending upon climate and plant variety. It contains palmitic acid, linoleic acid and stearic acid [52].

Calophyllum inophyllum (Polonga)

Calophyllum inophyllum is also called as polonga or hone; it is a medium to large evergreen tree belonging to the family *Clusiaceae.* It is non-edible oil seed found in India, Southeast Asia, East Africa and Australia. Polonga seeds contain 50-70% oil content that includes various saturated and unsaturated fatty acids. Calophyllum inophyllum oil is used in cosmetics, medicine, soap, and lamp oil in various parts of the world.

Schleichera oleosa (Kusum)

Schleichera oleosa is a well-recognised deciduous or evergreen tree which belongs to the family Sapindaceae, also known as kusum, found in Southeast Asia. Parts such as leaves, fruits, seeds are edible and are used in various medicines and dyes. Kusum oil contains 40% unsaturated and 53% saturated fatty acids [53].

Cascabela thevetia (Yellow Oleandar)

It is a poisonous, drought resistant non-edible shrub native throughout Mexico and in Central America. The height of this shrub is approximately

10-18 feet and its seed contains 60-65% oil content; it is and cultivated widely as an ornamental.

10.5 Properties of Non-Edible Oils for Biodiesel as a Future Fuel

FLASH POINT

Storage of non-edible biodiesel is a very important parameter and flash point is the deciding factor. The minimum temperature at which fuel can vaporise to form an ignitable mixture on application of ignition source under given conditions is the flash point of a fuel. It is inversely proportional to the fuel volatility. A lower value of flash point indicates higher flammability whereas a higher value of flash point makes them safe to transport and reduces the risk of fire [54, 55]. Since methyl esters obtained from these non-edible oils have a high value of flash point this makes them less volatile and safer to handle and use. It was found that the flash point of *Jatropha curcas* is 158.8°C whereas the flash point of *Mesua ferrea* is 112°C. Table 10.5 gives the flash point of different non-edible oils.

THE CETANE NUMBER

It is another important parameter that decides the quality of fuel, especially ignition quality. Higher value of cetane number is an indication of better ignition properties. The cetane number basically tells how easily ignition can occur and how smooth combustion can be. The higher value of cetane number the better. Since biodiesel has a higher value of cetane number when compared with petroleum it results in smooth burning, with high thermal efficiency and less consumption of fuel, and it can easily replace petrol. It was observed that *Moringa oleifera* is 66 whereas *Vernicia montana* has the lowest cetane number as raw oils have more unsaturated fatty acids (Table 10.5).

DENSITY

It is an engine performance characteristic which plays a crucial role in efficiency of fuel. It is defined as mass per unit volume of a liquid at a given temperature. Due to the presence of higher molecular weight triglycerides, the methyl esters of non-edible oils have higher density. *Simmondsia chinesis* has the highest density, i.e., 920kg/m3 while *Scleichera oleosa* has a lowest value 865.5 920kg/m3.

Table 10.5 Properties of biodiesel.

Biodiesel	Density Kgm^{-3}	Cetane number	Viscosity mm^2/s	Flash point (K)	Pour point (K)	Cloud point (K)	Iodine number mg g-1
Jatropha	880	51-52	4.8	3	279	0	135
Castor	886	53	12	43	247	28	84
Karanja	889	41.7	6.88	461	288.8	283	87
Mahua	916	51	3.98	423	279	-	71
Neem	890	-	5.53	402	276	278	85
Rubber	892	66.2	5.81	423	265	279	135

VISCOSITY

It is a property of a liquid to resist to flow due to internal friction. Fuel with high viscosity leads to incomplete combustion, fuel atomization is poor and carbon deposition. So, biodiesel must have low viscosity. Viscosity of jatropha curcas was found to be 4.8mm2/s, whereas viscosity of Azadirachta indica is 5.1 8mm2/s. Ricinus communis methyl esters are highly viscous 9.2 mm2/s.

IODINE NUMBER

The degree of unsaturation in fatty acids can be determined by iodine number. The higher value of iodine number means greater unsaturation and thus the lower the oxidative stability of biodiesel. Non-edible oil has high unsaturation and thus higher iodine value. Iodine numbers of non-edible oils are listed in Table 10.4.

CLOUD POINT AND POUR POINT

Cloud point and pour point are two important parameters for low temperature applications of a fuel. The temperature at which fuel starts to form crystals and further decrease in temperature increases the size of crystals, and fuel gels is the cloud point whereas pour point is the lowest temperature at which fuel will flow. Biodiesel fuel and methyl esters of these non-edible oils have quite similar cloud point and pour point. The compositions of the fatty acid of non-edible oils will have varied cloud and pour point (Table 10.5).

10.6 Extraction of Non-Edible Oil

Extensive research is taking place these days to replace the edible oils with non-edible oils but high FFA, less harvest of few non-edible crops, high content of polyunsaturated fatty acids posed an obstacle in commercial production of biodiesel [49]. A diesel which contains lower amount of polyunsaturated and saturated fatty acids and more amount of monosaturated fatty acid gives a better quality of biodiesel. To increase the yield of seed oil, many scientists and biotechnologists have now started utilising the technology like genetic engineering to modify plants to increase seed oil yield and improve the composition of fatty acids. For example, Roesler obtained increase 5% rapeseed oil by modifying chloroplast of rapeseed by using cytosolic version of acetyl-CoA carboxylase enzymes [56].

Different technologies have been used to extract non-edible oils from a variety of oil seeds. These can be classified as:

1. Soxhlet extraction
2. Enzymatic extractions
3. Mechanical screw press
4. Extraction using microwave

The techniques which are more versatile are soxhlet extraction techniques and mechanical screw press techniques which are widely used because of higher yield and ease of operation. scCO2 extraction technique offers various advantages over other extraction methods and thus is extensively used these days. Mechanical screw press is also familiar as hydraulic press is the most accepted method for the extraction of oils having high oil content such as Jatropha, Candle nut and olive [57–59]. This technique utilizes rotating worm shaft which increases the pressure, thus reducing the space and volume in the extraction chamber which helps in squeezing oil from various oil seeds. However, this technique has various limitations such as high maintenance cost, and tedious pre-treatments are also required such as particle size reduction, drying, dehulling, and this method also requires further oil purification. Jatropha seed oil is extracted using mechanical extraction method as moisture plays a very important role, whereas in case of extraction of hazel nut oil microwave radiations are used in pre-treatment. Soxhlet extraction technique is also very frequently used for the extraction of oil from various oil seeds. Various solvents can be used such as acetone, chloroform, ethanol, methanol and hexane which is most widely used [60, 61]. In this technique particle size plays a vital role since small particle size is suitable as it has greater interfacial area between solvent and solid matrices [62]. Increase in temperature increases the solubility and also influences the extraction rates. Soxhlet extraction method is used to extract *Catharanthus roseus* seed oil using methanol as a solvent which gives the highest yield, i.e., 31.50%. The combination of solvent extraction and microwave heating has many benefits such as high extraction rate and quality, less use of solvent and short extraction time [63, 64].

With the rising environmental pollution and concern about using various organic solvents in conventional oil extraction techniques which are volatile and toxic, Super critical CO_2 technology can be an attractive alternative as it is a green technology. It involves non-volatile and non-toxic solvents. This method requires low temperature and since it is a greenhouse gas, the CO_2 which is used in supercritical extraction method returns to the atmosphere after extraction. Thus, it does not cause greenhouse effect [65]. The scCO_2 technology has several advantages over existing methods such as high solubility, selectivity and not standing in need of any refining process. For example: due to the fact that candlenut oil is quite prone to oxidation because the amount of unsaturated fatty acids are quite high it requires an extraction method which works at low temperature. $scCO_2$ offers a very promising technique in the above case [66].

10.7 Emissions Characteristics of Non-Edible Vegetable Oils

Carbon monoxide emissions are substantially reduced when diesel is replaced by biodiesel. Emissions from *Jatopha curcas* biodiesel, *Madhuca indica* biodiesel, and *Hevea brasilensis* biodiesel decreased by 20-25%, 26-58% and 41-60%, respectively. Carbon dioxide is a main greenhouse gas responsible for global warming leading to change in climatic condition. Extensive studies are being focussed on reducing carbon dioxide emissions. Carbon dioxide emissions are reduced in case of *Jatopha curcas* biodiesel, *Hevea brasilensis* biodiesel and *Pongamia pinnata* biodiesel due to lower carbon to hydrocarbon ratio as compared to petroleum. Also, high emissions of carbon dioxide are not of much concern because usage of biodiesel maintains a closed carbon cycle which does not increase its level in the atmosphere. Fatty acid composition plays a crucial role in nitrogen dioxide emissions. The higher content of unsaturated fatty acids is responsible for a large number of emissions of nitrogen dioxide. The emissions from *Pongamia pinnata* biodiesel, *Madhuca indica* biodiesel, *Simmondisa chinesis* biodiesel are 4.15-14.18%, 6-16%, and 14-16% higher than that of petrodiesel.

10.8 Conclusion

The positive impact on the environment and the economy has made biodiesel quite engaging as an alternate fuel as compared to fossil fuels which are non-renewable and are at the verge of exhaustion. There are growing concerns, especially the struggle between fuel and food due to extensive use of edible oil crops and high feedstock cost leading to production cost. So, the need and the insistence for using non-edible crops as a suitable feedstock for the economic manufacturing of biodiesel is increasing substantially. Non-edible crops to be used as conceivable feedstock is also because they are not fit for consumption as they contain many toxic components. Not only that, these crops can be easily grown in harsh and non-fertile lands where moisture content is quite low and such lands are not fit for the growth for the production of edible crops. Fatty acid compositions and various properties of non-edible oils are quite similar to edible oil, thus can be easily replaced though high FFA are the main obstacles in the use of non-edible oils as a potential feedstock. The technology for extraction like $scCO_2$ has a great advantage as it gives non-edible oil with least possible FFAs content and moisture content. The properties of non-edible oils and subsequently biodiesel give ample evidence to exploit these non-edible crops to extract non-edible oil for mass-scale production of biodiesel synthesis without disrupting the comprehensive economy of food.

References

1. Shahsavari, A., and Akbari, M., (2018). *Renewable Sustainable Energy Rev.*, 90, 275–291.
2. Agarwal, K., Gupta, J. G., and A. Dhar., (2017). *Prog. Energy Combust. Sci.*, 61, 113–149.
3. Trutnevyte, E., McDowall, W., Tomei, J. and I. Keppo. (2016) *Renewable Sustainable Energy Rev.*, 55, 326–337.
4. Staffell, I., Scamman, D., Abad, A. V., Balcombe, P., Dodds, P. E., Ekins, P., Shah, N.and Ward, K. R. (2019). *Energy Environ. Sci.*, 12, 463–491.
5. Salvucci, R., Petrovíc, S., Karlsson, K., Wr˚ake, M., Uteng, T. P. and Balyk, O. (2019). *Energies*,12, 2232.
6. Karmakar, A., and Mukherjee, S. (2017) *Indian J. Agric. Res.*,51, 529–535.
7. Czarny, R. M., (2020). *The Nordic Dimension of Energy Security*, Springer, pp. 67– 99.
8. Farid, M. A. A., Roslan, A. M., Hassan, M. A., Hasan, M. Y., Othman M. R. and Shirai, Y. (2020). *Sustain. Energy Technol. Assess.*, 39, 100700.
9. Munir, M., Ahmad, M., Saeed, M., Waseem, A., Rehan, M., Nizami, A.-S, Zafar, M., Arshad, M. and Sultana, S. (2019). *Renewable Sustainable Energy Rev.*, 109, 321–332.
10. Pan, H., Li, H., Zhang, H., Wang, A., Jin, D. and Yang, S. (2018). *Energy Convers. Manage.*, 166, 534–544.
11. Mercuri, E., Manca, D., Abderazag, O., Patel, R. and Mujtaba, I. (2016) *Computer Aided Chemical Engineering*, Elsevier, vol. 38, pp. 1713–1718.
12. Rajak, U. and Verma, T. N. (2018). *Energy Convers. Manage.*, 166, 704–718.
13. Rodrigues, J., Miranda, I., Gominho, J., Vasconcelos, M., Barradas, G., Pereira, H., Bianchi-de-Aguiar, F. and Ferreira-Dias, S. (2016) *Ind. Crops Prod.*, 83, 614–619.
14. Correa, D. F., Beyer, H. L., Possingham, H. P., Thomas Hall, S. R. and Schenk. (2017). *Renewable Sustainable Energy Rev.*, 74, 1131–1146.

15. Folayan, A. J., Anawe, P. A. L., Aladejare, A. E. and Ayeni, A. (2019) *Energy Rep.*, 5, 793–806.
16. Sajjadi, A., Raman, A. A. A. and Arandiyan, H. (2016) *Renewable Sustainable Energy Rev.*, 63, 62–92.
17 Adewale, P., Dumont, M.-J. and Ngadi, M. (2015) *Renewable Sustainable Energy Rev.*, 45, 574–588.
18. Dong, T., Knoshaug, E. P., Davis, R., Laurens, L. M., Van Wychen, S., Pienkos. P. T. and Nagle, N. (2016). *Algal Res.*, 19, 316–323.
19. Atabani, A., Silitonga, A., Ong, H., Mahlia, T., Masjuki, H., Badruddin, and H. Fayaz, I. A. (2013) *Renewable Sustainable Energy Rev.*, 18, 211–245.
20. Kumar, A. and Sharma, S. (2008) *Ind. Crops Prod.*, 28, 1–10.
21. Abdullah, M., Salimon, J., Yousif, E. and Salih, N., J. (2013) *Assoc. Arab Univ. Basic Appl. Sci.*, 14, 83–86.
22. Ferrero, G. O., Faba, E. M. S., Rickert. A. A. and Eimer, G. A. (2020). *Renewable Energy*, 150, 128–135.
23. Demirbas, A., Bafail, A., Ahmad, W., and Sheikh, M., (2016). *Energy Explor. Exploit.*, 34, 290–318.
24. Pham,L. N., Van Luu, B., Phuoc, H. D., Le, H. N. T., Truong, H. T., Luu, P. D., Furuta, M., Imamura, K., and Maeda, Y., (2018). *J. Oleo Sci.*, ess17220.
25. Khurana, S. P., and Gaur, R. K., (2019 *Plant Biotechnology: Progress in Genomic Era*, Springer.
26. Ghadge, S.V., and Raheman, H., (2005) Biodiesel production from mahua (Madhuca indica) oil having high free fatty acids. *Biomass Bioenergy*, 28(6), 601–605.
27. Thangaraj, B., Ramachandran, K.B., and Raj, S.B., (2014). Homogeneous catalytic transesterification of renewable Azadirachta indica (Neem) oil and its derivatives to biodiesel fuel via acid/alkaline esterification processes. *International Journal of Renewable Energy and Biofuels*, 11, 1–16.
28. Atabani, A.E., Silitonga AS, H.C. Ong, *et al.* (2013) non-edible vegetable oils: A critical evaluation of oil extraction, fatty acid compositions, biodiesel production, characteristics, engine performance and emissions production. *Renewable and Sustainable Energy Reviews*, 18, 211–245.
29. M. Balat, *Energy Convers. Manage.*, 2011, 52, 1479–1492.
30. G.U. Chidrewar, J.H. Tanavade, S.H. Deshpande, *et al.* Anti-ulcer and anti-oxidant activity of leaves of Madhuca indica in rats. *Oriental Pharmacy and Experimental Medicine* 2010, 10(1), 13–20.
31. Acharya, N., Nanda, P., Panda, S., and Acharya, S., (2017). *Eng. Sci. Technol.*, 20, 511–517, 3536.
32. Mahalingam, A., Devarajan, Y., Radhakrishnan, S., Vellaiyan, S, and Nagappan, B., (2018). *Alexandria Eng. J.*, 57, 2627–2631.
33. Makama, B. Y., Wilson, O., and Claver, J. H., (2018). *Am. J. Environ. Prot.*, 6, 35–38.
34. Baroutian, S., Aroua, M.K., Raman, *et al.* (2010). Viscosities and densities of binary and ternary blends of palm oil+ palm biodiesel + diesel fuel at

different temperatures. *Journal of Chemical and Engineering Data*, 55(1), 504–507.
35. Kant, P., Shuirong, W., Chaliha, S., (2011). Going beyond Jatropha: Can an expanded land and feedstock base help India meet its ambitious biodiesel target? *Institute of Green Economy*, IGREC working paper IGREC-22.
36. Chhetri, A.B., Tango, M.S., Budge, S.M. *et al.* (2008). Non-edible plant oils as new sources for biodiesel production. *International Journal of Molecular Sciences*, 9(2), 169–180.
37. Mathiarasi, R., and Partha, N., (2017). *Saudi Chem. Soc.*, 21, 11–17.
38. Naik, P.L., and Katpatal, D.C., (2013). Optimization of effective parameters of Pongamia Pinnata (Karanja) biodiesel using taguchi method. *International Journal of Engineering Research and Applications*, 3(4), 701–706.
39. Gravalos, I., Ziakas, N., Loutridis, S., and Gialamas, T., (2019). *Comput. Electron. Agric.*, 166, 104986.
40. Patel, R. L., and Sankhavara, C., (2017). *Renewable Sustainable Energy Rev.*, 71, 464- 474
41. Ortiz-Martínez, V., Salar-García, M., Palacios-Nereo, F., Olivares-Carrillo, P., (2016). J. Quesada-Medina, A. P. De Los Ríos and F. Hernández-Fernández, *J. Supercrit. Fluids*, 113, 23–30.
42. Khan, T.M.Y., Atabani, A.E., Badruddin, I.A., *et al.*, (2014). Recent scenario and technologies to utilize nonedible oils.
43. Ni, Z.-h., Li, F.-s., Wang, H., Wang, S., Gao, S.-y., and Zhou, L., (2020). *Renewable Energy*, 145, 93–98.
44. Subroto, E., Manurung, R., Heeres, H. J., and Broekhuis, A. A., (2015). *Ind. Crops Prod.*, 63, 294–302.
45. Moser, B.R., (2011). Biodiesel production, properties, and feedstocks. In *Vitro Cellular and Developmental Biology-Plant*, 45, 229–266.
46. Sandouqa, A., and Al-Hamamre, Z., (2019). *Renewable Energy*, 130, 831–842.
47. Azad, A., Rasul, M., and Bhatt, C., (2019). *Energy Procedia*, 156, 159–165.
48. Aransiola, E., Ehinmitola, E., Adebimpe, A., Shittu, T., and Solomon, B., (2019). *Advances in Eco-Fuels for a Sustainable Environment*, pp. 53–87.
49. Gadhave, S. L., and Ragit, S. S., (2020). *Biofuels*, 11, 49–55.
50. Liu, Z., Li, J., Knothe, G., Sharma, B. K., and Jiang, J., (2019). *Energy Fuels*, 33, 5110–5115.
51. Boukandoul, S., Casal, S., and Zaidi, F., (2018). *Agriculture*, 2018, 8, 150.
52. V. Karthickeyan, *Fuel*, 2019, 235, 538–550.
53. Banka, S., and Parikh, S. P., (2019). *Asia-Pac. J. Chem. Eng.*, 14, e2310.
54. Sanjay, B., (2013). Non-Conventional Seed Oils as Potential Feedstocks for Future Biodiesel Industries: A Brief Review. *Research Journal of Chemical Sciences*, 3(5), 99-103.
55. Vuppaladadiyam, V.K. Sangeetha, C.J., Sowmya, V., (2013). Transesterification of Pongamia pinnata Oil Using Base Catalysts: A Laboratory Scale Study. Universal *Journal of Environmental Research and Technology*, 3, (1), 113-118.

56. Roesler, K., Shintani, D., Savage, L., Boddupalli, S., and Ohlrogge, J. (1997) *J. Plant Physiol.*, 113(1), 75–81.
57. Martín, C., Moure, A., Martíın, G., Carrillo, E., Domínguez., H., and Parajó, J. C., (2010). *Biomass Bioenergy*, 2010, 34, 533–538.
58. Tambunan, A. H., Situmorang, J., Silip, J., Joelianingsih, A., and Araki, T., (2012). *Biomass Bioenergy*, 43, 12–17.
59. Veneziani, G., Esposto, S., Taticchi, A., Selvaggini, R., Urbani, S., Maio, I. Di., Sordini, B., and Servili, M. (2015). *J. Agric. Food Chem.*, 63, 6066–6074.
60. Uquiche, E., Jeréz, M., and Ortíz, J., (2008). *Innovative Food Sci. Emerging Technol.*, 2008, 9, 495–500.
61. Qin, S., Sun, Y., Meng, X., and Zhang, S., (2010). *Energy Explor. Exploit.*, 28, 37–46.
62. Karthikeyan, M., Renganathan, S., Baskar, G., and Nambirajan, S., (2017). *Energy Sources, Part A*, 39, 1746–1753.
63. Liew, S. Q., Ngoh, G. C., Yusoff, R., and Teoh, W. H., (2016). *Int. J. Biol. Macromol*, 93, 426–435.
64. Wu, L., Hu, M., Li, Z., Song, Y., Yu, C., Zhang, H., Yu, A., Ma, Q., and Wang, Z., (2016). *Food Chem*, 92, 596–602.
65. Hossain, S., Balakrishnan, V., Rahman, N. N. N. Ab., Rajion, Z. A., and Kadir, M. O. Ab., (2013). *J. Supercrit. Fluids*, 83, 47–56.
66. Subroto, E., Widjojokusumo, E., Veriansyah, B., and Tjandrawinata, R. R., (2017). *J. Food Sci. Technol*, 54, 1286.

11

Ecological and Economic Aspects of Oil and Fats

Shivang Dhoundiyal, Awaneet Kaur* and Md. Aftab Alam

Department of Pharmacy, School of Medical and Allied Sciences, Galgotias University, Greater Noida, Uttar Pradesh, India

Abstract
Fats and oils are fundamental for healthy nutrition as they provide energy, insulate and safeguard the body's vital organs, provide essential fatty acids, and are necessary for the absorption of the fat-soluble vitamins A, D, E, and K. The vast quantities of oilseed and fruit grown around the world attest to the vital role fats and oils play in the international economy. Billions of people produce and process oils and fats around the world, contributing to their economic significance. The importance of the animal feed generated from processing of most oil plants gives a special importance to the production of oils and fats. Their value could further improve if oil-rich crops were made available to the chemical industries as a renewable resource.

***Keywords*:** Palm oil, ecologically sustainable, economic aspects, price disparities

11.1 Introduction

Due to our high fat-requirements, our predecessors have been called "fat hunters" [1]. Despite the fact that some parts of the world consume too many oils and fats, it is estimated that 720 to 811 million people will be undernourished globally by 2020 [2, 3]. It is crucial to ensure that malnourished persons have access to fats because they provide 25-30% of the daily calorie needs in the average healthy diet [4]. To achieve current recommendations for fat consumption, an additional 45 million tonnes of

**Corresponding author*: kaurawaneet@gmail.com

Divya Bajpai Tripathy, Anjali Gupta, Pooja Agarwal, Anuradha Mishra, and Arvind Kumar Jain (eds.) Oils and Fats as Raw Materials for Industry, (285–306) © 2024 Scrivener Publishing LLC

dietary fats and oils per year is needed [5]. Both areas of high fat consumption, notably from animal sources, and those of low-fat consumption will see changes in their fat intake. An extra 88–139 Mt is needed if the "fat gap" is expected to persist until 2050. In 2019, soybean oil produced 9.88 grams of fats per person worldwide, followed by palm oil at 7.17 g, sunflower oil at 4.35 g, butter and ghee at 3.62 g, as well as rapeseed and mustard oil at 3.51 g [6]. As a group, these five supply routes account for somewhere around 62% of total worldwide consumption. Roughly 208 million tons of oil is generated worldwide annually from oil crops, which provide the vast majority (around 80%) of all edible fats consumed [7]. In 2019, total production of animal fats (including "dairy," which includes butter, ghee, cheese, milk, etc.) was 46 million tons, with an additional 6 million tons produced in pigs and tallow (7.3 million tons) [8]. The question of how to enhance plant oil production, as well as animal fat production from plant-fed animals, has implications for the health of both humans and the planet. Figure 11.1 shows the contribution of various crops in worldwide oil production. To generate the relatively similar amount of oil, for instance, various oil crops require different amounts of land and have varying yields [9, 10]. Palm oil is a common crop in the tropics, while soybeans thrive in the subtropics and the temperate zones of the planet. Since vegetable oils can be substituted for one another in many recipes, changes in the amount of land devoted to oil production will reflect changes in the availability of

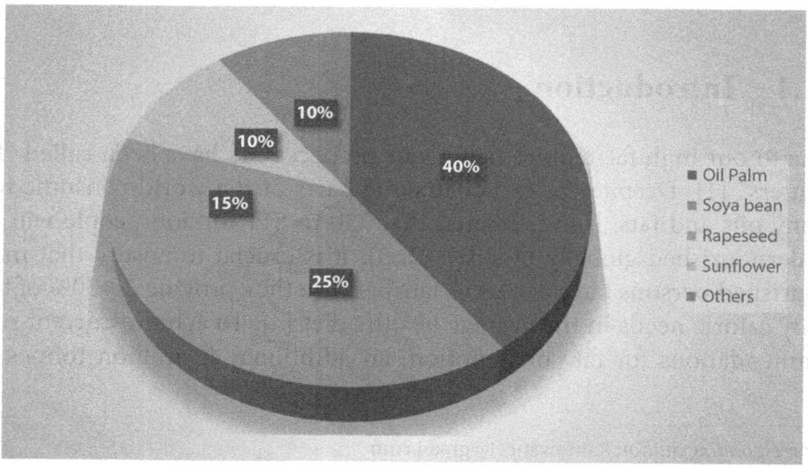

Figure 11.1 The percentage contribution of various crops to global vegetable oil production.

different types of oil [11]. Here we will discuss the ecological and economic aspects of oils and fats and how that affects people worldwide.

11.1.1 Nutritionally Valued Fats

Fats (also known as lipids) are essential to the function and structure of cellular membrane and are also utilized as a source of energy [12]. Additionally, fats aid in the digestion and absorption of fat-soluble vitamins, providing important fatty acids (A, D, E, and K) [13]. Oils are a form of fat, and the term "oil" is sometimes used interchangeably with "fat" to describe fats which are liquid at room temperature. Dietary fats are categorized as monounsaturated, polyunsaturated, or saturated depending on the amount of double bonds in their fatty acid chains. Trans fatty acids represent a unique fourth category of dietary fat that can be created through the industrial partial hydrogenation of vegetable oils (as in margarines) or found in animal products such as beef, lamb, and dairy. Foods like fish, various oils from plants, nuts, and seeds are excellent sources of monounsaturated fats [14]. Saturated fats are most commonly found in animal products and palm oil. Most fats are complicated chemical combinations of many different fatty and non-fatty acids, including the majority of saturated fatty acids. When it comes to one's health, each of these various fats might have either favourable or harmful effects. In addition to their effect on cardiovascular disease, these eating habits have an influence on cognitive function, digestive wellness, and weight management.

Carbohydrates, which have been linked to health issues, tend to fill in the void left when fats are cut from the diet.

As a result, there are now established norms for how people should be instructed to eat. Mediterranean diet, which contains vegetables, fruits, legumes, and healthy grains, is now widely recommended [15]. In addition to their effect on cardiovascular disease, these eating habits have an influence on cognitive function, digestive wellness, and weight management. Carbohydrates, which have been linked to health issues, tend to fill the void left when fats are cut from the diet.

11.1.2 Affordability of High Nutritional Fats

When contrasted with other categories of food, fats are typically sold at a lower price [16]. The total cost of fats for an individual on a daily basis is less than US$0.20, and they account for approximately 4% of the global average price of a healthy diet. Protein-rich foods can cost anywhere from 60 cents to $1, whereas dairy products, fruits, and vegetables cost around

70 cents. The cost of staples that are based on starch is $0.40. When compared to the international poverty line, which is established at US$1.90 per day for low-income countries, these costs appear to be quite high when viewed in this context [17]. The price of fat varies depending on the type and location. Although fat prices may vary based on the type of fat and the market in which it is supplied, customers can often save money and reduce waste by purchasing goods made in their own regions. The price of fat varies depending on the type and location [18].

11.1.3 Influence of Palm Oil on Human Health

Saturated fatty acids make about 40–50% of palm oil [19], making it the most saturated vegetable oil (other than coconut oil) utilized in food preparation and cooking [20]. Complex relationships exist between dietary oils, saturated fat, and health, but it is generally accepted that a larger intake of saturated fat leads to an increase blood cholesterol [21] and increases the risk of death from cardiovascular disease [22]. A switch to unsaturated fats such as polyunsaturated or monounsaturated has been demonstrated in long-term research to substantially and statistically lower cardiovascular risks from stroke and myocardial infarction [23]. A 6-8% reduction in cardiovascular disease is observed when saturated lauric, myristic, and stearic acids are substituted with palmitic acids, while a 10-12% reduction is seen when palmitic acids are substituted for saturated fatty acids. Large-scale longitudinal investigations conducted over a period of 24–28 years provide support for this [24]. Therefore, the current long-term results provide compelling data for the medical benefits of swapping out palm oil's saturated fats for polyunsaturated and monounsaturated fats.

11.1.4 Consequences of Palm Oil on the Economy and Human Health

A lack of data prevents researchers from accurately estimating palm oil's negative impact on health as well as on the associated expenses. Logistic regressions have been used to show that replacing palm oil with soybeans would reduce the incidence of myocardial infarction [25], and linear panel regressions have shown a correlation between rising palm oil consumption and higher death rates from ischemic heart disease in developing nations [26]. Effects of food tax on palm oil in India have been simulated using an economic-epidemiologic model [27]. The World Health Organization has called the evidence regarding palmitic acid intake and cardiovascular disease "convincing," [28] albeit the exact number of cases that can be attributed to palm oil consumption is still unknown.

Although health has been left out of economic analyses, palm oil production has been related to lower poverty and higher gross regional product [29], as well as to beneficial effects on smallholder income and greater inequality in palm oil farmer contracts [30]. Additionally, it has been demonstrated that the development of oil palm improves farmers' livelihoods significantly by raising their consumption, enhancing the productivity of their agricultural activities, and boosting their secondary income as a result of the time saved [31].

11.1.5 Consequences of Fats Production on Environment

The total area of farmland across the globe increased by 9% between 2003 and 2019, with the yearly expansion rate nearly doubling as a direct result of growth in South America and Africa [32]. Since native plant life is threatened by both increased crop production and environmental protection efforts, it is evident that these two aims are in conflict with one another and dense forests have been replaced on half of the newly cultivated land. This expansion puts the habitability of Earth in jeopardy [33]. Oil crops occupy 23% of the world's total farmland, which amounts to 331 million acres, but their area increased by 41% between 2003 and 2019. Overly abundant nitrogen fluxes are concentrated in places where soybean and dairy cultivation is more advantageous. Soybean, sunflower, dairy, olive oil, and peanut-producing areas all share concerns regarding the integrity of the biosphere. In the palearctic regions where crops like sunflowers and tallow are cultivated, and in the humid tropics, where crops like oil palm and soybeans are grown, worries about land-system change predominate [34].

11.1.6 The Prospects of Ecologically Sustainable Oil and Fat Industries Enterprises

The manufacturing of oil, fats, margarine, edible oil, and sales all function as part of the same complex, making them all part of the food sector. Seeds including sunflower, soybean, rapeseed, flax, peanuts, hemp, etc., are used as raw materials in the oil and fat business. On average, these plants yield 35–40% oil in their seeds, and the best cultivars yield over 50% oil [35]. It is important to encourage the mutual active progress of the state and business for the sake of the proper and dynamic growth of the agro-industrial complex. One of the primary responsibilities of both organizations should be to streamline the process of conforming to international standards of corporate governance by strengthening the laws that are currently on the books. Second, the implementation of international standards should be made mandatory rather than voluntary, and compliance should be strictly policed. Institutional support, agrarian enterprise, information and consulting help to farmland producers and rural population, small-scale agricultural production development, and agricultural collaboration to assure personal farm market integration are all areas that need special attention [36].

Some of the greatest difficulties now facing the oil and fats sector are listed below. A handful of corporations have significant influence over the market. There are approximately 1,200 businesses operating within this sector of the economy; however, the top nine companies were responsible for 85% of the output of unrefined sunflower oil in 2013, 66% of the production of refined sunflower oil, and 82% of the manufacturing of margarine products [37]. The oil and fat industry has become increasingly concentrated, which has led to the underutilization of production capacity among various businesses. This is because monopolistic corporations have a natural tendency to maximize profits in order to maintain their dominant market position. Lack of innovation and investment in the oil and fat business is another major cause for concern.

11.2 Disparities in Price

Oils and fats are considered to be commodities, which means that they are traded on international raw materials exchanges such as the ones that are located in Chicago and Kuala Lumpur. As a result, the prices of these products are subject to frequent and significant fluctuations. The practices of material purchasing and stockpiling that are utilized in the industry of chemical processing are intended to help level out these fluctuations. Coconut oil (2.20%/year) and tallow (2.15%/year) have experienced a slower average yearly rate of growth than the overall rate of inflation and a significantly slower rate of increase than the average for petrochemical products over the past 25 years, despite the fact that there have been some significant price spikes during this time period [38]. This is because the overall rate of inflation has been higher than the average rate of increase for petrochemical products. There is a very good chance that these patterns will remain constant.

11.2.1 The Monetary Impact of Essential Oils

As the editor of the *Journal of Essential Oil Research*, Dr. Brian Lawrence estimates that the world produces roughly 104,000 tonnes of the 20 most widely used essential oils. Since this number includes sectors outside of the cosmetics industry, such as the food, pharmaceutical, and home supply sectors, it should be used only as a guide. Large multinational corporations also have a stranglehold on the pharmaceuticals sector, with the United States, Europe, and Japan constituting the industry's three largest markets [39]. The essential oil market demands appear to be stable. The 18 most valuable essential oils account for roughly 75% of the world's total output value. Since high-value goods are traded in relatively small quantities, the concentration in terms of tons is considerably greater (e.g., rose, jasmine). Many countries import essential oils, including USA, France, UK, Japan, and Germany. In 2005, imports into the United States totalled US$390.9 m from the United States, or 14% of the total. The top five exporters of essential oils are the United States, France, India, the United Kingdom, and Brazil. Together, they exported essential oil goods worth $936.3 m (or around 40% of global exports) while the United States is the world's leading exporter. Its $351.7 billion in exports last year accounted for 17% of the world total; the United Kingdom, France, India, and Brazil follow behind the United States. India contributes only 4% of global output, but its value share is far higher at 21-22%. However, India's contribution of worldwide essential oil and fragrance material exports is only 0.4%, indicating that the country has a lot of space to grow in the rapidly expanding natural raw material sector of the global economy [40].

11.2.2 Economical Effects of Oils

Both huge corporations and individual farmers engage in oil palm cultivation. Many investigators have looked at how increasing the number of oil palms grown affects GDP and other measures of human prosperity. For many tropical nations, palm oil is a significant source of GDP and export revenue. International trade in palm oil was worth a total of $30 billion in 2018, with the largest exporters being Indonesia and Malaysia. Nearly 10% of Indonesia's total exports are palm oil [41]. Palm oil exports also make up about 5% of overall exports in numerous smaller countries like Honduras, Papua New Guinea, Solomon Islands, and Guatemala. Farmers, workers, and other supply chain participants have all benefited economically from the growth of oil palm [42]. Increased farm earnings, new jobs, and better rural infrastructure are all boons for rural families and towns. Numerous studies conducted in Southeast Asia found that oil palm planting led to increased household income, capital accumulation, and spending on food, healthcare, and education among smallholder farmers [43]. Farming households in Africa benefit economically when oil palm is cultivated. Oil palm farmers in Guinea were shown to have greater income stability and greater food security than those who did not grow oil palm or other cash crops. After accounting for any confounding factors, a study conducted in Ghana indicated that oil palm farmers enjoy greater incomes and experience less multidimensional poverty than other farmers [44]. There is a high demand for workers because most oil palm operations are done manually [45]. Both large-scale corporate plantations and smaller-scale family farms have contributed to the growth of the oil palm sector, which has provided employment and economic benefits to many people who had previously been landless. Many rural families rely on income from oil palm work, and oil palm workers tend to be better off than their counterparts in other agricultural niches [46].

11.2.3 Biodiesel Feedstock Price Comparison

The current significant challenge to biodiesel commercialization is the product's high price. When compared to diesel fuel made from petroleum, the cost of producing biodiesel is astronomical. While diesel fuel made from petroleum typically costs around $0.35 per litre, biodiesel typically costs more than $0.5 per litre [47]. The economic performance of a biodiesel plant can be determined if certain factors are specified. These factors include plant capacities, process technologies, raw material prices, and chemical costs [48]. Several factors, including base stock, location, seasonal fluctuations in crop output, the costs of crude petroleum, and others, affect the overall cost of biodiesel fuels [49]. The cost of feedstock has a major influence on the economy of biodiesel production. Depending on the source, raw material costs can make up as much as 75% of the total price of producing biodiesel. An estimated US$0.158 per litre goes into the production of biodiesel [50, 51], an estimated feedstock cost of US$0.539 per litre for refined soy oil, and the estimated cost of US$0.70 per litre of soy-based biodiesel production [52]. Feedstock price is a major cost in biodiesel production, hence the type of feedstock used has an impact on the industry's bottom line. Some plant oils, such as soybean and palm oils, have a very erratic price, and in years when worldwide prices are particularly high, their use may be redirected from the production of biodiesel to the food market. The increased demand for rapeseed oil has a major impact on oil prices, which are roughly 25% higher in Europe than in the US. The rising demand for biodiesel in the European Union and the falling cost of rapeseed oil have followed a similar trajectory. Rapeseed prices increased by 63% between 2002 and 2007 [53]. Soybean oil is the major oil utilized in manufacturing of biodiesel across the US [54]. In the United States, the price of one litre of biodiesel made from crude soybean oil has been selling for between $0.40 and $0.48 per litre [55]. Palm oil is the most widely sold oil in the world and has import potential. Beef tallow and yellow grease are two alternative feedstocks that can be used to create biodiesel, both of which are affordable and abundant because of their use in other sectors [56]. Producing biodiesel from animal fat is the most cost-effective option (US$0.4-US$0.5 per litre), while the standard method of transesterification of vegetable oil costs about US$0.6-US$0.8 per litre [57].

As production volumes increase, the price of each unit of output should decrease by between $0.01 and $0.03 per litre due to economic scales. More than US$0.9 per litre (feedstock US$3.6/GJ) is spent on biomass-to-liquids (BTL) diesel made from lignocellulose, although this price has the potential to be reduced to US$0.7-US$0.8 per litre [58]. The price of producing biodiesel decreases with increasing output. Although the cost of feedstock is higher for larger facilities, production costs are lower, and co-product utilization is higher [59].

11.3 Environmental Effects of Oils

Although the oil palm boom has helped the economy flourish, it has also had detrimental impacts on the ecology. Many studies have linked oil palm to biodiversity loss, ecosystem dysfunction, and deforestation [60]. Oil palm is typically grown in monocultures, despite the fact that a typical rainforest can support >470 different tree species per hectare [61]. Due to their single canopy layer rather than multiple forest strata, lack of complex and rich vegetation, and almost complete absence of leaf litter and woody debris, all of which are crucial to the high biodiversity of tropical forests, these monocultures are structurally much less complex than the forests they replace [62]. Furthermore, oil palm plantations are inhospitable to the vast majority of forest species due to insecticides, fertilisers, and frequent human disturbance. There is a wide variation in the amount of carbon stored in a tropical rainforest's trees, from 175 to 272 tons per hectare [63]. Oil palm plantations have standing carbon reserves ranging from 2 to 60 tons per hectare, based on age and planting density [64].

11.3.1 Alternative Diesel Fuels: Plant Oils and Oleochemicals

Plant oils were utilized in diesel engines for quite some time, even before the diesel engine was invented. In 1900, Rudolf Diesel, the inventor of the diesel engine, demonstrated his innovation by fuelling it with peanut oil [65]. In the 1930s and 1940s, researchers explored the feasibility of substituting plant oils for diesel fuel in automobiles. Motivation to go beyond conventional, petroleum-based fuels came from the fuel and energy crises of the late 1970s and early 1980s, as well as associated concerns about the depletion of the world's non-renewable resources [66]. Properties of palm oil-based biofuels compared to mineral diesel are listed in Table 11.1. They have recently emerged as a leader in the field of developing sustainable energy sources. Alternative diesel fuels derived from plant oils, such as biodiesel, have been the subject of countless academic researches and reports from around the world [67]. They are now in the early phases of commercialisation, having progressed from the strictly experimental fuels stage. However, there are a number of technological and economic considerations that need to be addressed before these fuels can be considered fully viable.

Table 11.1 Comparing the psychochemical properties of palm biodiesel with mineral diesel.

S.no	Properties	Palm oil based biodiesel	Mineral diesel	Unit	Comparison of palm diesel and mineral diesel
1	Density	880	835	(kg/m^3)	Higher (↑)
2	Kinematic viscocity	4.61	2.95	(mm^2/s)	Higher (↑)
3	Heating value	38.5	44.8	MJ/kg	Lower (↑)
4	Flash value	175	70	°C	Higher (↑)
5	Cloud point	10	-12	°C	Higher (↑)
6	Acid value	0.49	0.16	mg KOH/g	Higher (↑)

Researchers have tested numerous plant oils in order to determine which one works best as biodiesel. Researchers focus on locally available oils due to the significant association between plant oil's abundance and its potential as a biodiesel fuel source. This resulted in discussions about using soybean oil as a biodiesel feedstock at the forefront in the USA while in Europe, rapeseed oil is a leading contender, and in tropical regions, coconut oil and palm oil are the preferred options. Sunflower and safflower oils, among others, have also been the subject of research. Other potential biodiesel sources include animal fats and wasted or discarded cooking oils [68].

11.3.2 Effect of Biofuels on the Environment

Biodiesel and diesel fuels have many similarities, which is why it is considered a viable diesel replacement [69]. Running the engine lean reduces temperature, while running it rich reduces oxygen supplies, burn time, and engine speed [70]. These are all techniques for lowering NO_x emissions from a biodiesel engine. Combustion temperature, combustion period length, and the availability of biodiesel can all influence the total amount of NOx released into the environment. Since the nineteenth century, alcohols have been used to power engines. For spark-ignition (SI) engines, ethanol is generally considered the most sustainable, bio-based, and ecologically acceptable fuel alternative. The major benefit of using ethanol as a fuel for SI engines is that it can be made from renewable energy sources such as sugar cane, cassava, a range of waste biomass materials, corn, and barley. Ethanol enhances engine performance and decreases exhaust emissions since it has a greater evaporation heat, octane number, and flammability temperature than gasoline. The results of the engine test revealed that when ethanol was added to unleaded gasoline, the engine performed better in terms of torque, power, fuel economy, and carbon monoxide (CO) and hydrocarbon emissions (HC) [71]. Different biodiesel formulations and conventional diesel formulations had different effects on tailpipe emissions. Diesel fuel blends containing up to 20% biodiesel are generally compatible with existing storage and distribution networks, and can be used in most diesel-powered equipment. Compared to running a conventional diesel engine on ordinary diesel fuel, running a biodiesel engine produces much less carbon monoxide, sulphates, polycyclic aromatic hydrocarbons, nitrated polycyclic aromatic hydrocarbons, and particulate matter. Higher levels of nitrogen oxide (NOx) are produced when using biodiesel as a fuel in a diesel engine. There are many positive outcomes that can result from using biofuels on a societal, economic, and individual level. Alternatives to petroleum as motor fuel include biodiesel, vegetable oil, ethanol, bio methanol, and biomass pyrolysis oil due to their liquid nature, abundance, renewability, high combustion efficiency, low sulphur and aromatic content, and biodegradability [72]. The high oxygen concentration of biofuels sets them apart from petroleum feedstocks. Oxygen concentrations in biofuels range from 10% to 45%, while oxygen levels in petroleum are negligible. This has a significant impact on the chemical properties of biofuels. Oxygenates are simply recycled hydrocarbons with the proper molecular structure to provide some knocking resistance. Furthermore, the oxygen in them improves fuel combustion efficiency, resulting in fewer hydrocarbons in the exhaust. The sole drawback is that oxygenated fuels have less energy concentration [73].

11.3.3 Effect of Biofuels on the Economy

Feedstock costs account for approximately 75%-80% of the total operational cost of biofuel production, making them the most important economic issue to address. The feedstock requires human labour, methanol, and a catalyst and all have substantial costs [74]. The process cost was estimated to be US$0.158/l, with the feedstock cost (including refining costs) estimated to be US$0.539/l [75]. Palm oil is the best choice among options that are traded globally and could be imported soon [76]. Feedstock, transition technique, production scale, and geographical location all play a role in determining the final price of biofuels. In terms of energy expenditure, ethanol production is more expensive than gasoline production. Only Brazilian ethanol comes close to being a practical alternative to standard gasoline. Compared to sugar cane ethanol produced in Brazil, the price of U.S.-made ethanol is far greater. Grain and sugar beet ethanol are even more expensive in Europe. Scale, process efficiency, feedstock costs, capital and labour expenses, co-product accounting, and the form of estimations are all factors that contribute to these gaps. Creating bio-based products at scale is currently quite expensive in developed countries. Even though the price of producing biofuels is three times that of petroleum fuels, there are many non-economic benefits that should be considered. In contrast to the OECD, the cost of manufacturing biofuels in developing nations is comparable with the worldwide market price of petroleum fuel [77]. The economic benefit of using biofuel will involve the following effects: the value of the feedstock would rise, creating more manufacturing jobs in rural areas; income taxes would rise; investments would be made in plant and equipment; greenhouse gas emissions would decrease; the country's reliance on imported crude oil would decrease; and agriculture would benefit from the availability of new consumer and labour markets. On-food crops have gained considerable importance in recent years. If the biofuel production is to be increased, the compulsory set-aside policy, which allows the cultivation of non-food crops, could be implemented. Renewable liquid fuels are increasingly being used as alternatives to petroleum. Greenhouse gas reduction, territorial expansion, social structure and agricultural production and food supply safety are some of the benefits that renewable liquid fuels are believed to offer [78].

11.3.4 Recent Developments in Palm Oil Harvesting

As the world's largest producer, grower, and exporter of crude palm oil (CPO) and palm kernel oil, Malaysia has tremendous potential to become a leading worldwide biodiesel player (PKO). A rise in palm oil exports from Malaysia resulted in a record of Malaysian Ringgit 59.77 billion in revenue, while the total area planted with oil palms increased by 3.47%, from 4.69 million hectares to 4.85 million hectares. Although just 12% of the world's oils and fats are produced, Malaysia accounts for 44% of its exports [79]. Biodiesels are reusable fuels obtained from organic sources that are used as an alternative to mineral diesel [80]. They consist of the long-chain fatty acid monoalkyl ester, which can be found in both plant and animal fats. Biodiesel is produced through a process that converts oils and lipids into long-chain mono alkyl esters. Fatty acid methyl esters (FAME) are the end products of transesterification. Transesterification of biodiesel made from organic vegetable oils and animal fats produces methyl and ethyl esters of fatty acids. It's important to have easy access as biodiesel production is dependent on the locally accessible feedstock. The majority of the world's biodiesel comes from the Asian and African continents. Biodiesel is produced primarily from palm oil in Southeast Asian countries like Malaysia and Indonesia, but rapeseed is used in Europe. Furthermore, in the United States, soybean output was the highest, followed by corn and sunflower [81].

11.3.5 Oils and Oleochemicals in Pesticidal Formulations

The chemical industry relies heavily on oils and fats, especially those derived from plants. Advantages of using plant oils and oleochemicals over mineral oils include their sustainability and abundance worldwide (often as by-products of other industries). In addition, they can be broken down in the environment, are non-flammable, and have a lower potential for triggering health issues and allergic reactions to the end users [82].

Inert components like surfactants, wetting agents, dispersion agents, emulsifying agents, adjuvants, solvents, and carrier/diluents for pesticide formulations are manufactured from plant oils such as coconut, soybean, corn, palm, and palm kernel oils and their oleochemicals [83]. Both active and inert chemicals are necessary for the successful production of pesticide formulations. The term "active ingredient" refers to the chemical or compounds in a product that are responsible for the product's intended effect, such as its ability to prevent, destroy, repel, or mitigate insects, fungi, weeds, rodents, or other pests. But supplements, often known as inert substances, have no pesticidal effect whatsoever and are therefore ineffective [84]. Emulsifiers, organic solvents, dispersion agents, and wetting agents are all examples of inert chemicals commonly used in pesticide formulation.

11.4 Global Trends

The oil palm tree is indigenous to Central and West Africa and thrives in the lowland wet tropics [85]. Palm oil was already in use in ancient Egypt some thousands of years ago, according to historical documents. It was during the colonial period that Europe took an interest in palm oil, albeit production was still primarily limited to Africa [86]. Small farmers in Africa are responsible for the vast majority of oil palm production [87]. Early in the twentieth century, primarily European corporations constructed a small number of massive oil palm farms [88]. There was a massive expansion in the area devoted to oil palms around the world, with much of that growth happening in Southeast Asia, particularly in Indonesia and Malaysia. In some Southeast Asian countries, governments have invested in and subsidised vegetable oil production in response to increased global demand [89]. Despite its potential economic benefits, oil palm output in Africa has barely risen over the past few decades due to unsteady regimes and a lack of private investment. Much of the land that was converted to oil palm cultivation in Southeast Asia was formerly used for food or cash crop production, or was degraded forest or fallow land. Large corporations and state agencies in Indonesia and Malaysia hold roughly 60% of oil palm land owing to long-term land concessions from the respective governments [90]. Figure 11.2 depicts the worldwide consumption of oils and fats. The future of palm oil production is bright because of the ever-increasing demand. It is true that there is still space in Southeast Asia that may be used for oil palm plantations, but there are growing environmental fears and competition from efforts to protect natural areas including the rainforest [91].

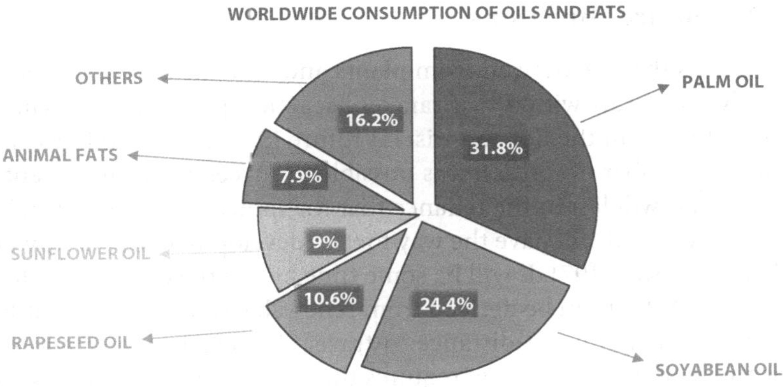

Figure 11.2 Global utilization of oils and fats.

11.4.1 Addressing Growing "Fats" Demands

As the global population continues to rise, there will be an increased demand for fat, which will likely result in an increase in the importance of oil crops. As a result of the metabolic costs involved in transforming plant material into animal material, more farmland would be required to produce the same quantity of animal fat that can be produced directly from plants. This would be the case even though the same quantity of fat can be produced directly from plants. Animals eat a wide range of plants, including oil crops as well as grasses and other plants that are inedible to humans. The extent to which animal fats and vegetable oils are produced in competition with one another, as well as the conditions under which this occurs, are debatable and highly dependent on the context [92]. It is possible to increase oil crop productivity in two ways: (1) by increasing the yield of existing crops, hence increasing oil production on a given area of land, and (2) by dedicating additional land towards the cultivation of oil. Both of these methods are effective in increasing oil crop output. As things stand, the vast majority of vegetable oil comes from just four different types of plants: oil palm, soybeans, sunflower, and rapeseed. Although this obviously is dependent on a variety of factors, average yields can vary quite a bit between different agricultural subsectors. The spread of oil palm and coconut plantations in Southeast Asia [93], soybean farms in South America [94], and the expansion of maize, groundnut, and cotton in the equatorial regions of Africa are just a few examples of the widespread effects that the proliferation of oil crops has had on the natural ecosystems of the planet [95]. There are many other examples.

11.4.2 Future Directions

Oils and fats that are derived from plants and animals have the capability to supply chemicals with a vast range of reaction products that will have significant value in the not-too-distant future [96]. These products can be useful in a number of ways. If it is possible to successfully implement this technology, it will lessen the reliance that the industry has on petrochemical feedstocks and will pave the way for the development of entirely novel synthetic pathways [97]. It will be some time before the chemical potential of oils and fats that can be derived from renewable sources is fully utilized. There is still a significant distance to travel. The expansion of successful applications of this technology requires the collaboration of scientists specializing in a variety of fields, including chemistry, biology, plant breeding, and agriculture, amongst others.

References

1. Ben-Dor M, Gopher A, Hershkovitz I, Barkai R. Man the fat hunter: the demise of Homo erectus and the emergence of a new hominin lineage in the Middle Pleistocene (ca. 400 kyr) Levant. *PLoS One*. 2011 Dec 9;6(12):e28689.
2. Blundell JE, Stubbs RJ, Golding C, Croden F, Alam R, Whybrow S, Le Noury J, Lawton CL. Resistance and susceptibility to weight gain: individual variability in response to a high-fat diet. *Physiology & Behavior*. 2005 Dec 15;86(5):614-22.
3. World Health Organization. The State of Food Security and Nutrition in the World 2021: Transforming food systems for food security, improved nutrition and affordable healthy diets for all. Food & Agriculture Org.; 2021 Jul 12.
4. Liu AG, Ford NA, Hu FB, Zelman KM, Mozaffarian D, Kris-Etherton PM. A healthy approach to dietary fats: understanding the science and taking action to reduce consumer confusion. *Nutrition Journal*. 2017 Dec;16(1):1-5.
5. Bajželj B, Laguzzi F, Röös E. The role of fats in the transition to sustainable diets. *The Lancet Planetary Health*. 2021 Sep 1;5(9):e644-53.
6. Horn B, Ferreira C, Kalantari Z. Links between food trade, climate change and food security in developed countries: A case study of Sweden. *Ambio*. 2022 Apr;51(4):943-54.
7. Mirpoor SF, Giosafatto CV, Porta R. Biorefining of seed oil cakes as industrial co-streams for production of innovative bioplastics. A review. *Trends in Food Science & Technology*. 2021 Mar 1;109:259-70.
8. Al Hasan SM, Saulam J, Mikami F, Kanda K, Ngatu NR, Yokoi H, Hirao T. Trends in per Capita Food and Protein Availability at the National Level of the Southeast Asian Countries: An Analysis of the FAO's Food Balance Sheet Data from 1961 to 2018. *Nutrients*. 2022 Jan 29;14(3):603.

9. Meijaard E, Brooks TM, Carlson KM, Slade EM, Garcia-Ulloa J, Gaveau DL, Lee JS, Santika T, Juffe-Bignoli D, Struebig MJ, Wich SA. The environmental impacts of palm oil in context. *Nature Plants*. 2020 Dec;6(12):1418-26.
10. 1. Ritchie, H., & Roser, M. (2020). Environmental Impacts of Food Production. - References - Scientific Research Publishing [Internet]. Scirp.org. 2022 [cited 1 October 2022]. Available from: https://www.scirp.org/(S(czeh2tfqyw2orz553k1w0r45))/reference/referencespapers.aspx?referenceid=2980887
11. Parsons S, Raikova S, Chuck CJ. The viability and desirability of replacing palm oil. *Nat Sustain*. (2020) 3:412–8. doi: 10.1038/s41893-020-0487-8
12. Cena H, Calder PC. Defining a healthy diet: evidence for the role of contemporary dietary patterns in health and disease. *Nutrients*. 2020 Jan 27;12(2):334.
13. Sanders TA. Introduction: the role of fats in human diet. *Functional Dietary Lipids*. 2016 Jan 1:1-20.
14. 1. Astrup A, Magkos F, Bier D, Brenna J, de Oliveira Otto M, Hill J. Saturated Fats and Health: A Reassessment and Proposal for Food-Based Recommendations. *Journal of the American College of Cardiology*. 2020;76(7):844-857.
15. Aleta A, Brighenti F, Jolliet O, Meijaard E, Shamir R, Moreno Y, Rasetti M. A need for a paradigm shift in healthy nutrition research. *Frontiers in Nutrition*. 2022;9.
16. Herforth A, Bai Y, Venkat A, Mahrt K, Ebel A, Masters WA. Cost and affordability of healthy diets across and within countries: Background paper for The State of Food Security and Nutrition in the World 2020. FAO Agricultural Development Economics Technical Study No. 9. Food & Agriculture Org.; 2020 Dec 12.
17. Jolliffe D, Prydz EB. Estimating international poverty lines from comparable national thresholds. *Journal of Economic Inequality*. 2016 Jun;14(2):185-98.
18. [Internet]. 2022 [cited 1 October 2022]. Available from: https://www.ers.usda.gov/data-products/oil-crops-yearbook.aspx
19. Sun Y, Neelakantan N, Wu Y, Lote-Oke R, Pan A, van Dam RM. Palm oil consumption increases LDL cholesterol compared with vegetable oils low in saturated fat in a meta-analysis of clinical trials. *Journal of Nutrition*. 2015 Jul 1;145(7):1549-58.
20. Foster R, Williamson CS, Lunn J. Briefing Paper: Culinary oils and their health effects. *Nutrition Bulletin*. 2009 Mar;34(1):4-7.
21. Mensink RP, Zock PL, Kester AD, Katan MB. Effects of dietary fatty acids and carbohydrates on the ratio of serum total to HDL cholesterol and on serum lipids and apolipoproteins: a meta-analysis of 60 controlled trials. *American Journal of Clinical Nutrition*. 2003 May 1;77(5):1146-55.
22. Verschuren W. Serum total cholesterol and long-term coronary heart disease mortality in different cultures. Twenty-five-year follow-up of the

seven countries study. *JAMA: Journal of the American Medical Association.* 1995;274(2):131-136.

23. Jakobsen MU, O'Reilly EJ, Heitmann BL, Pereira MA, Bälter K, Fraser GE, Goldbourt U, Hallmans G, Knekt P, Liu S, Pietinen P. Major types of dietary fat and risk of coronary heart disease: a pooled analysis of 11 cohort studies. *American Journal of Clinical Nutrition.* 2009 May 1;89(5):1425-32.
24. Zong G, Li Y, Wanders AJ, Alssema M, Zock PL, Willett WC, Hu FB, Sun Q. Intake of individual saturated fatty acids and risk of coronary heart disease in US men and women: two prospective longitudinal cohort studies. *BMJ.* 2016 Nov 23;355.
25. Kabagambe EK, Baylin A, Ascherio A, Campos H. The type of oil used for cooking is associated with the risk of nonfatal acute myocardial infarction in Costa Rica. *Journal of Nutrition.* 2005 Nov 1;135(11):2674-9.
26. Chen BK, Seligman B, Farquhar JW, Goldhaber-Fiebert JD. Multi-Country analysis of palm oil consumption and cardiovascular disease mortality for countries at different stages of economic development: 1980-1997. *Globalization and Health.* 2011 Dec;7(1):1-0.
27. Basu S, Babiarz KS, Ebrahim S, Vellakkal S, Stuckler D, Goldhaber-Fiebert JD. Palm oil taxes and cardiovascular disease mortality in India: economic-epidemiologic model. *BMJ.* 2013 Oct 22;347.
28. Who J, Consultation FE. Diet, nutrition and the prevention of chronic diseases. *World Health Organ Tech Rep Ser.* 2003 Feb 27;916(i-viii):1-49.
29. Shahputra MA, Zen Z. Positive and negative impacts of oil palm expansion in Indonesia and the prospect to achieve sustainable palm oil. In *IOP Conference Series: Earth and Environmental Science* 2018 Feb 1 (Vol. 122, No. 1, p. 012008). IOP Publishing.
30. Gatto M, Wollni M, Asnawi R, Qaim M. Oil palm boom, contract farming, and rural economic development: Village-level evidence from Indonesia. *World Development.* 2017 Jul 1;95:127-40.
31. Kubitza C, Krishna VV, Alamsyah Z, Qaim M. The economics behind an ecological crisis: livelihood effects of oil palm expansion in Sumatra, Indonesia. *Human Ecology.* 2018 Feb;46(1):107-16.
32. Potapov P, Turubanova S, Hansen MC, Tyukavina A, Zalles V, Khan A, Song XP, Pickens A, Shen Q, Cortez J. Global maps of cropland extent and change show accelerated cropland expansion in the twenty-first century. *Nature Food.* 2022 Jan;3(1):19-28.
33. Gerten D, Heck V, Jägermeyr J, Bodirsky BL, Fetzer I, Jalava M, Kummu M, Lucht W, Rockström J, Schaphoff S, Schellnhuber HJ. Feeding ten billion people is possible within four terrestrial planetary boundaries. *Nature Sustainability.* 2020 Mar;3(3):200-8.
34. Meijaard E, Abrams JF, Slavin JL, Sheil D. Dietary Fats, Human Nutrition and the Environment: Balance and Sustainability. *Frontiers in Nutrition.* 2022:439.

35. Agegenehu M, Honermeier B. Effects of seeding rates and nitrogen fertilization on seed yield, seed quality and yield components of false flax. *Die Bodenkultur.* 1997;48(1):15.
36. Живко ЗБ, Родченко СС, Руда II, Боруцька ЮЗ, Бойко ОТ. Management, finance, economics: modern problems and ways of their solutions: collective monograph.
37. Laudan R. *Cuisine and Empire: Cooking in World History.* Univ. of California Press; 2013.
38. Baumann H, Bühler M, Fochem H, Hirsinger F, Zoebelein H, Falbe J. Natural fats and oils—renewable raw materials for the chemical industry. *Angewandte Chemie International Edition in English.* 1988 Jan;27(1):41-62.
39. Raina VK, Srivastava SK, Aggarwal KK, Syamasundar KV, Khanuja SP. Essential oil composition of Cymbopogon martinii from different places in India. *Flavour and Fragrance Journal.* 2003 Jul;18(4):312-5.
40. Devi MP, Chakrabarty S, Ghosh SK, Bhowmick N. Essential oil: its economic aspect, extraction, importance, uses, hazards and quality. In *Value Addition of Horticultural Crops: Recent Trends and Future Directions.* 2015:269-78.
41. Feintrenie L, Chong WK, Levang P. Why do farmers prefer oil palm? Lessons learnt from Bungo district, Indonesia. *Small-Scale Forestry.* 2010 Sep;9(3):379-96.
42. Naylor RL, Higgins MM, Edwards RB, Falcon WP. Decentralization and the environment: Assessing smallholder oil palm development in Indonesia. *Ambio.* 2019 Oct;48(10):1195-208.
43. Obidzinski K, Andriani R, Komarudin H, Andrianto A. Environmental and social impacts of oil palm plantations and their implications for biofuel production in Indonesia. *Ecology and Society.* 2012 Mar 1;17(1).
44. Ahmed A, Dompreh E, Gasparatos A. Human wellbeing outcomes of involvement in industrial crop production: Evidence from sugarcane, oil palm and jatropha sites in Ghana. *Plos one.* 2019 Apr 25;14(4):e0215433.
45. van Noordwijk M, Susswein PM, Tomich TP, Diaw C, Vosti S. *Land use practices in the humid tropics and introduction to ASB benchmark areas.* Bogor, Indones: ICRAF; 2001.
46. Mingorría S, Gamboa G, Martín-López B, Corbera E. The oil palm boom: socio-economic implications for Q'eqchi'households in the Polochic valley, Guatemala. *Environment, Development and Sustainability.* 2014 Aug;16(4):841-71.
47. Zhang Y, Dube MA, McLean DD, Kates M. Biodiesel production from waste cooking oil: 1. Process design and technological assessment. *Bioresource Technology.* 2003 Aug 1;89(1):1-6.
48. Zhang Y, Dubé MA, McLean DD, Kates M. Biodiesel production from waste cooking oil: 2. Economic assessment and sensitivity analysis. *Bioresource Technology.* 2003 Dec 1;90(3):229-40.
49. Demirbas A, Karslioglu S. Biodiesel production facilities from vegetable oils and animal fats. *Energy Sources, Part A.* 2007 Feb 1;29(2):133-41.

50. Krawczyk T. Biodiesel-alternative fuel makes inroads but hurdles remain. *Inform*. 1996;7:801-15.
51. Costa AO, Oliveira LB, Lins MP, Silva AC, Araujo MS, Pereira Jr AO, Rosa LP. Sustainability analysis of biodiesel production: A review on different resources in Brazil. *Renewable and Sustainable Energy Reviews*. 2013 Nov 1;27:407-12.
52. Haas MJ, McAloon AJ, Yee WC, Foglia TA. A process model to estimate biodiesel production costs. *Bioresource Technology*. 2006 Mar 1;97(4):671-8.
53. Jank MJ, Kutas G, do Amaral LF, Nassar AM. EU and US policies on biofuels: potential impacts on developing countries.
54. Haas MJ. Improving the economics of biodiesel production through the use of low value lipids as feedstocks: vegetable oil soapstock. *Fuel Processing Technology*. 2005 Jun 25;86(10):1087-96.
55. Gire A, Jeffery N, Schulte J, Snow J, VanDyk R. Waste cooking oil to biodiesel automated fuel pump project.
56. Demirbas AH. Inexpensive oil and fats feedstocks for production of biodiesel. *Energy Educ Sci Technol Part A*. 2009 Apr 1;23(1):1-13.
57. Balat M. Potential alternatives to edible oils for biodiesel production–A review of current work. *Energy Conversion and Management*. 2011 Feb 1;52(2):1479-92.
58. Hajjari M, Tabatabaei M, Aghbashlo M, Ghanavati H. A review on the prospects of sustainable biodiesel production: A global scenario with an emphasis on waste-oil biodiesel utilization. *Renewable and Sustainable Energy Reviews*. 2017 May 1;72:445-64.
59. Miranowski JA. Economic drivers of biofuels expansion. University Extension, Iowa State University. Retrieved October. 2006 Nov;8:2008.
60. Qaim M, Sibhatu KT, Siregar H, Grass I. Environmental, economic, and social consequences of the oil palm boom. *Annual Review of Resource Economics*. 2020 Jan 1;12(1):321-44.
61. Valencia R, Balslev H. High tree alpha-diversity in Amazonian Ecuador. *Biodiversity & Conservation*. 1994 Feb;3(1):21-8.
62. Luskin MS, Albert WR, Tobler MW. Sumatran tiger survival threatened by deforestation despite increasing densities in parks. *Nature Communications*. 2017 Dec 5;8(1):1-9.
63. Sayer J, Ghazoul J, Nelson P, Boedhihartono AK. Oil palm expansion transforms tropical landscapes and livelihoods. *Global Food Security*. 2012 Dec 1;1(2):114-9.
64. Kho LK, Jepsen MR. Carbon stock of oil palm plantations and tropical forests in M alaysia: a review. *Singapore Journal of Tropical Geography*. 2015 Jul;36(2):249-66.
65. Kirakosyan A, Kaufman PB. *Recent Advances in Plant Biotechnology*. Dordrecht: Springer; 2009.

66. Salimon J, Salih N, Yousif E. Industrial development and applications of plant oils and their bio based oleochemicals. *Arabian Journal of Chemistry.* 2012 Apr 1;5(2):135-45.
67. Tripathy, D. B., & Mishra, A. (2016). Renewable plant-based raw materials for industry. *Sustainable Inorganic Chemistry*, 353.
68. Demirbaş A. Biodiesel fuels from vegetable oils via catalytic and non-catalytic supercritical alcohol transesterification and other methods: a survey. *Energy Conversion and Management.* 2003 Aug 1;44(13):2093-109.
69. Krahl J, Knothe G, Munack A, Ruschel Y, Schröder O, Hallier E, Westphal G, Bünger J. Comparison of exhaust emissions and their mutagenicity from the combustion of biodiesel, vegetable oil, gas-to-liquid and petrodiesel fuels. *Fuel.* 2009 Jun 1;88(6):1064-9.
70. Demirbas A. Economic and environmental impacts of the liquid biofuels. *Energy Education Science and Technology.* 2008 Dec 1;22(1):37-58.
71. Najafi G, Ghobadian B, Tavakoli T, Buttsworth DR, Yusaf TF, Faizollahnejad MJ. Performance and exhaust emissions of a gasoline engine with ethanol blended gasoline fuels using artificial neural network. *Applied Energy.* 2009 May 1;86(5):630-9.
72. Demirbas A. Theoretical heating values and impacts of pure compounds and fuels. *Energy Sources, Part A.* 2006 Feb 1;28(5):459-67.
73. Demirbas A. Political, economic and environmental impacts of biofuels: A review. *Applied Energy.* 2009 Nov 1;86:S108-17.
74. Haas MJ, McAloon AJ, Yee WC, Foglia TA. A process model to estimate biodiesel production costs. *Bioresource Technology.* 2006 Mar 1;97(4):671-8.
75. Dene T, Hole J. *Enabling Biofuels: Biofuel Economics.* Final report. Auckland, New Zealand: Minister of Transport, Covec Ltd. 2006 Jun.
76. Wu M. Analysis of the Efficiency of the US Ethanol Industry 2007. Center for Transportation Research, Argonne National Laboratory. 2008 Mar 27.
77. UN (United Nations). The emerging biofuels market: regulatory, trade and development implications. United Nations Conference on Trade and Development. 2006.
78. Demirbas A, Dincer K. Sustainable green diesel: a futuristic view. *Energy Sources, Part A.* 2008 May 22;30(13):1233-41.
79. Sumathi S, Chai SP, Mohamed AR. Utilization of oil palm as a source of renewable energy in Malaysia. *Renewable and Sustainable Energy Reviews.* 2008 Dec 1;12(9):2404-21.
80. Janaun J, Ellis N. Perspectives on biodiesel as a sustainable fuel. *Renewable and Sustainable Energy Reviews.* 2010 May 1;14(4):1312-20.
81. Yusuf NN, Kamarudin SK, Yaakub Z. Overview on the current trends in biodiesel production. *Energy Conversion and Management.* 2011 Jul 1;52(7):2741-51.
82. Fukuyama T, Tajima Y, Ueda H, Hayashi K, Shutoh Y, Harada T, Kosaka T. Allergic reaction induced by dermal and/or respiratory exposure to

low-dose phenoxyacetic acid, organophosphorus, and carbamate pesticides. *Toxicology*. 2009 Jul 10;261(3):152-61.

83. Tripathy, D. B., Yadav, A., & Mishra, A. (2017). Applications of petrochemicals: A mini review. *Recent Adv. Petrochem. Sci, 2*, 67-69.
84. Abhilash PC, Singh N. Pesticide use and application: an Indian scenario. *Journal of Hazardous Materials*. 2009 Jun 15;165(1-3):1-2.
85. Corley R, Tinker P. The Oil Palm. 5th ed. Chichester, UK: Wiley Blackwell; 2016.
86. Carrere R. Oil palm in Africa: Past, present and future scenarios. *WRM Series on Tree Plantations*. 2010 Dec;15(111):1-78.
87. Ordway EM, Naylor RL, Nkongho RN, Lambin EF. Oil palm expansion and deforestation in Southwest Cameroon associated with proliferation of informal mills. *Nature Communications*. 2019 Jan 10;10(1):1-1.
88. Byerlee D, Falcon WP, Naylor R. *The Tropical Oil Crop Revolution: Food, Feed, Fuel, and Forests*. Oxford University Press; 2017.
89. Cramb R, Curry G. Oil palm and rural livelihoods in the Asia-Pacific region. Special issue. *Asia Pacific Viewpoint* (Print). 2012;53(3).
90. Byerlee D, Falcon WP, Naylor R. *The Tropical Oil Crop Revolution: Food, Feed, Fuel, and Forests*. Oxford University Press; 2017.
91. Krishna V, Euler M, Siregar H, Qaim M. Differential livelihood impacts of oil palm expansion in Indonesia. *Agricultural Economics*. 2017 Sep;48(5):639-53.
92. Mottet A, de Haan C, Falcucci A, Tempio G, Opio C, Gerber P. Livestock: On our plates or eating at our table? A new analysis of the feed/food debate. *Global Food Security*. 2017 Sep 1;14:1-8.
93. Meijaard E, Abrams JF, Juffe-Bignoli D, Voigt M, Sheil D. Coconut oil, conservation and the conscientious consumer. *Current Biology*. 2020 Jul 6;30(13):R757-8.
94. Song XP, Hansen MC, Potapov P, Adusei B, Pickering J, Adami M, Lima A, Zalles V, Stehman SV, Di Bella CM, Conde MC. Massive soybean expansion in South America since 2000 and implications for conservation. *Nature Sustainability*. 2021 Sep;4(9):784-92.
95. Meijaard E, Abrams JF, Slavin JL, Sheil D. Dietary Fats, Human Nutrition and the Environment: Balance and Sustainability. *Frontiers in Nutrition*. 2022:439.
96. de Jong E, Higson A, Walsh P, Wellisch M. Bio-based chemicals value added products from biorefineries. *IEA Bioenergy, Task42 Biorefinery*. 2012;34.
97. Mowery DC, Rosenberg N. *Paths of Innovation: Technological Change in 20th-Century America*. Cambridge University Press; 1999 Oct 28.

12

Oils and Fats: Raw Materials for Corrosion Inhibitor

Smriti Dwivedi[1]* and Anita Kushwaha[2]

[1]Department of Applied Science (Chemistry), Galgotias College of Engineering and Technology, Greater Noida, U.P, India
[2]Department of Applied Chemistry, Gautam Buddha University, Greater Noida, U.P, India

Abstract
Corrosion is controlled by various organic and inorganic compounds, which are responsible for contamination and pollution in the environment. Strong environmental concerns and growing regulations have increased the need for vegetable oil and plant extracts that can be used as corrosion inhibitors naturally. Fatty acids which are extracted from plants and plant products like essential oils are acceptable to the environment. They are easily available and also renewable sources for inhibitors of corrosion. This chapter gives a brief review of various oils and fats which are used as corrosion inhibitors and will become alternatives to organic and inorganic corrosion inhibitors.

Keywords: Corrosion inhibitor, fatty acids, vegetable oil, essential oil, environment

12.1 Introduction

Corrosion is a material's degradation due to its interaction with the environment chemically and electrochemically. The major problems due to corrosion are that it affects metals' performance and safety. Corrosion inhibitors can control the above problems practically and economically

**Corresponding author*: s.dwivedi@galgotiacollege.edu
S. Dwivedi: ORCID: https://orcid.org/0000-0001-9193-2935
A. Kushwaha: ORCID: https://orcid.org/0000-0002-7583-9199

Divya Bajpai Tripathy, Anjali Gupta, Pooja Agarwal, Anuradha Mishra, and Arvind Kumar Jain (eds.)
Oils and Fats as Raw Materials for Industry, (307–332) © 2024 Scrivener Publishing LLC

with continuous corrosion resistance [1]. Various heterocyclic compounds of Nitrogen and Sulphur and their substituents are effective corrosion inhibitors. It was shown [2–5] that research is going on to find corrosion inhibitors of organic origin in different types of corrosive media. Organic substances like Pyrrole, thiophene, hydrazine and derivatives while borates, dichromates, silicates, phosphates, chromates, tungstates, molybdates, and arsenates are inorganic substances and both types of inhibitor prevent metal corrosion in acid solutions. These corrosion inhibitors are also useful in the formulation of anticorrosive coatings and primers. Sulphur containing aldehydes. i.e., thioaldehydes, different alkaloids like strychnine, papaverine, nicotine, quinine and acetylenic compounds are used as corrosion inhibitors in acid medium while nitrite, phosphate, benzoate and chromate, act as corrosion inhibitors in neutral media. Corrosion is prevented when inhibitors inhibit the reaction of the metal with any acid or alkaline or neutral medium. In spite of its various advantages, toxicity is a major disadvantage of organic and inorganic corrosion inhibitors so that their use has affected the environment and come under severe criticism from people.

Anticorrosive activity is shown by various synthetic compounds but most of them are highly toxic and harmful [6] to the environment and for human beings also. It was found that synthetic compounds are non-biodegradable and very expensive, so the natural products become the key area of research for use as inhibitors for corrosion.

Green corrosion inhibitors do not contain heavy metals or other toxic compounds so they are biodegradable and safe to the environment. Metal Corrosion is inhibited by various naturally occurring substances in acidic and alkaline environments. Oil obtained from plants such as castor, linseed and sunflower oil inhibit the corrosive action on metallic surfaces successfully by modifying their behavior electrochemically [7–9]. Oils adsorbed on the metal surface due to their lubricating properties and also depends on their boundary regime property. Long-chain aliphatic organic molecules which are polar in nature, adsorbed on metallic surface of positively charged metal, by creating barriers or films which are self-assembled and increase resistance against corrosion and increase properties of oil like viscosity, intermolecular interaction and surface tension which is suitable for metal and inhibitor interaction [10, 11].

These advantages have prompted many researchers to examine natural substances as corrosion inhibitors by doing laboratory work on various oil-producing plants like fennel oil [12], pear seed oil [13–16], argan oil [13, 16], rosemary oil [17–19], Thymus oil [20, 21], pennyroyal mint oil [22], lavender oil [23], Artemisia and jojoba oil [24]. Artemisia and Jojoba oil has been investigated for its anticorrosive properties [25, 26]. So this paper highlights the role of various oils and fatty acids as green corrosion inhibitors.

12.2 Essential Oil as Corrosion Inhibitor

12.2.1 Essential Oil

Essential oils have multiple origins. i.e., extracted from any part of a plant and are generally present in all the vegetable species. They are aromatic liquid organic substances and compared to other known plant extracts they are highly volatile. Apart from this they are non-oily and under high temperature they are susceptible to decomposition [27]. As we discussed, they have multiple origins; they are found in seeds (nutmeg, coriander), fruits (pepper), wood (cedar, sandalwood), leaves (eucalyptus), flowering tops (lavender), mint roots or rhizome (ginger), etc. [28].

12.2.2 Corrosion Inhibition Mechanism by Essential Oils

The corrosion inhibition mechanism of essential oils and plant extracts is based on an electrochemical process which is applicable for all corrosion inhibitors. The inhibitory action of the essential oils is usually carried out at the surface of solid phases at the time of adsorption of species, at the transfer of electric charges or at the time when species is transported in solution, or at the formation of superficial intermediaries which are elementary reaction stages. The action of an inhibitor occurs near the surface in their immediate vicinity because intervention of inhibitor within solution does not take place in the process of transport of oxygen, protons and reaction products which are electro-active species. Essential oils' mechanism of action and their inhibitory effect is based on the presence of active constituents. Corrosion is reduced by the formation of a protecting layer on the metal surface or by improvement of passivation which previously exists by adsorption mechanism [29]. The adsorption process in metals is affected by essential oils' molecules structures, the nature of the metal, and charge distribution [30]. Either by chemisorption mechanism or by physisorption mechanism or as a combination of both mechanisms essential oils are adsorbed at the metal or solution interface.

In Chemisorption mechanism chemical interaction takes place by chemical bond formation or by electron exchange. between metal surface atoms and inhibitor atoms of essential oil. In the physisorption mechanism it takes place by van der Waals forces, which is physical interaction between the surface of metal and ions contained in the solution.

The following types of action of the inhibitor can be seen:

1. In acidic media, there is creation of a barrier between the corrosive medium and metal by surface adsorption.
2. In neutral media, barrier formation is by interaction between the corrosive medium and the inhibitor.
3. Molecules with polar groups at both ends form a layer on the metal surface by chemical bonding which is a continuous layer.
4. Metal surface passivation is done by increasing the pH near the metal surface by addition of buffers.
5. The rate of corrosion is also reduced by oxidizing inhibitors and metal is passivated.
6. When mineral salts precipitate and form surface films, corrosion rate reduces.

12.2.3 Various Essential Oils for Corrosion Inhibition

Corrosion inhibition is done by use of various essential oils. Essential oils are extracted from different plants and their parts like from their flowers, fruits, leaves and seeds [31–41].

12.2.3.1 Tea-Tree Oil

Tea-tree extract contains antioxidants, and has various applications in the cosmetic, pharmaceutical food industries but its oil shows anticorrosive properties also. Tea-tree essential oil had tested for their inhibitory action against steel in 1 M hydrochloric acid. At a temperature of 298 K, the inhibitory efficiency of oil increases to 96% at 2 g/L of essential oil with the increase in concentration [42].

12.2.3.2 Chamomile Essential Oil

Chamomile has a fruity odor and its oil is widely used in the flavour and fragrance industry. They have anti-allergic, antidepressant and anti-inflammatory properties. Ben Hmamou *et al.* analysed chamomile oil on C38 steel in 1 M hydrochloric for corrosion efficiency, where he found that the efficiency is 90% with 2 g/L of chamomile essential oil while there is no effect of temperature on the efficiency rate [43].

12.2.3.3 *Mentha Oil*

Mentha spicata (Spearmint) is a creeping and perennial herb with a strong aromatic odor. The essential oil of mentha spicata has both insecticidal and mutagenic activity. The oil of mentha spicata has a characteristic spearmint odor and is rich in carvone. Znini *et al.* [44] studied the inhibitory action of Mentha spicata on 1 M HCl medium on steel corrosion. The inhibitory effect of this oil at 298 K is 97% at 2 g/L of inhibitor. But on higher temperatures (from 303 to 333 K) there is a decrease in the efficiency observed.

12.2.3.4 *White Sagebrush (Artemisia Herba-Alba) Oil*

White sagebrush oil inhibits corrosion; it has been examined by various authors with the help of gravimetric measurements, electrochemical impedance methods and by combination of polarization and on the corroded steel in three different acids. Sagebrush essential oil in 0.5 M H_2SO_4 was studied by Ouachikh *et al.* [45] and their results showed that corrosion rate is reduced by this oil through cathodic action. Oil concentration reached 74% at 1 g/L and their inhibition efficiency increases with temperature. Artemisia herba-alba essential obtained from aerial parts was studied by Boudiala *et al.* [46] and tested against stainless steel for its corrosion inhibition efficiency in 1M HCL acid by using potentiodynamic polarization measurements, electrochemical impedance spectroscopy and scanning electron microscopy. Rate of corrosion is reduced and inhibition efficiency increased with Artemisia's essential oil inhibitor and concentration reaches 88% with 1 g/L of substance.

12.2.3.5 *Mint Fowl Essential Oil*

Bouyanzer *et al.* [47] evaluated the corrosion behavior of mint fowl essential oil, also known as Menthapulégium, on steel in 1 M HCl solution by using electrochemical impedance and gravimetric measurements. When oil content increases, corrosion inhibition action increases and at 2.76 g/L it reaches 80%, which shows that essential oils of mint fowl are a good corrosion inhibitor.

12.2.3.6 *Eucalyptus Oil*

Rekkab *et al.* [48] investigated the role of oil obtained from Eucalyptus globulus, botanical name Myrtaceae. Corrosion of steel C38 in H_2SO_4 solution is inhibited by using electrochemical impedance spectroscopy, potentiodynamic polarization technique and weight loss techniques. When the concentration of essential oil increases, the inhibition efficiency increases to 75.8% at 6 g/L. The role of essential oils and their effect on temperature in the interval of 298–328 K was studied in 0.5 M H_2SO_4 on the behavior of C38 steel for corrosion. It was found that with increase in temperature there is a decrease in the inhibitory efficiency of essential oil.

12.2.3.7 Atlas Cedar (Cedrus Atlantica) Essential Oil

El-miziani *et al.* [49] studied corrosion inhibition by Atlas cedar (Cedrus atlantica) essential oil in 0.1 M Na_2CO_3 solution by using the impedance technique and polarization curve technique. As oil concentration increases, the inhibition efficiency increases and inhibition efficiency reaches 70% at 2000 ppm of concentration, and thus corrosion rate decreases.

12.2.3.8 Salvia Aucheri Mesatlantica Essential Oil

Znini *et al.* [50] studied the inhibitory action of Salvia aucheri mesatlantica in 0.5 M H_2SO_4 solution on the corrosion of steel by using electrochemical techniques and weight loss method technique. The results showed that the inhibitory efficiency was 86.12% at 2g/L and increased with the increase in essential oil concentration.

12.2.3.9 Fennel Essential Oil

Lahhit *et al.* [51] studied inhibitory action on corrosion of steel in 1 M HCl by fennel essential oil (foeniculum vulgaris) with the help of transient electrochemical techniques and gravimetric methods. The results showed that when concentration of Fennel essential oil increases, the inhibitory efficiency increases and at 3 ml/L a it reaches 76% and corrosion rate decreases.

12.2.3.10 Vervain Essential Oil

Ben Hmamou *et al.* [52] investigated C38 steel in HCL 1M for inhibitory action of vervain essential oil by using potentiodynamic polarization method, electrochemical techniques and weight loss method technique. Results of Vervain oil were investigated in temperature range of 298–328 K and found that the inhibitory effect decreases with increase in temperature whereas inhibitory effect increases with increase in inhibitor Vervain oil concentration.

12.2.3.11 *Rosmarinus Officinalis Essential Oil*

Essential oil of Rosmarinus officinalis was tested in 0.5 MH_2SO_4 on the corrosion of steel for its inhibitory action by El Ouariachi *et al.* [53]. Inhibitory action was studied by using electrochemical polarization and weight loss measurements method. Results showed that at 1 g/L, corrosion inhibition reached 61%, which means it increases with increase in oil concentration but there is no change with increase in temperature.

12.2.3.12 *Helichrysum Italicum Oil*

L. Majidi *et al.* [54] studied the inhibitory effect of essential oils obtained from aerial parts of Helichrysum italicum, which increases with concentration of the essential oil and at 2 g/L it reached 82.33%. They also studied the effect of temperature without and with essential oil in HCl solution of 1 M concentration on the behavior of mild steel for corrosion inhibition.

12.2.3.13 *Combined Effect of Clove, Basil and Atlas Cedar Essential Oils*

R-T Loto *et al.* [55] studied the combined effect of essential oils of clove, atlas, cedar, basil. These mixed oils were studied in H_2SO_4 and HCl solutions of 0.5 M on carbon steel of low grade to find the inhibition properties against corrosion. The results showed that inhibition efficiencies values are 95.48% and 95.32% in H_2SO_4 solution, and in HCl solution it was 92.7% and 97.98% efficiency of these mixed oils at various concentrations.

12.2.3.14 *Combined Effect of Salvia Officinalis and Simmondsia Chinensis Essential Oils*

Essential oils of salvia officinalis and simmondsia chinensis in combination was studied and conducted by R-T Loto *et al.* [56] to assess the corrosion inhibition properties in 1 M of H_2SO_4 on mild steel. The electrochemical test of this mixture showed optimal inhibition efficiency of 86.58% to 6% and 83.29% to 5%, respectively, in volumetric concentration.

12.2.3.15 *Citrus Sinensis (CS) Essential Oil*

R-T Loto *et al.* [57] evaluated the Citrus sinensis (CS) essential oil. The inhibition properties of these oils were evaluated for ordinary carbon steel in HCl solution and H_2SO_4 solution of 0.5 M concentration. On assessment through weight loss method, it was revealed that there is optimal inhibition of 76.95% in H_2SO_4 and 81.61% in HCl. The result showed inhibition efficiency in both acids increases with increased concentration and decreases with respect to exposure time.

12.2.3.16 Ptychotis Verticillata Essential Oil

E. El Ouariachi *et al.* [58] investigated essential oil of Ptychotis verticillata and in HCl (1M) used as a mild steel corrosion inhibitor. Corrosion inhibition was evaluated by electrochemical impedance spectroscopy, weight loss measurement and by potentiodynamic bias curves. The inhibition effect increases with the increase in corrosion inhibitor concentration and in HCl of 1 M concentration at 1.08 g/L the inhibition effect was 74.6% and it is through adsorption of inhibitory molecules essential oil on the metallic surface.

Various other sources of vegetable oils can also be used as corrosion inhibitors, such as biopolymers and biofuels [59–61]. Unsaturated fatty acid bioproducts and Fatty acid amides obtained from rice bran oil is best for the corrosion inhibitors synthesis. For the optimum inhibitor concentration, the inhibition efficiency was at least 95% [62]. The process of on electrochemical evaluations of fatty amides at all test temperatures showed that it is used as corrosion inhibitors.

12.3 Fatty Acids as Corrosion Inhibitors

The negative impact due to conventional anticorrosive products decreases by the use of the green inhibitor that is extracted from plants [63]. Synthesization of many molecules of corrosion inhibitor, an extract of vegetable oils, was used for the process of fatty acid derivatization. In aggressive media for mild steel, derived fatty acids such as ethyl ester, hydrazides, sulfate and sulfate-amine salt, oxadiazoles, triazoles, imidazoline, thiosemicarbazide of phenyl, ethoxylate, and hydrazides of phenyl were very important and effective corrosion inhibitors. For some metals like steel, aluminum, zinc and other metals, substances like phenols, saponins, tannins, oils and fats are used as inhibitors [64]. These compounds can be found in many plants and other natural resources [65]. Amino acids and biopolymers like several natural products [66] have very high inhibition efficiency due to rich sources of fatty acids obtained from plant extract. Several fatty acids with some structural modification in various heterocyclic derivatives, due to the presence of more active centers in their structures which are easily adsorbed on metal surfaces [67, 68], such types of compounds for environmental regulations which can also replace the conventional toxic chemicals [69] from the atmosphere.

12.4 Copper Corrosion Inhibitor by Fatty Amidine

The derivatives of fatty acids produce sustainable and green chemicals as precursor materials [70, 71] and are ideal candidates, inexpensive and abundant. One type of compound of derivatization oleochemicals fatty acids which can be obtained from the palm oil industry is a renewable source such as an inhibitor of copper. Due to the presence of long alkyl chains, it can be easily bound on metal surfaces which can help to form the terminal moiety and a protective layer on the surface of metal. There are many fatty acid derivatives–based corrosion inhibitors like acyclic and heterocyclic compounds for metal surfaces are present [72–82]. The common structure of the RnC(=NR) NR_2 in fatty amidine is a derivative of oxoacid, act as a corrosion inhibitor, wherein an amino and imine group have substituted oxo group and the hydroxyl group respectively [83]. Stearamidine hydrochloride salt is used as a corrosion inhibition compound on a copper surface due to the presence of fatty amidine. The inhibition efficiency of stearamidine in 3.0 wt.% of NaCl determined by the use of electrochemical, thermodynamic parameters and morphological analysis.

12.5 Palm Oil as Corrosion Inhibitor

The most common saturated fatty acid which is derived from plants and animals [84] is Palmitic acid or hexadecanoic acid with molecular formula is $CH_3(CH_2)14COOH$ or $C_{16}H_{32}O_2$ in IUPAC nomenclature is a major component of the oil; its name indicates oil comes from palm that is palm kernel oil, palm oil, and oil of coconut. For corrosion inhibition of metals, raw material of palm oil–based fatty acids and monoethanolamin can be good for "environmentally friendly" in acidic media like amides [85] and surfactants [86]. Additional heteroatom groups from monoethanolamine should cause additional interactions with metal surfaces compared to fatty acids or normal fatty esters, thus increasing their corrosion prevention properties [87]. Palm oil is obtained from fatty acids which achieve inhibitor properties of the products [88]; it has high inhibition efficiency against the mild steel corrosion in air and 1M NaOH, especially on corrosion of

Table 12.1 Typical fatty acid composition (%) of palm oil.

Chain length of fatty acid	Mean	Observed range	Standard deviation
12	0.3	0 – 1	0.12
14	1.1	0.9 – 1.5	0.08
16	43.5	39.2 – 45.8	0.95
16.1	0.2	0 – 0.4	0.05
18	4.3	3.7 – 5.1	0.18
18.1	39.8	37.4 – 44.1	0.94

ductile iron and protective shield. Table 12.1 shows the typical fatty acid composition of palm oil. Saturated and unsaturated fatty acids in approximately equal amounts are present in palm oil [89].

12.6 Flower Extracts as Corrosion Inhibitor

The use of inhibitors in acidic solutions for protection and prevention of unexpected metal dissolution and acid consumption against corrosion is one of the most important practical methods [90, 91]. Naturally occurring products as renewable sources of materials [92, 93] are used by scientists as corrosion inhibitors due to decreased hazardous effects and the fact that they are inexpensive, readily available, ecologically acceptable and environmentally friendly. In the present time, for mild steel the leaves of certain plants like Terminalia belivia, Terminalia chebula, Murraya koenigii, Embilica officianilis, Accacia conicianna, Swertia angustifolia, Sapindus trifolianus. Eucalyptus, anis, Telferia occidentalis, Annona squamosa, Hibiscus, Azadirachta indica Vernonia amydalina, Accacia Arabica, Ricimus communis coriander, Nypa fructicans wurmb, Pongamia glabra, Carica papaya, Eugenia jambolans, Occimum viridis, garden cress and black cumin have been used as the corrosion inhibitor [94–100] in acidic medium. A few of the fruits like lignin along with Papaia, castor oil fruits, acacia gum and Poinciana pulcherrima, Tobacco, Datura stramonmium and Cassia occidentalis have also been used [101–105]. In acidic media, the good anticorrosion activity showed by onion, garlic and bitter gourd for mild steel and also oil extracts of Eugenol, Acetyl-eugenol,

Ginger, Artemisia oil, Mentha pulegium, and Jojoba are used for inhibition [106] of steel in acidic medium [107, 108]. Nowadays some plants such as Azydracta indica, Auforpio turkiale and Calotropis procera are very useful as acid corrosion inhibitors. The effect of anticorrosive on carbon steel in 1 M HCl used Quinine has been used. An inhibition effect of Zenthoxylum alatum extract was investigated [109–111] in aqueous Orthophosphonic acid. By the use of potentiodynamic polarization, weight loss process and electrochemical impedance spectroscopy (EIS) methods the development of green inhibitors [112, 113] of mild steel in H2SO4 solution are also reported [114].

12.7 Fatty Amide Derivatives Used as Corrosion Inhibition of Carbon Steel

Fatty amides, a distillate-based solution of fatty acid of palm, was used as a corrosion inhibitor. However, very few natural products such as pharmaceutically active compounds (antibiotics and antibacterials), extracts of various parts of plants and so on [115–119] are environmentally acceptable. Carbon steel in 3% NaCl is inhibited corrosion studied and found by Hashim, Saleh & Toff [120]. However, the increase of concentration of inhibitors and the performance of corrosion inhibitors fluctuates.

12.8 Unsaturated Fatty Acid Derived by Microalgae as Corrosion Inhibitor

For effective inhibition of mild steel, the molecules like 9,12–octadecadienoic acid, hexadecanoic acid (C16:0) and 9,12,15-octadecatrienoic acid are used due to presence of active functional groups in the fatty acid having the capability of fast electron transfer and having a biodegradable, renewable, and economically feasible environment [121–123]. According to the reviewers, the organic compounds potentially have properties to develop an adhesive layer on surfaces of metal due to increasing the rate of reaction of chemisorption and physisorption [124, 125]. The plant extracts as green inhibitors under an acidic environment were utilized [126–130] and it is possible to replace synthetic products by the plant-based inhibitors. The work of M. Hiri *et al.* (2016) synthesized integrating orange peel extracts having protective ability and forming a thin film over the surface of carbon steel [131].

Nowadays, as an anodic inhibitor extracts of brassica campestris to COR-TEN steel, contributing 80–84% inhibition [132] are used in chloride solution. In non-arable as well as fragile land and in the environment, the microalgae are present. These microalgae are photosynthetic, eukaryotic, and oxygenic microorganisms widespread [133]. Microalgae are an abundant source of oil accumulation. It is the first time the researcher has introduced microalgal oil as anti-corrosion proficiency in acidic medium for mild steel. The algal extracts carrying active molecules are usually rich in heteroatoms; their excellence have already proven toward corrosion inhibition [134–136] in acidic environments. The transfer of electrons toward a metal surface is promoted by available nonbonded (lone pair) and p-electrons in organic inhibitor molecules, it increases the inhibitor properties [137]. In Microalgal oil, saturated or unsaturated double bonds in free fatty acids, oleochemicals in the form of triglycerides, methyl esters fatty acids, amines, and fatty alcohols are present [138]. Due to the unsaturated bonds, cations and also in fatty acids of carboxylate anion, the presence of active functional groups enhance the surface chemistry and increase the adsorption characteristics [139]. The anti-corrosion behavior of nove, fatty acid-rich chlorophycean microalgae strain Scenedesmus sp. enhanced effectiveness under an acidic environment. The green approach to maximize the inhibition efficiency by creating adsorptives act as green inhibitors.

12.9 Other Green Inhibitors

The use of natural products as inhibitors shown by a number of works, these natural materials contain $-NH_2$, $-NR_2$, -SR and -COOH and phytochemicals (tannins, amino acids, alkaloids, saponin phenols, flavonoids, etc.) as functional groups which provide adsorption by these substances on the metal surface. Black pepper, a rich source for piperine alkaloid, collected from the Piper nigrum plant which belongs to the Piperaceae family, using weight loss process, its anticorrosion effect obtained and electrochemical impedance, polarization and SEM studies show their green inhibition [140]. A use of rape-cake extracts, macerating in the 2-propanol alcohol [141] and from Boraginaceae family the part of plant Trichodesma indicum (Linn) R. Br. [142] having the alkaloid extract showed the corrosion inhibition effect. Corrosion inhibition effects on various metals have been investigated from extracts of Funtumia elastica, Hibiscus sabdariffa, Chamaemelum mixtum, Wrightia tinctoria, Telfairia occidentalis, Piper guineense, Azadirachta indica, Tridax procumbens, Ocimum gratissimum,

Nigella sativa, Lupinousalbus, Ipomoea triloba and Chromolaena odorata and so on [143–153]. The above work shows their inhibition properties in highly corrosive media and obtains nitrogen-containing substances from plant and animal raw materials. As corrosion inhibitors, animal fats can also be used [154].

Wherever possible, local raw materials should be used to decrease the cost of obtaining inhibitors. Based on knowledge, there are a number of sources of raw materials in Uzbekistan, to protect equipment from corrosion developing in the oil and gas industry by their effective composite inhibiting materials that are based on vacuum distillation waste - cotton soap stock - gossypol resin. Gossypol resin, liquid ammonia, amino alcohols-monoetonolamine ($H_2N_2H_2CHOH$), diethanolamine ($HN(CH_2CH_2OH)_3$), triethanolamine ($N(CH_2CH_2OH))_3$, β(N,N diethylamine ($(C_2H)_2NCH_2CH_2OH$), as a secondary raw material is formed from oil and fat plants in large number of quantities. According to chemical composition the gossypol resin [155, 156] is heterogeneous in nature, due to its variability has not been fully established. The hexahedral dialdehyde phenol that is homologous of 2,2-binaphthyl are present in gossypol molecules, in most organic solvents their substances are soluble and practically in water it is insoluble. The amino acids, carbohydrates and other components present in cottonseed, are the products of its thermal and chemical transformation due to polycondensation. The basis of gossypol resin is free fatty acids, then polymerized fatty acids, as well as gossypol.

Due to hydroxyl group, the capability of gossypol produces derivatives in large numbers like polyhydric phenol. When treating gossypol with the help of soluble alkalis, which are destroyed under the action of mineral acids [155] with the release of free gossypol after that soluble of water gossypolates (such as sodium, potassium, ammonium) are formed.

12.10 Conclusions

From the study it is concluded that a variety of green corrosion inhibitors are there and some of those green corrosion inhibitors showed almost the same corrosion inhibition efficiency in acid medium as synthetic inhibitors. These natural inhibitors, especially essential oils and fatty acids, that are obtained from plant extracts are environmentally friendly [157]. The use of these corrosion inhibitors lessens the toxicity of synthetic inhibitors and they are environment friendly. Although plant-derived oils and fats are nonrenewable and nontoxic in nature for which it should be used for future corrosion inhibition, we should understand that these oils are for food for human beings and their excessive use should not disturb the ecological balance which will be affected partially. But the major problems with these oils is that they are seasonal plants and take a long time to grow and for cultivation. Therefore, for mild steel corrosion inhibition and for other metal corrosion inhibition we have to look further for a more economical and more sustainable source of oil towards metal corrosion inhibition.

References

1. M. Heydari and M. Javidi, Corrosion inhibition and adsorption behavior of an amido-imidazoline derivative on API 5L X52 steel in CO2-saturated solution and synergistic effect of iodide ions, *Corrosion Science*, vol. 61, pp. 148–155, 2012.
2. M. Bouklah, B. Hammouti, T. Benhadda, and M. Benkadour, Thiophene derivatives as effective inhibitors for the corrosion of steel in 0.5 M H2SO4, *Journal of Applied Electrochemistry*, vol. 35, no. 11, pp. 1095–1101, 2005.
3. A. S. Fouda, A. A. Al-Sarawy, and E. E. El-Katori, Pyrazolone derivatives as corrosion inhibitors for C-steel HCl solution, *Desalination*, vol. 201, pp. 1–13, 2006.
4. A. Fiala, A. Chibani, A. Darchen, A. Boulkamh, and K. Djebbar, Investigations of the inhibition of copper corrosion in nitric acid solutions by ketene dithioacetal derivatives, *Applied Surface Science*, vol. 253, no. 24, pp. 9347–9356, 2007.
5. U. R. Evans, *The Corrosion and Oxidation of Metals*, Hodder Arnold, 1976.
6. Ostovari, A.; Hoseinieh, S. M.; Peikari, M.; Shadizadeh, S. R.; Hashemi, S. J. *Corros. Sci.* 2009, 51, 1935–1949.
7. Godavarthi, S.; Porcayo-Calderon, J.; Casales-Diaz, M.; Vazkuez-Velez, E.; Neri, A.; Martinez-Gomez, L. Electrochemical analysis and quantum chemical behaviour of castor oil-based inhibitors. *Curr. Anal. Chem.* 2016, 12, 476–488.
8. Yang, Z. N.; Liu, Y. W.; Chen, Y. Linseed Oil Based Amide as Corrosion Inhibitor for Mild Steel in Hydrochloric Acid. *Int. J. Electrochem. Sci.* 2018, 13, 514–529.
9. Balakrishnan, T.; Sathiyanarayanan, S.; Mayavan, S. Advanced anticorrosion coating materials derived from sunflower oil with bifunctional properties. *ACS. Appl. Mater. Interfaces* 2015, 7, 19781–19788.
10. Tian, H.; Liu, W.; Liu, A.; Gao, X.; Han, P.; Ding, R.; Yang, C.; Wang, D. Controlled delivery of multi-substituted triazole by metalorganic framework for efficient inhibition of mild steel corrosion in neutral chloride solution. *Corros. Sci.* 2018, 131, 1–6.

11. Han, P.; He, Y.; Chen, C.; Yu, H.; Liu, F.; Yang, H.; Ma, Y.; Zheng, Y. Study on Synergistic mechanism of Inhibitor Mixture Based on Electron Transfer Behavior. *Sci. Rep.* 2016, 27, No. 33252.
12. Lahhit N, Bouyanzer A, Desjobert JM, Hammouti B, Salghi R, Costa J, Jama C, Bentiss F, Majidi L, Fennel (Foeniculum Vulgare) Essential Oil as Green Corrosion Inhibitor of Carbon Steel in Hydrochloric Acid Solution, *Port Electrochim Acta* 29:127, 2011.
13. Ben Hmamou D, Salghi R, Bazzi L, Hammouti B, Al-Deyab SS, Bammou L, Bazzi L, Bouyanzer A (2012) *Int J Electrochem Sci* 7:1303.
14. Afia L, Salghi R, El B, Bazzi L, Errami M, Jbara O, Al-Deyab SS, Hammouti B, Corrosion Inhibition of Steel in 1 M Hydrochloric Acid Medium by Chamomile Essential Oils, *Int J Electrochem Sci* 6:5918, 2011.
15. Afia L, Salghi R, Bammou L, Bazzi E, Hammouti B, Bazzi L, Application of Argan plant extract as green corrosion inhibitor for steel in 1 mol/L HCl, *Acta Metall Sin* 25:10, 2012.
16. Afia L, Salghi R, Bazzi E, Zarrouk A, Hammouti B, Bouri M, Zarrouk H, Bazzi L, Bammou L, Argan hulls extract: green inhibitor of mild steel corrosion in 1 M HCl solution, *Res Chem Intermed.* 38:1707–1717, 2012.
17. Afia L, Salghi R, Bammou L, El B, Hammouti B, Bazzi L, Bouyanzer A, Testing Natural compounds: Argania spinosa Kernels Extract and Cosmetic Oil as Ecofriendly Inhibitors for Steel Corrosion in 1 M HCl, *Int. J. Electrochem. Sci.*, 6, 5918–5939, 2011.
18. Chetouani A, Hammouti B, Benkaddour M, A study of anti-corrosive effects of Artemisia oil on steel, *Res Pig Tech* 33:26, 2004.
19. El Ouariachi E, Paolini J, Bouklah M, Elidrissi A, Bouyanzer A, Hammouti B, Desjobert J-M, Adsorption properties of Rosmarinus of ficinalis oil as green corrosion inhibitors on C38 steel in 0.5 M H2SO4, *J. Costa, Acta Metall Sin* 23:13, 2010.
20. Bendahou M, Benabdallah M, Hammouti B, A study of rosemary oil as a green corrosion inhibitor for steel in 2 M H_3PO_4, *Pigm Res Techn* 35:95, 2006.
21. Bouyanzer A, Hammouti B, Naturally occuring ginger as corrosion inhibitor for steel in molar hydrochloric acid at 353 K, *Bull Electrochem* 20:63, 2004.
22. Bammou L, Chebli B, Salghi R, Bazzi L, Hammouti B, Mihit M, El Idrissi H, Thermodynamic properties of Thymus satureioides essential oils as corrosion inhibitor of tinplate in 0.5 M HCL: Chemical characterization and electrochemical study, *Green Chem Lett Rev* 3:173, 2010.
23. Bouyanzer A, Hammouti B, Majidi L, Pennyroyal oil from Mentha pulegium as corrosion inhibitor for steel in 1 MHCl., *Mater Lett* 60:2840, 2006.
24. Zerga B, Sfaira M, Rais Z, Ebn Touhami M, Taleb M, Hammouti B, Imelouane B, Elbachiri A, Lavender oil as an eco-friendly inhibitor for mild steel in 1 M HCl, *Mater. Tech* 97:297, 2009.
25. Bouyanzer A, Hammouti B., A study of anti-corrosive effects of Artemisia oil on steel, *Pigm. Resin Techn.* 33:287, 2004.

26. P. C. Okafor, M. E. Ikpi, I. E. Uwah, E. E. Ebenso, U. J. Ekpe, and S. A. Umoren, "Inhibitory action of Phyllanthus amarus extracts on the corrosion of mild steel in acidic media," *Corrosion Science*, vol. 50, no. 8, pp. 2310–2317, 2008.
27. A. Y. El-Etre and M. Abdallah, Natural honey as corrosion inhibitor for metals and alloys. II. C-steel in high saline water," *Corrosion Science*, vol. 42, no. 4, pp. 731–738, 2000.
28. Turgeon, M., "Profil des produits forestiers première transformation des huiles essentielles" 2001–3086 p. 1., Déc 2001.
29. Belakhdar, J., "La pharmacopée marocaine traditionnelle". Idis Press (Ed.). Paris, p. 764, 1997.
30. O. L. Riggs. Jr., *Corrosion Inhibitors*, 2nd Ed., C. C. Nathan, Houston, 1973.
31. A. El-Etre., Inhibition of acid corrosion of carbon steel using aqueous extract of olive leaves. *Journal of Colloid and Interface Science*, 314 (2), 578–583, 2007.
32. J. C. Da Rocha, J. A. d. C. P. Gomes, and E. D'Elia, Corrosion inhibition of carbon steel in hydrochloric acid solution by fruit peel aqueous extracts. *Corrosion Science*, 52 (7), 2341–2348, 2010.
33. A. Zaabar, R. Aitout, L. Makhloufi *et al.*, Inhibition of acid corrosion of mild steel by aqueous nettle extracts. *Pigment & Resin Technology*, 43 (3), 127–138, 2014.
34. A. A. Koffi, S. Muralidharan, and A. Trokourey. Corrosion inhibition of carbon steel using extract of Mussaenda erythrophylla leaves: interfacial action mode in sulfuric acid medium/Inhibition de la corrosion de l'acier à l'aide d'extrait de feuilles de Mussaenda erythrophylla: mode d'action interfaciale en milieu acide sulfurique, *Journal de la Société Ouest-Africaine de Chimie*, 40, 31, 2015.
35. K. Shalabi, and A. A. Nazeer., Adsorption and inhibitive effect of Schinus terebinthifolius extract as a green corrosion inhibitor for carbon steel in acidic solution. *Protection of Metals and Physical Chemistry of Surfaces*, 51 (5), 908–917, 2015.
36. M. Prabakaran, S.-H. Kim, K. Kalaiselvi *et al.*, Highly efficient Ligularia fischeri green extract for the protection against corrosion of mild steel in acidic medium: Electrochemical and spectroscopic investigations. *Journal of the Taiwan Institute of Chemical Engineers*, 59, 553–562, 2016.
37. A. Fouda, A. Abousalem, and G. El-ewady. Mitigation of corrosion of carbon steel in acidic solutions using an aqueous extract of Tilia cordata as green corrosion inhibitor. *International Journal of Industrial Chemistry*, 8 (1), 61–73, 2017.
38. P. Muthukrishnan, B. Jeyaprabha, and P. Prakash., Adsorption and corrosion inhibiting behavior of Lannea coromandelica leaf extract on mild steel corrosion. *Arabian Journal of Chemistry*, 10 S2343-S2354, 2017.
39. T. Bhuvaneswari, V. Vasantha, and C. Jeyaprabha., Pongamia Pinnata as a green corrosion inhibitor for mild steel in 1N sulfuric acid medium. *Silicon*, 10 (5), 1793–1807, 2018.

40. M. Prabakaran, S.-H. Kim, A. Sasireka *et al.*, Polygonatum odaratum extract as an eco-friendly inhibitor for aluminum corrosion in acidic medium. *Journal of Adhesion Science and Technology*, 32 (18), 2054–2069, 2018.
41. Y. Qiang, S. Zhang, B. Tan *et al.*, Evaluation of Ginkgo leaf extract as an eco-friendly corrosion inhibitor of X70 steel in HCl solution. *Corrosion Science*, 133, 6–16, 2018.
42. Nasser Otaifah, Yaaqob. Etude de l'inhibition de l'acier C38 par de l'huile essentielle en milieu acide (HCl) 1 M, 2016.
43. Ben Hmamou, D., Salghi, R., Zarrouk, A., Hammouti, B., AlDeyab, S. S., Bazzi, Lh., Zarrok, H., Chakir, A., Bammou, L., The Inhibited effect of Phenolphthalein towards the corrosion of C38 Steel in Hydrochloric Acid, *Int. J. Electrochem. Sci.*, 7, 2361–2373, 2012.
44. Znini, M. & Bouklah, M. & Majidi, Lhou & Kharchouf, S & Aouniti, A & Bouyanzer, A. & Hammouti, B. & Costa, Jean & Al-Salem, Sultan. (2011). Chemical Composition and Inhibitory Effect of Mentha Spicata Essential Oil on the Corrosion of Steel in Molar Hydrochloric Acid. Int. *J. Electrochem. Sci. International Journal*. 6. 691–704.
45. O. Ouachikh, A. Bouyanzer, M. Bouklah, J. M. Dexjobert, J. Costa, B. Hammouti, L. Majidi. *Surf. Rev. Lett.* 16 (01), 2009: 49–54.
46. M. Boudalia, R-M. Fernández-Domene, M. Tabyaoui, A. Bellaouchou, A. Guenbour, J. García-Antón Green approach to corrosion inhibition of stainless steel in phosphoric acid of Artemesia herba albamedium using plant extract, *Journal of Materials Research and Technology* Volume 8, Issue 6, November–December 2019, pp. 5763–5773.
47. Bouyanzer, A., Hammouti, B., Majidi, L., Pennyroyal oil from Mentha pulegium as corrosion inhibitor for steel in 1 M HCl, *Materials Letters*, 60, 2840–2843, 2006.
48. S. Rekkab, H. Zarrok, R. Salghi, A. Zarrouk, Lh. Bazzi, B. Hammouti, Z. Kabouche, R. Touzani, M. Zougagh (2012) Green Corrosion Inhibitor from Essential Oil of Eucalyptus globulus (Myrtaceae) for C38 Steel in Sulfuric Acid Solution. *J. Mater. Environ. Sci.* 3 (4) 613–627, 2012.
49. I. Elmiziani, S. Houbairi, M. Essahli, A. Lamiri, S. Lhaloui, A. Lamiri, Lead corrosion inhibition by cedrus atlantica as a green inhibitor in 0.1M Na2CO3 solution, *International Journal of Advanced Chemistry*. 5, 1–7, 2017.
50. M. Znini, L. Majidi, A. Bouyanzer, J. Paolini, J-M. Desjobert, J. Costa, B. Hammouti: Essential oil of Salvia aucheri mesatlantica as a green inhibitor for the corrosion of steel in 0.5 M H2SO4 10.1016/j.arabjc.2010.09.017.
51. Lahhit, N. and al. Fennel (Foeniculum Vulgare) Essential Oil as Green Corrosion Inhibitor of Carbon Steel in Hydrochloric Acid Solution. *Port. Electrochim. Acta* [online]. 2011, vol. 29, n. 2, pp. 127–138. ISSN 0872–1904.
52. Dris Ben Hmamou, Rachid Salghi, Abdelkader Zarrouk, Hassan Zarrouk, Mohamed Errami, Belkheir Hammouti, Laila Afia, Lhoucine Bazzi, Lahcen Bazzi. *Acta Metall. Sin. (Engl. Lett.)* Vol. 23 No. 1 pp. 13–20 February 2010.

53. E. El Ouariachi, J. Paolini, M. Bouklah, A. Elidrissi, A. Bouyanzer 1), B. Hammouti, J-M. Desjobert and J. Costa Adsorption properties of Rosmarinus officinalis oil as green corrosion inhibitors on C38 steel in 0.5 M H2SO4. *Acta Metall. Sin. (Engl. Lett.)* Vol. 23 No. 1 pp. 13–20 February 2010.
54. Cristofari, G & Znini, M. & Majidi, Lhou & Costa, Jean & Hammouti, B. & Paolini, Julien. Helichrysum italicum subsp. italicum Essential Oil as Environmentally Friendly Inhibitor on the Corrosion of Mild Steel in Hydrochloric Acid. *International Journal of Electrochemical Science*. 7, 2012.
55. R-T. Loto, T. Olukeye, E. Okorie, Synergistic combination effect of clove essential oil extract with basil and atlas cedar oil on the corrosion inhibition of low carbon steel, *South African Journal of Chemical Engineering*, Volume 30, October 2019, pp. 28–41.
56. R-T. Loto, O. Olowoyo, Corrosion inhibition properties of the combined admixture of essential oil extracts on mild steel in the presence of SO4 2– anions Volume 26, December 2018, pp. 35–41.
57. R-T. Loto, E-H. Mbah, J-I. Ugada, Corrosion inhibition effect of citrus sinensis essential oil extract on plain carbon steel in dilute acid media. *South African Journal of Chemical Engineering*, Vol. 35, January 2021, pp. 159–64.
58. Ouariachi, El & Bouyanzer, A. & Salghi, Rachid & Hammouti, B. & Desjobert, J.-M & Costa, Jean & Paolini, Julien & Majidi, Lhou. Inhibition of corrosion of mild steel in 1M HCl by essential oils or solvent extracts of Ptychotis Verticillata. *Research on Chemical Intermediates*. 41. 10.1007/s11164-013-1246-5, 2015.
59. J. Porcayo-Calderon, L. M. Martínez de la Escalera, J. Canto, and M. Casales-Diaz, Imidazoline derivatives based on coffee oil as CO2 corrosion inhibitor, *International Journal of Electrochemical Science*, vol. 10, pp. 3160–3176, 2015.
60. S. Godavarthi, J. Porcayo-Calderon, M. Casales-Diaz, E. Vazquez-Velez, A. Neri, and L. Martinez-Gomez, Electrochemical analysis and quantum chemistry of castor oil-based corrosion inhibitors, *Current Analytical Chemistry*, vol. 12, pp. 476–488, 2016.
61. M. Alam, D. Akram, E. Sharmin, F. Zafar, and S. Ahmad, Vegetable oil based eco-friendly coating materials: a review article, *Arabian Journal of Chemistry*, vol. 7, no. 4, pp. 469–479, 2014.
62. E. Reyes-Dorantes, J. Zuñiga-Díaz, A. Quinto-Hernandez, J. Porcayo-Calderon, J. G. Gonzalez-Rodriguez, and L. Martinez-Gomez, Fatty Amides from Crude Rice Bran Oil as Green Corrosion Inhibitors, *Journal of Chemistry* Vol. 2017, Article ID 2871034, https://doi.org/10.1155/2017/2871034
63. Faustin, M., Maciuk, A., Salvin, P., Roos, C. & Lebrini, M. Corrosion inhibition of C38 steel by alkaloids extract of *Geissospermum* leave in 1 M hydrochloric acid: Electrochemical and phytochemical studies. *Corros. Sci.*, 92: 287–300, 2015.
64. Rajalakshmi, R., Prithiba, A. & Leelavathi, S. An overview of emerging scenario in the frontiers of eco-friendly corrosion inhibitors of plant origin for Mild Steel. *J. Chem. Acta*, 1(1), 6–13, 2012.

65. Yoo, S. H., Kim, Y. W., Chung, K., Baik, S. Y. & Kim, J. S. Synthesis and corrosion inhibition behavior of imidazoline derivatives based on vegetable oil. *Corros. Sci.*, 59: 42–54, 2012.
66. P. B. Raja and M. G. Sethuraman, Natural products as corrosion inhibitor for metals in corrosive media — A review, *Materials Letters,* vol. 62, pp. 113–116, 2008.
67. M. M. Osman and M. N. Shalaby, Some ethoxylated fatty acids as corrosion inhibitors for low carbon steel in formation water, *Materials Chemistry and Physics,* vol. 77, pp. 261–269, 2003.
68. T. Szauer and A. Brandt, On the role of fatty acid in adsorption and corrosion inhibition of iron by amine - Fatty acid salts in acidic solution, *Electrochimica Acta,* vol. 26, pp. 1257–1260, 1981.
69. B. E. Amitha Rani and B. B. J. Basu, Green inhibitors for corrosion protection of metals and alloys: An overview, *International Journal of Corrosion,* vol. 2012, p. 380217, 2012.
70. U. Biermann, U. Bornscheuer, R. A. M. Meier, O. J. Metzger, and J. H. Schafer, Oils and fats as renewable raw materials in chemistry, *Angewandte Chemie International Edition*, vol. 50, pp. 3854–3871, 2011.
71. R. N. P. Vennestrøm, M. C. Osmundsen, H. C. Christensen, and E. Taarning, Beyond petrochemicals: the renewable chemicals industry, *Angewandte Chemie International Edition*, vol. 50, pp. 10502–10509, 2011.
72. M. B. Badran, A. A. Abdel Fattah, and A. A. Abdul Azim, New corrosion inhibitors based on fatty materials-I. epoxidized fatty materials modified with aliphatic amines, *Corrosion Science*, vol. 22, pp. 513–523, 1982.
73. A. M. Quraishi, D. Jamal, and N. R. Singh, Inhibition of mild steel corrosion in the presence of fatty acid thiosemicarbazides, *Corrosion*, vol. 58, pp. 201–207, 2002.
74. R. Martínez-Palou, J. Rivera, G. L. Zepeda *et al.*, Evaluation of corrosion inhibitors synthesized from fatty acids and fatty alcohols isolated from sugar cane wax, *Corrosion*, vol. 60, pp. 465–470, 2004.
75. A. M. Quraishi and D. Jamal, Fatty acid triazoles: novel corrosion inhibitors for oil well steel (N-80) and mild steel, *Journal of the American Oil Chemists' Society*, vol. 77, pp. 1107–1111, 2000.
76. H. S. Yoo, W. Y. Kim, K. Chung, Y. S. Baik, and S. J. Kim, Synthesis and corrosion inhibition behavior of imidazoline derivatives based on vegetable oil, *Corrosion Science*, vol. 59, pp. 42–54, 2012.
77. A. Yildirim and M. Çetin, Synthesis of undecanoic acid phenylamides as corrosion inhibitors, *European Journal of Lipid Science and Technology*, vol. 110, pp. 570–575, 2008.
78. S. Ghareba and S. Omanovic, Interaction of 12-aminododecanoic acid with a carbon steel surface: towards the development of "green" corrosion inhibitors, *Corrosion Science*, vol. 52, pp. 2104–2113, 2010.
79. N. Ochoa, F. Moran, N. Pébère, and N. Tribollet, Influence of flow on the corrosion inhibition of carbon steel by fatty amines in association with

phosphonocarboxylic acid salts, *Corrosion Science*, vol. 47, pp. 593–604, 2005.
80. A. M. Quraishi and A. F. Ansari, Fatty acid oxadiazoles as corrosion inhibitors for mild steel in formic acid, *Journal of Applied Electrochemistry*, vol. 36, pp. 309–314, 2006.
81. A. Yildirim and M. Çetin, Synthesis and evaluation of new long alkyl side chain acetamide, isoxazolidine and isoxazoline derivatives as corrosion inhibitors, *Corrosion Science*, vol. 50, pp. 155–165, 2008.
82. S. Öztürk, A. Yildirim, M. Çetin, and M. Tavasli, Synthesis of quaternary, long-chain N-alkyl amides and their corrosion inhibition in acidic media, *Journal of Surfactants and Detergents*, vol. 17, pp. 471–481, 2014.
83. P. G. Moss, S. A. P. Smith, and D. Tavernier, Glossary of class names of organic compounds and reactive intermediates based on structure (IUPAC recommendations 1995), *Pure and Applied Chemistry*, vol. 67, pp. 1307–1375, 1995.
84. Beare-Rogers, J., Dieffenbacher, A., and Holm, J.V. Lexicon of Lipid Nutrition (IUPAC Technical Report). *Pure and Applied Chemistry.* 73(4): 685–744, 2001.
85. Porcayo-Calderon, J., Rivera-Muñoz, E.M., PezaLedesma, C., Casales-Diaz, M., Martínez de la Escalera, L.M., Canto, J., Martinez-Gomez, L., Sustainable development of palm oil: synthesis and electrochemical performance of corrosion inhibitors. *J. Electrochem. Sci. Technol.* 8(2): 133-145, 2017.
86. Ismayilov, I.T., Abd El-Lateef, H.M., Abbasov, V.M., Efremenko, E.N., Aliyeva, L.I., Akhmadbeyova, S.F., Novel synthesized surfactants based on palm oil and monoethanolamine as corrosion inhibitors for mild steel in CO2 environments. *Am. J. Chem.* 4(5): 155-165, 2014.
87. Effect of Palm Oil as Corrosion Inhibitor on Ductile Iron and Mild Steel, *Pacific Journal of Science and Technology*, Vol. 12, No. 2, November 2011.
88. Chen, Yun, Yang, Wenzhong, Formulation of Corrosion Inhibitors, March 28th, 2019 Reviewed: July 11th, 2019; Published: August 20th, 2019, DOI: 10.5772/intechopen.88533
89. Sundram. Kalyana. Palm Oil: Chemistry and Nutrition Updates. Malaysian Palm Oil Board (MPOB): Kuala Lumpur, Malaysia. 2010.
90. Ashassi-Sorkhabi, H., Seifzade, D., Hosseini, M.G., EN, EIS and polarization studies to evaluate the inhibition effect of 3H-phenothiazin-3-one, 7-dimethylamin on mild steel corrosion in 1 M HCl solution, *Corros. Sci.* 50, 3363, 2008.
91. Satapathy, A.K., Gunasekaran, G., Sahoo, S.C., Kumar, A., Rodrigues, P.V., Corrosion inhibition by Justicia gendarussa plant extract in hydrochloric acid solution, *Corros. Sci.* 51, 2848, 2009.
92. Abdel-Gaber, A.M., Abd-El-Nabey, B.A., Saadawy, M., The role of acid anion on the inhibition of the acidic corrosion of steel by lupine extract, *Corros. Sci.* 51 (2009)1038-1042.

93. Raja, P.B., Sethuraman, M.G., Instantly Investigating the Adsorption of Polymeric Corrosion Inhibitors on Magnesium Alloys by Surface Analysis under Ambient Conditions, *Mater. Lett.* 62, 113, 2008.
94. Noor, E.A., Comparative Study on the Corrosion Inhibition of Mild Steel by Aqueous Extract of Fenugreek Seeds and Leaves in Acidic Solutions, *J. Engg. Appl.* Sci. 3, 23, 2008.
95. Buchweishaija, J., Mhinzi, G.S., Natural Products as a Source of Environmentally Friendly Corrosion Inhibitors: The Case of Gum Exudate from Acacia seyal var. seyal, *Port. Electrochim. Acta* 26, 257, 2008.
96. Oguzie, E.E., Evaluation of the inhibitive effect of some plant extracts on the acid corrosion of mild steel, *Corros. Sci.* 50, 2993, 2008.
97. Okafor, P.C., Ikpi, M.E., Uwaha, I.E., Ebenso, E.E., Ekpe, U.J., Umoren, S.A., Synergistic Effect of Allium cepa-Zn2+ System on the Corrosion of Carbon Steel in Ground Water. *Corros. Sci.* 50, 2310, 2008.
98. Valek, L., Martinez, S., Copper corrosion inhibition by Azadirachta indica leaves extract in 0.5 M sulphuric acid, *Mater. Lett.* 61, 148, 2007.
99. Oguzie, E.E., Evaluation of the inhibitive effect of some plant extracts on the acid corrosion of mild steel, *Corros. Sci.* 50, 2993, 2008.
100. Singh, Ambrish, Ahamad, I., Singh, Vinod Kumar, Quraishi, M.A., Effect of fruit extracts of some environmentally benign green corrosion inhibitors on corrosion of mild steel in hydrochloric acid solution, *Journal of Materials and Environmental Science*, 1, 13, 2010.
101. Noor, E.A., Potential of aqueous extract of Hibiscus sabdariffa leaves for inhibiting the corrosion of aluminum in alkaline solutions, *J. Appl. Electrochem.* 39, 1465, 2009.
102. De Souza, F.S., Spinelli, A., Caffeic acid as a green corrosion inhibitor for mild steel, *Corros. Sci.* 51, 642, 2009.
103. Raja, P.B., Sethuraman, M.G., Instantly Investigating the Adsorption of Polymeric Corrosion Inhibitors on Magnesium Alloys by Surface Analysis under Ambient Conditions, *Mater. Lett.* 62, 1602, 2008.
104. El-Etre, A.Y., Inhibition of aluminum corrosion using Opuntia extract, *Corros. Sci.* 45, 2485, 2003.
105. Badiea, A.M., Mohana, K.N., Corrosion Mechanism of Low-Carbon Steel in Industrial Water and Adsorption Thermodynamics in the Presence of Some Plant Extracts, *J. Mater. Eng. Perform.* 18, 1264, 2009.
106. Singh, A., Singh, V. K., Quraishi, M. A., Effect of fruit extracts of some environmentally benign green corrosion inhibitors on corrosion of mild steel in hydrochloric acid solution, *J. Mater. Environ. Sci.* 1 (3) 162–174, 2010.
107. Chauhan, L.R., Gunasekaran, G., Investigation of Aloe lateritia Gel as Corrosion Inhibitor for Mild Steel in 2 M HNO3 and 1 M H2SO4 Media, *Corros. Sci.* 49, 1143, 2007.
108. El-Etre, A.Y., Abdallah, M., El-Tantawy, Z.E., Performance of Tannin/Glycerol-Chromate Hybrid Conversion Coating on Aluminium, *Corros. Sci.* 47, 385, 2005.

109. Orubite, K.O., Oforka, N.C., Inhibition of corrosion of mild steel in HCl solution by extracts of leaves of Mypa fruticans wurmb, *Mater. Lett.* 58, 1768, 2004.
110. Evic Grassino, A.N., Grabaric, Z., Pezzani, A., Fasanaro, G., Voi, A.L., Influence of essential onion oil on tin and chromium dissolution from tin-plate, *Food Chem. Toxicol* 47, 1556, 2009.
111. Torres-Acosta, A.A., Opuntia-Ficus-Indica (Nopal) mucilage as a steel corrosion inhibitor in alkaline media, *J. Appl. Electrochem.* 37, 835, 2007.
112. Quraishi, M.A., Singh, Ambrish, Singh, Vinod Kumar, Yadav, Dileep Kumar, Singh, Ashish Kumar, Green approach to corrosion inhibition of mild steel in hydrochloric acid and sulphuric acid solutions by the extract of *Murraya koenigii* leaves, *Mater. Chem. Phy.* 122, 114, 2010.
113. Quraishi, M.A., Yadav, Dileep Kumar, Ishtiaque, A., Green Approach to Corrosion Inhibition by Black Pepper Extract in Hydrochloric Acid Solution, *Open Corrosion Journal* 2, 56, 2009.
114. Kumar, S. Ananth, Sankar, A., Vijayan, M., Rameshkumar, S. "Magnolia Champaca-Flower Extracts as Corrosion Inhibitor for Mild SteeliIn Acid Medium". *IOSR Journal of Engineering (IOSRJEN)* Vol. 3, Issue 8, pp. 10–14, August 2013.
115. S. Gudić, L. Vrsalović, M. Kliškić, I. Jerković, A. Radonić, and M. Zekić, Corrosion inhibition of AA 5052 aluminum alloy in NaCl solution by different types of honey, *Int. J. Electrochem. Sci.*, vol. 11, pp. 998–1011, 2016.
116. S. K. Shukla and E. E. Ebenso, Corrosion inhibition, adsorption behavior and thermodynamic properties of streptomycin on mild steel in hydrochloric acid medium, *Int. J. Electrochem. Sci.*, vol. 6, no. 8, pp. 3277–3291, 2011.
117. M. A. Bedair, M. M. B. El-Sabbah, A. S. Fouda, and H. M. Elaryian, "Synthesis, electrochemical and quantum chemical studies of some prepared surfactants based on azodye and Schiff base as corrosion inhibitors for steel in acid medium," *Corros. Sci.*, vol. 128, no. April, pp. 54–72, 2017.
118. C. Rahal *et al.*, Olive leaf extract as natural corrosion inhibitor for pure copper in 0.5M NaCl solution: A study by voltammetry around OCP, *J. Electroanal. Chem.*, vol. 769, pp. 53–61, 2016.
119. K. Boumhara *et al.*, Use of Artemisia Mesatlantica essential oil as green corrosion inhibitor for mild steel in 1 M hydrochloric acid solution, *Int. J. Electrochem. Sci.*, vol. 9, no. 3, pp. 1187–1206, 2014.
120. M. A. Hashim, S. N. M. Saleh, and M. R. M. Toff, Surface adsorption mechanism: A molecular self-assembly phenomenon with application to surface protection, 2010, vol. 1217, pp. 279–286.
121. Azzaoui, K.; Mejdoubi, E.; Jodeh, S.; Lamhamdi, A.; RodriguezCastellon, E.; Algarra, M.; Zarrouk, A.; Errich, A.; Salghi, R.; Lgaz, H. Eco Friendly Green Inhibitor Gum Arabic (GA) for the Corrosion Control of Mild Steel in Hydrochloric Acid Medium. *Corros. Sci.* 2017, 129, 70–81.
122. Raja, P. B.; Fadaeinasab, M.; Quereshi, A. K.; Rahim, A. A.; Osman, H.; Litaudon, M.; Awang, K. Evaluation of green corrosion inhibition by alkaloid

extracts of Ochrosia oppositifolia and isoreserpiline against mild steel in 1 M HCl medium. *Ind. Eng. Chem. Res.* 2013, 52, 10582–10593.

123. Krishnegowda, P. M.; Venkatesha, V. T.; Krishnegowda, P. K. M.; Shivayogiraju, S. B. Acalyphatorta Leaf Extract as Green Corrosion Inhibitor for Mild Steel in Hydrochloric Acid Solution. *Ind. Eng. Chem. Res.* 2013, 52, 722–728.
124. Ji, G.; Anjum, S.; Sundaram, S.; Prakash, R. Musa paradisica peel extract as green corrosion inhibitor for mild steel in HCl solution. *Corros. Sci.* 2015, 90, 107–117.
125. Srivastava, M.; Tiwari, P.; Srivastava, S. K.; Prakash, R.; Ji, G. Electrochemical investigation of Irbesartan drug molecules as an inhibitor of mild steel corrosion in 1 M HCl and 0.5 M H2SO4 solutions. *J. Mol. Liq.* 2017, 236, 184–197.
126. Mobin, M.; Rizvi, M.; Olasunkanmi, L. O.; Ebenso, E. E. Biopolymer as Tragacanth gum as a green corrosion inhibitor for carbon steel in 1 M HCl solution. *ACS Omega* 2017, 2, 3997–4008.
127. Singh, A.; Ahamad, I.; Singh, V. K.; Quraishi, M. A. Inhibition effect of environmentally benign Karanj (Pongamia pinnata) seed extract on corrosion of mild steel in hydrochloric acid solution. *J. Solid State Electrochem.* 2011, 15, 1087–1097.
128. Abiola, O. K.; James, A. O. The effects of Aloe vera extract on corrosion and kinetics of the corrosion process of zinc in HCl solution. *Corros. Sci.* 2010, 52, 661–664.
129. Loto, C. A.; Joseph, O. O.; Loto, R. T.; Popoola, A. P. I. Corrosion Inhibitive Behaviour of Camellia Sinensis on Aluminium Alloy in H2SO4. *Int. J. Electrochem. Sci.* 2014, 9, 1221–1231.
130. Chaubey, N.; Singh, V. K.; Savita; Quraishi, M. A.; Ebenso, E. E. Corrosion Inhibition of Aluminium Alloy in Alkaline Media by Neolamarkia Cadamba Bark Extract as a Green Inhibitor. *Int. J. Electrochem. Sci.* 2015, 10, 504–518.
131. M'hiri, N.; Veys-Renaux, D.; Rocca, E.; Ioannou, I.; Mihoubi Boudhrioua, N.; Ghoul, M. Corrosion inhibition of carbon steel in acidic medium by orange peel extract and its main antioxidant compounds. *Corros. Sci.* 2016, 102, 55–62.
132. Casaletto, M. P.; Figa, V.; Privitera, A.; Bruno, M.; Napolitano, A.; Piacente, S. Inhibition of COR-TEN steel corrosion by green extracts of Brassica campestris. *Corros. Sci.* 2018, 136, 91–105.
133. Vidyashankar, S.; Gopal, V. S. K.; Swarnalatha, G. V.; Kavitha, M. D.; Chauhan, V. S.; Ravi, R.; Bansal, A. K.; Singh, R.; Pande, A.; Ravishankar, G. A.; Sarada, R. Characterization of fatty acids and hydrocarbons of chlorophycean microalgae towards their use as biofuel source. *Biomass Bioenergy* 2015, 77, 75–91.
134. Kamal, C.; Sethuraman, G. M. Spirulina platensis – A novel green inhibitor for acid corrosion of mild steel. *Arabian J. Chem.* 2012, 5, 155–161.
135. Kamal, C.; Sethuraman, G. M. Hydroclathrus clathratus marine alga as a green inhibitor of acid corrosion of mild steel. *Res. Chem. Intermed.* 2013, 39, 3813–3828.

136. Rai, M. P.; Khanra, A.; Rai, S.; Srivastava, M.; Prakash, R. Pivotal role of levoglucosenone and hexadecanoic acid from microalgae Chlorococcum sp. for corrosion resistance on mild steel: electrochemical, microstructural and theoretical analysis. *J. Mol. Liq.* 2018, 266, 279–290.
137. Kamal, C.; Sethuraman, G. M. Caulerpin: A bis-Indole Alkaloid As a Green Inhibitor for the Corrosion of Mild Steel in 1 M HCl Solution from the Marine Alga Caulerpa racemosa. *Ind. Eng. Chem. Res.* 2012, 51, 10399–10407.
138. Alam, M.; Akram, D.; Sharmin, E.; Zafar, F.; Ahmad, S. Vegetable oil based eco-friendly coating materials: A review article. *Arabian J. Chem.* 2014, 7, 469–479.
139. Robertson, E. J.; Beaman, D. K.; Richmond, G. L. Designated Drivers: The Differing Roles of Divalent Metal Ions in Surfactant Adsorption at the Oil-Water Interface. *Langmuir* 2013, 29, 15511–15520.
140. Raja, Pandian B., Sethuraman, Gopalakrishnan M., Inhibitive effect of black pepper extract on the sulphuric acid corrosion of mild steel, *Materials Letters* 62, 2977–297,9 2008.
141. Chygyrynets' E., Vorobyova V., A Study of Rape-Cake Extract as Eco-Friendly Vapor Phase Corrosion Inhibitor, *Chem. Chem. Technol.*, 8, 235, 2014.
142. Mangai A., Ravi S.: Comparative Corrosion Inhibition Effect of Imidazole Compounds and of Trichodesma indicum (Linn) R. Br. on C38 Steel in 1 M HCl Medium. *J. Chem.*, 2013, 4, 2013.
143. Raghavendra N., Ishwara Bhat J.: Green approach to inhibition of corrosion of aluminum in 0.5 M HCl medium by tender arecanut seed extract: insight from gravimetric and electrochemical studies. *Res. Chem. Intermed.*, 42, 635, 2016.
144. Adindu, C. B., Oguzie, Emeka E., Ogukwe, Cynthia E., Corrosion Inhibition of Mild Steel in 0.5 M H2SO4 Using Ethanol Extract of Funtumia elastica, *Int. Lett. Chem., Phys., Astron.*, 68, 24. 2016.
145. Njoku D., Ukaga I., Ikenna O. *et al.* Natural products for materials protection: Corrosion protection of aluminium in hydrochloric acid by Kola nitida extract, *J. Mol. Liq.*, 219, 417, 2016.
146. Okafor P., Ebenso E.: Inhibitive action of Carica papaya extracts on the corrosion of mild steel in acidic media and their adsorption characteristics, *Pigm. Resin Technol.*, 36, 134, 2007.
147. Eddy N., Ebenso E.: Corrosion inhibition and adsorption properties of ethanol extract of Gongronema latifolium on mild steel in H2SO4, *Pigm. Resin Technol.*, 39, 77, 2010.
148. Patel N., Jauhari S., Mehta G.: Inhibitive Effect by Acid Extract of Ficus Exasperata Leaves on the Sulphuric Acid Corrosion of Mild Steel. *J. Chem.*, 6, S189, 2009.
149. Oguzie E., Iheabunike Z., Oguzie K. *et al.*: Corrosion Inhibiting Effect of Aframomum melegueta Extracts and Adsorption Characteristics of the Active Constituents on Mild Steel in Acidic Media, *J. Dispers. Sci. Technol.*, 34, 516, 2013.

150. Oguzie E., Ogukwe C., Ogbulie J. *et al.*: Broad spectrum corrosion inhibition: corrosion and microbial (SRB) growth inhibiting effects of Piper guineense extract, *J. Mater. Sci.*, 47, 3592, 2012.
151. Oguzie E., Evaluation of the inhibitive effect of some plant extracts on the acid corrosion of mild steel, *Corros. Sci.*, 50, 2993, 2008.
152. Oyedeji O., Adeniyi B., Ajayi O., Konig W., Essential Oil Composition of Piper guineense and its Antimicrobial Activity. Another Chemotype from Nigeria, *Phytother. Res.*, 19, 362, 2005.
153. Aribo S., Olusegun S., Ibhadiyi L. *et al.*, Green inhibitors for corrosion protection in acidizing oilfield environments, J. Assoc. *Arab Univ. Basic Appl. Sci.*, 24, 34, 2017.
154. Pujar M., Miller A., Proroka M. *et al.*, A novel corrosion inhibitor for aluminum alloys using the beef lipids, *J. Mater. Sci. Lett.*, 19, 1797, 2000.
155. Rzhekhin V.P. Gossypol and its derivatives. L., Ed. VNIIZh, 1955.
156. Slozina G.Z., Rzhekhin V.P. and other Proceedings of VNIIZh, vol. XXV., L., 1965.
157. Tripathy, D. B., & Mishra, A. (2016). Renewable plant-based raw materials for industry. *Sustainable Inorganic Chemistry*, 353.

Index

Also of Interest

Books in the series, "Bioprocessing in Food Science"

NUTRITIONAL SCIENCE AND TECHNOLOGY: Concept to Application, Edited by Tejpal Dhewa, Anil Kumar Puniya, and Anil Panghal, ISBN: 97811198089. This new volume in the groundbreaking series, "Bioprocessing in Food Science," Nutritional Science and Technology, is a comprehensive resource covering topics such as human nutrition, food safety concerns, and emerging technologies like nanotechnology, and it also addresses the need for innovative food products, contamination, and regulatory challenges in the global food market.

HARVESTING FOOD FROM WEEDS, Edited by Prerna Gupta, Navnidhi Chhikara, and Anil Panghal, ISBN: 9781119791973. An exciting new volume in the groundbreaking new series, "Bioprocessing in Food Science," Harvesting Food from Weeds focuses on the latest processes, industrial applications, and leading research on this important, evolving new area, for engineers, scientists, students, and other industry professionals.

MICROBES IN THE FOOD INDUSTRY, Edited by Navnidhi Chhikara, Anil Panghal, and Gaurav Chaudhary, ISBN: 9781119775584. This newest volume in the groundbreaking new series, "Bioprocessing in Food Science," focuses on the latest processes, industrial applications, and leading research on microbes in the food industry, for engineers, scientists, students, and other industry professionals.

NOVEL TECHNOLOGIES IN FOOD SCIENCE, Edited by Navnidhi Chhikara, Anil Panghal, and Gaurav Chaudhary, ISBN: 9781119775584. The third book in the series, "Bioprocessing in Food Science," covers the latest and greatest new technologies in food science.

FUNCTIONAL FOODS, Edited by Navnidhi Chhikara, Anil Panghal, and Gaurav Chaudhary, ISBN: 9781119775560. Presenting cutting-edge information on new and emerging food engineering processes, *Functional Foods*, the second volume in the groundbreaking new series, "Bioprocessing in Food Science," is an essential reference on the modeling, quality, safety, and technologies associated with food processing operations today.

THERMAL FOOD ENGINEERING OPERATIONS, Edited by Nitin Kumar, Anil Panghal, and M. K. Garg, ISBN: 9781119775591. Presenting cutting-edge information on new and emerging food engineering processes, Thermal Food Engineering Operations, the first volume in the new series, "Bioprocessing in Food Science," is an essential reference on the modeling, quality, safety, and technologies associated with food processing operations today.

Other related titles from Scrivener Publishing

ESSENTIAL OILS: Extraction Methods and Applications, Edited by Inamuddin, Tariq Altalhi, and Jorddy Neves Cruz, ISBN: 9781119829355. Written and edited by some of the world's top experts in the field, this exciting new volume provides up-to-date information on essential oils' chemical structure, therapeutic, and biological actions, and defines their functional uses and evaluates the advantages and disadvantages of their use in a variety of sectors.

FUNCTIONAL POLYMERS IN FOOD SCIENCE, Part 1: Food Packaging, Edited by Giuseppe Cirillo, Umile Gianfranco Spizzirri, and Francesca Iemma, ISBN: 9781118594896. Provides an extensive overview of the link between polymer properties and food quality. *NOW AVAILABLE!*

FUNCTIONAL POLYMERS IN FOOD SCIENCE, Part 2: Food Processing, Edited by Giuseppe Cirillo, Umile Gianfranco Spizzirri, and Francesca Iemma, ISBN: 9781118595183. Provides an extensive overview of the link between polymer properties and food quality. *NOW AVAILABLE!*

ADVANCES IN FOOD SCIENCE AND TECHNOLOGY Volume 1, Edited by Visakh, P.M., Sabu Thomas, Laura B. Iturriaga, and Pablo Daniel Ribotta, ISBN: 9781118121023. Written in a systematic and comprehensive manner the book reports recent advances in the development of food science and food technology areas. *NOW AVAILABLE!*

ADVANCES IN FOOD SCIENCE AND NUTRITION Volume 2, Edited by Visakh P.M., Laura B. Iturriaga, and Pablo Daniel Ribotta, ISBN: 9781118137093. This important book comprehensively reviews research on new developments in all areas of food chemistry/science and nutrition. *NOW AVAILABLE!*

Applied Water Science Volume 1: Fundamentals and Applications, Edited by Inamuddin, Mohd Imran Ahamed, Rajender Boddula and Tauseef Ahmad Rangreez, ISBN 9781119724766. Edited by one of the most well-respected and prolific engineers in the world and his team, this is the first volume in a two-volume set that is the most thorough, up-to-date, and comprehensive volume on applied water science available today. *COMING IN SUMMER 2021*

Sustainable Water Purification, by Safiur Rahman and M. R. Islam, ISBN 9781119650997. This is the only book that takes a zero-waste approach to propose 100% sustainable water purification techniques. *NOW AVAILABLE!*

Printed in the USA/Agawam, MA
September 18, 2026

910661.119